D1452351

LABORATORIES

LABORATORIES

A Guide to Master Planning, Programming, Procurement, and Design

Fernand Dahan, FAIA, AICP

W. W. Norton & Company
New York • London

A NORTON PROFESSIONAL BOOK

Printed in the United States of America
First Edition

The text of this book is composed in Bembo
with the display set in Akzidenz Grotesk
Manufacturing by Hamilton Printing Company
Book design and composition by Silvers Design

Library of Congress Cataloging-in-Publication Data

Dahan, Fernand.
Laboratories : a guide to planning, programming, and design / Fernand Dahan.
p. cm.
Includes index.
ISBN 0-393-73058-1
TH4652 .D35 2000
727'.5—dc21

00-058405

ISBN 0-393-73058-1

W. W. Norton & Company, Inc., 500 Fifth Avenue, New York, N.Y. 10110
www.wwnorton.com

W. W. Norton & Company Ltd., 10 Coptic Street, London WC1A 1PU

0 9 8 7 6 5 4 3 2 1

Contents

Preface

THIS BOOK is not intended as an engineering or an architectural precis on laboratory design. Its aim instead is to explain specific laboratory activities and provide guidelines for identifying the steps needed for each activity. It should also help all professionals involved in the design of laboratories to understand each others' roles. It is geared first toward architects but also addresses industrial hygienists and mechanical, electrical, fire, structural, and civil engineers involved in the programming, conceiving, designing, or master planning of most types of laboratories. It is also addressed to corporate and governmental laboratory owners and managers and to personnel involved in managing the facilities.

The information contained herein is not only the result of my experience in all of the above phases of programming, design, and master planning, but also derives from my observations of the functionality of completed, occupied projects. I may say, therefore, that this is a personal book, as many of the suggested guidelines resulted from my own experiences. I was involved in the different phases of acquiring a laboratory and, in so doing, I monitored and observed what needed to be done and what was accomplished. Once the laboratory became functional, I continued to monitor what went right and what needed correction; throughout this process my colleagues and I discussed how to improve the procedures involved in laboratory programming, design, and master planning.

By nature, I share the philosophy of the sixteenth-century French moralist Montaigne, who wrote: "There is more ado to interpret interpretations than to interpret things, and more books upon books than upon all other subjects. We do nothing but comment upon one another." In my research, I made a point of not reading the publications of my contemporaries and recent predecessors whose field of studies matched my own until I formed my own opinion. I preferred to examine the facts directly, analyze them, and if needed discuss them with colleagues involved in the same kind of work. I did not want to use facts that had been filtered through the minds of others and then interpret them without knowing the author's rationale or which facts which were relevant or irrelevant. My aim was to remain unfettered by the influence of other thinkers with whom I could not discuss my views as I did with my colleagues. And so this book is the result of my observations and experience in this particular field. Thus again I could quote Montaigne: "Je suis moi-meme la matière de mon livre" (I am myself the substance of my book).

No author, however, can claim to be the only source of the information provided in his book. This book is no exception. Undeniably, the men and women with whom I discussed my experiences and the interpretation of codes and regulations significantly influenced my way of thinking and helped me be who I am.

Further, a writer's thoughts, once laid on paper, are never in themselves sufficient to comprehensively accomplish the writer's design. Yet another process is required—one that calls for deliberate and extensive preparation, research, analysis, and review. This book is also about that process: I studied, analyzed, and learned new facts about laboratory programming, design, and master planning. I exchanged ideas with other knowledgeable colleagues, and, more often than not, sought their advice and followed it. Much in this book resulted from the exchange of ideas with others and the seeking of advice. I am lucky to have been surrounded by immensely talented colleagues throughout my twenty-eight years as a government architect. Many of these colleagues were with the U.S. Environmental Protection Agency (USEPA). Others were not. Some of them have retired; others are still active. I am honored to have been associated with all my friends and colleagues at the USEPA and elsewhere who throughout the years have helped me mold my thinking and set me straight when I strayed. The following individuals fall in this category; they had an impact on my formation as a laboratory architect and deserve special acknowledgment.

Alan M. Clark, a real estate specialist now retired, and the former head of the real estate branch at the USEPA. Throughout my early years as a laboratory architect, Alan patiently familiarized me with most real estate–related aspects of acquiring a laboratory facility. After these early years, my friend and colleague Robert Garrison, also a real estate specialist and still active with the USEPA, kept my knowledge up to date and kindly reviewed the chapter dealing with this subject.

Carl Sandine, who is presently retired, was my first guide on how to program and design a laboratory facility. Since then, my discussions and interactions with Joseph Gillian and William H. Ridge helped me sharpen my views on laboratory components design.

Thomas Ashmore, Tony Castro, Stephanie James, Russell Kulp, Ray Lum, and Rolando Santos, all of whom are talented mechanical engineers, and Hadi Janbakhsh, Gary Hanson, and Mario Sanches, who are also talented electrical engineers, helped me throughout the years to shape my thoughts on design of laboratory mechanical and electrical systems and on how to refine such systems to respond to very specific requirements. I also thank Andre Wagner of Trier, Germany, for helping me convert the American measurements to metric equivalents.

I owe my consciousness of and sensitivity to health and safety requirements and to prudent practice in laboratory design to Larry Gaffney, who is presently retired, David Wiseman, Sheldon Rabinovitz, Howard Wilson, Patricia Weggel, Leo Stein, and Thomas Oberholzer.

Bobby Carrol, Charles Hooper, and Russell Wright, all three of Athens, Georgia; Brenda Bettencourt of San Francisco, California; Robert Menzer of Gulf Breeze, Florida; Michael Johnston of Manchester, Washington; Joseph Pernice of Edison, New Jersey; Marvin Wood of Ada, Oklahoma; and Jon Yeagley of Denver, Colorado, deserve thanks. Among others, each of them in their respective fields of expertise helped me better understand laboratory users' needs. Through their criticism of my work in procuring laboratories for them, they provided me with invaluable information.

I want to thank Charles Porter, former director of the USEPA's National Air and Radiation Laboratory in Montgomery, Alabama; Thomas Horsh, P.E., who was head of the Planning and Real Estate Department of the Atlantic division of the Naval Facilities Engineering Command in Norfolk, Virginia; and

Luther Mellen, chief of the Architectural, Engineering and Real Estate Branch at the USEPA, for helping me seeing the big picture. Thanks also to Philip Wirdzek of the USEPA, for raising my consciousness of and sensitivity to the energy conservation and pollution-prevention efforts related to laboratories, and to Lance Swanshost, civil engineer, for helping me understand the importance of site selection and planning at the very early stages of the planning process.

I want to express my appreciation to William Ridge, architect, USEPA, for his help in reviewing chapter 13, and to John Birri, special project coordinator for the USEPA Region 2 Laboratory, for helping me edit and revise the chapter on instruments used in laboratories. I also want to express a special appreciation to Patricia Weggel, of the EPA's Technical Support and Evaluation Branch of the Safety, Health, and Environmental Division, for reviewing the chapter on special laboratory rooms.

I believe that there is no substitute for the discerning eyes of others, especially those whose knowledge and experience make them experts in their field and who can articulate their knowledge in practical terms. I am honored to thank the following individuals, who acted as my critical peers and submitted comments on various draft chapters. While their suggestions improved the manuscript, their assistance in no way relieves me of the sole responsibility for the content of this book.

Paul Jarvis, P.E. was the chief of Facilities Engineering and Real Estate and my first boss when I started working at the USEPA as a laboratory architect. A few years later, Paul moved to the position of chief engineer for the Science and Education Administration of the U.S. Department of Agriculture. From there he was offered and accepted the position of director of the Division of Engineering Services for the U.S. National Institutes of Health. When he retired a few years later, Paul had more than thirty years of experience in laboratory design and was recognized worldwide as a laboratory expert. He serves as a consultant in this field to the World Health Organization and the United Nation Development Program. Paul, thanks for reviewing my work.

Arthur G. Garikes, AIA, is the executive vice president of Garikes, Wilson, and Karisberger, in Birmingham, Alabama. Arthur developed the open laboratory concept in clinical laboratory design and established the "STAT RING" concept for accessioning and primary processing of laboratory specimens. He is recognized as one of the innovators and first users of flexible laboratory casework. He authored chapters in four clinical laboratory planning manuals for the College of American Pathologists and the American Public Health Association. Arthur had more than forty years of experience in laboratory design and is recognized worldwide as a laboratory expert. He has been an adjunct professor at the School of Public Health at the University of Alabama in Birmingham and is a member of the Subcommittee for Laboratory Design of the National Committee for Clinical Laboratory Standards. He is also a frequent lecturer for the American Hospital Association. Art, thanks for reviewing my work.

Howard Wilson is the chief of the EPA's Technical Support and Evaluation Branch of the Safety, Health, and Environmental Division of the USEPA. For the last eleven years his responsibility has been to assure that his agency laboratories are designed, constructed, and operated in accordance with safety, health, and environmental considerations published in regulations and national consensus standards. Howard is also responsible for the update and rewriting of the USEPA Facility Safety Manual, a document encompassing all standards and guidelines of the USEPA laboratories. He has a degree in civil engineering, is known nationally as a laboratory safety expert, and has addressed large governmental and nongovernmental audiences on such issues. Howard, thanks for reviewing my work.

Christopher Rousseau is a partner with the firm of Newcomb and Boyd consulting engineering in Atlanta, Georgia, and a national authority on the design of mechanical systems for laboratories and other high-technology buildings. He is a member of the American Society of Heating, Refrigerating, Air-conditioning Engineers' technical committee on laboratory systems. Chris is responsible for the design of numerous mechanical systems for laboratory facilities and for other types of advanced-technology facilities. Among the most recent projects are laboratories for the USEPA, Emory University, the Coca Cola company and the University of North Carolina. Chris holds a bachelor's degree in mechanical engineering from Georgia Tech University in Atlanta, Georgia, and is often a guest lecturer in his area of expertise. Chris, thanks for reviewing my work.

Ulrich M. Linder, principal and cofounder of the firm of Earl Walls Associates, dedicated his career to the planning and design of state-of-the-art laboratory facilities. Today he is recognized all over the world as an authority on laboratory design. In one of his early projects, he worked closely with the late Louis I. Kahn to complete the construction documents of the now-famous Salk Institute. Since then, Ulli, as he is known to his friends and colleagues, participated in the design of over a million square feet of all types of technical and advanced technology facilities. Ulli received a Diplom Ingenieur in architecture from the Akademie Für Angewandte Technik, in Germany; he often is a guest lecturer in his area of expertise and has received many prestigious awards and high honors for his projects. Ulli, thanks for reviewing my work.

In conclusion, it is only fitting to thank the people at W. W. Norton, particularly my editor, Nancy Green, and copy editor Casey Ruble, for their judicious suggestions, which improved and made this book what it is, and also my patient wife, who throughout the years encouraged me to write this book and experienced long lonely hours when I was writing it.

Fernand W. Dahan
Rockville, Maryland
September 2000

Introduction

THIS book defines, categorizes, and explains most types of laboratories. It discusses their planning and programming as well as the procurement of space for laboratory activities. It gives information on site selection, on the design of laboratory rooms, laboratory wings, and the facility as a whole, and on strategic master planning of laboratory facilities.

This book also describes the function and requirements of certain highly specialized laboratory rooms. It recommends guidelines for the relationship between architects and engineers designing laboratories and provides guidelines for the design of the architectural, mechanical, and electrical systems servicing the laboratory. Finally, it provides a partial list and brief description of functions of the instruments used in a laboratory.

This books guidelines should always be adapted to the situation at hand by the project architect, the engineers, and the industrial hygienist. This is because the design of a new laboratory or the alteration of an existing laboratory usually has to address numerous complex problems with many components that often conflict with each other. For instance, a suggested guideline initially may appear to be an answer to the problem at hand; however, after all the problem's components have been considered, the guideline may be proved inadequate. Further, sometimes a group of problems can be individually solved by particular guidelines, but the combination of those solutions causes new problems. Therefore, I highly recommend that the design team (including the architect, mechanical and electrical engineer, and industrial hygienist) examine all the components of the situation and collectively decide which guideline or combined set of guidelines to follow.

The design guidelines provided in this book incorporate the principles of sustainable design. They were conceived to lead building designers—architects, engineers, industrial hygienists, and others—to come up with green laboratory buildings. By definition, these buildings have minimal impact on the environment over their lifetime and promote their occupants' health and productivity. Methods and means to reduce waste and prevent pollution are suggested and indoor air quality is addressed. This book also promotes the use of natural light and provides designs that incorporate natural lighting in laboratory rooms as well as related offices.

The Laboratory: Definition and Categorization

A laboratory is a place for analysis, education, production, experimentation, and research. It is a place where chemicals, drugs, and other materials, living or not, are handled, tested, analyzed, and studied, and where related operations are demonstrated or performed.

Laboratories can be as varied as the functions and operations undertaken in them. Therefore, it is very important to know, at the inception, the purpose or purposes for which the laboratory is needed and to identify the operations that will take place in it. Additionally, the chemicals or other materials that will be handled, analyzed, tested, or stored in the laboratory, as well as the quantities in which they will be used, must be known. It is also important to know how many operators will be in each space and how many people will be within or near the space in which operations are being conducted. This information helps determine the type and class of the laboratory being considered and the applicable standards and design requirements.

Laboratories can be categorized in several ways depending on the intent of the categorization. For our purposes, we will classify them into five major groups:

- chemical
- biological
- biochemical
- physical
- psychological

Our concern is with the first four groups; the space requirements, equipment, and services of psychological laboratories are different from those of the other groups.

A *chemical laboratory* is one in which the tasks and operations involve use of chemicals in the context of analytical chemistry.

A *biological laboratory* deals with plants and animals. This type of laboratory may have thousands of varieties. They may deal with the origin, history, physical characteristics, life processes, and habits of plants and animals and include botany and zoology and their subdivisions.

Biological laboratories that are concerned with chemical processes and involve chemistry with plants, animals, and their life processes are known as biochemical laboratories. Biochemical laboratories are grouped in the same way as the chemical laboratories: these laboratories are subcategorized according to the quantities of chemicals used or stored in them and to the characterization of these chemicals or of the operations in which they are used. They fall into the same hazard and risk categories as the chemical laboratories.

Biological and biochemical laboratories may be subjected to biohazards. If they are, these biohazards must be identified and classified, which will help determine the appropriate safeguards to prevent laboratory infections and to protect the validity of the experiments conducted. Laboratory-acquired infections fall into one of the following types: bacterial, fungal, parasitic, rickettsial, and viral. Biohazardous agents have been divided into five classes.*

- Class 1 agents are of no (or minimal) hazard under ordinary conditions.
- Class 2 agents are ordinary and potentially hazardous.
- Class 3 agents involve special hazards.
- Class 4 agents require the most stringent containment conditions.
- Class 5 comprises animal agents that are prohibited by law in the United States.

Biosafety laboratories involving use of the above-named hazardous agents are grouped in four categories:

Level I biosafety laboratories are suitable for agents of minimal potential hazard to the environment and to laboratory personnel working with the agents.

Level II biosafety laboratories are suitable for agents of moderate potential hazard to the environment and to the laboratory personnel working with the agents.

Level III biosafety laboratories are used for clinical diagnostics, teaching, research, and production that involve indigenous or exotic agents that may cause serious or potential lethal diseases from inhalation exposure.

Level IV biosafety laboratories are required for work with dangerous and exotic agents that pose a high risk of life-threatening disease.

In a *physical laboratory*, tasks and operations deal with physics and physical elements and do not use chemicals in research and experimentation.

The purpose of this book is to acquaint all the people involved in the design and procurement of laboratories or of laboratory space with their various roles. While the book offers sets of guidelines and some obvious code requirements, the final specifications, criteria, and designs should be the responsibility of the professionals dealing with the specific project.

The first of this book's fourteen chapters is entitled "Predesign Activities: Planning to Program the Requirements." This chapter presents and discusses the factors of planning and programming a laboratory facility. It identifies the people who should be involved and explains their roles, how they should be organized, and what their expected outputs are. It deals with the assessment of the tasks and operations considered for a facility and suggests development of an operation plan. This chapter also identifies options for structure acquisition and

* The United States Public Health Services (USPHS) and the United States Department of Agriculture (USDA) have formulated the standard to evaluate hazards associated with biohazardous agents. See "Classification of Etiologic Agents on the Basis of Hazards," Center for Disease Control, USPHS, 3rd ed., June 1972. This classification does not include genetic hybrid viral agents. These agents, because of their potential hazardous nature, require specific precautions to reduce or eliminate the chances that these hybrids become part of the pathogens or agents that may infect the general population.

proceeds with the analysis and prioritization of these options to determine their level of efficiency and convenience. It explains what a "program of requirements" for a laboratory facility is, specifies the information it should contain, and suggests ways of preparing it so that the facility responds to the needs of the operation plan. Finally, it deals with the selection and roles of consultants, architects, and engineers and identifies the contributions of each professional in the context of the planning and programming function.

The second chapter, "Laboratory Space: Renovated Versus New, Owned Versus Leased," states and explains the different options available for acquisition of laboratory space and proposes a method for analysis and prioritization of these options. It also discusses how to undertake the investigation of existing physical conditions and of other (e.g., financial or time) factors. Also addressed is how these factors relate to each other and to the results of the investigation and analysis, and how they influence or are influenced by financial and budgetary considerations. It also shows how to use this information to establish a baseline for use in life-cycle cost analyses. Additionally, the professional determines the most economical acquisition approach by using a life-cycle cost analysis and a "conceived" model facility that includes all the factors.

The third chapter, "Selecting a Site for the Laboratory," states the criteria by which a site being considered for a laboratory must be evaluated. These criteria include health, safety, and environmental considerations, the engineering and operation plans, and the public perception. It discusses the possible impact of the site and surrounding environment on the laboratory operations, as well as the impact of laboratory operations, if any, on the site and surrounding environment. The chapter explores the possibility of accidental chemical contamination and discusses laboratory control technologies that diminish such risks. It identifies the different types of laboratories and recommends the preferred location for each type. It looks at the lot size and topography and deals with zoning and covenants' requirements and restrictions. It discusses the public perception of the laboratory location and suggests ways for alleviating public apprehension and anxiety about having a laboratory as a neighbor. The chapter also identifies criteria for evaluation of the site: its characteristics, location, size, and accessibility. It discusses the services and utilities available on or around the site and how these should be provided, as well as the importance of the site's proximity to major suppliers of laboratory needs.

The fourth chapter, "Designing the Laboratory Room," provides information on how to prepare "room data sheets" and describes how to show the "preferred room arrangement." It explains how the information common to many rooms can be grouped and incorporated in the "standard requirements as used in the room data sheets." It explains the factors used to size the laboratory rooms in terms of modules and provides typical layouts for laboratory rooms of different sizes. The chapter discusses hood requirements and their effect on the layout of the laboratory rooms and gives guidelines on hood locations in one-, two-, three-, and four-module laboratory rooms. It covers exit doors and the function of eye washes and emergency showers, and it recommends the types of doors, eye washes, and emergency showers that should be used for various locations. The chapter discusses how to prepare the room data sheet and shows how to provide coordinated information to assist the reviewers and the designer in determining the fire- and explosive-hazard levels of each room in order to provide a safe and efficient design.

The fifth chapter, "Designing the Laboratory Wings," provides information on how to arrange working groups of rooms by using information provided by the room data sheets. It indicates how to organize rooms into blocks (a group of rooms having similar functions and requirements) according to the rooms' adjacency, separation, functional, and organizational requirements. Also discussed is how these blocks can be arranged into laboratory wings. The chapter identifies and explains the six major considerations that must be resolved prior to the grouping of blocks and rooms into wings (number of floors, adjacencies and separations, flexibility and expansibility, windows, services and utilities, wings' location and orientation, and the location of fume-hood exhaust stacks). It shows different wing arrangements, and discusses the design advantages and disadvantages of each.

The sixth chapter, "Designing the Laboratory Facility," deals with the principles that govern the design of a laboratory facility. It recommends ways to group and/or segregate different laboratory activities according to function, hazards, and risks. It looks at such grouping in terms of wind direction, cross-contamination between rooms, blocks, and wings and considers the relationship between exhaust stacks and fresh-air intakes. This chapter also recommends ways to address module selection and providing services and utilities to laboratory wings and rooms. It offers guidelines for vertical and horizontal circulation within the laboratory and within the building that houses the facility. It also suggests requirements for the corridors, building framing, doors, windows, and exterior and interior walls of laboratory wings. This chapter provides information on furniture and casework for the laboratory wing; on steel, wood, and polypropylene cabinets; and on the different ways of setting cabinets in laboratory rooms. It discusses the advantages and disadvantages of fixed and flexible types of cabinet arrangements and provides information on different types of countertops and on when to use each type. It describes how to measure and respond to the storage needs of laboratory wings. It also states requirements for wings in laboratory support space, how to determine the amount of space required for laboratory operations, and how to efficiently subdivide and equip this space. The chapter also explains different ways of storing chemicals and gases and describes the requirements for chemical storage in cabinets, refrigerators, and freezers. It also defines stock rooms and describes stock rooms' requirements.

The seventh chapter, "Storage of Chemicals and Chemical Wastes in a Laboratory Facility," deals with the requirements for storage of large quantities of chemicals and of chemical waste in a laboratory facility. It refers to the National Fire Protection Association (NFPA) classification that subdivides laboratories into classes A, B, and C and sets rules for the maximum quantities of flammable and combustible chemicals and gases stored in each class of laboratories. The chapter also identifies chemical classifications and what each group consists of. Understanding these classifications will help designers and laboratory management determine how chemicals and gases should be segregated and stored and how a facility storing such materials should be designed.

This chapter also addresses the location and accessibility requirements for the hazardous materials storage building. It states where the building should be located in relation to property lines and to other structures and dictates the requirements for vehicular and pedestrian accessibility to the building and protection of pedestrians and air intakes from the vehicles accessing the building. The chapter also explains the handling, transfer, and storage requirements of the different classes of liquids as well as the general building protection requirements. Chemical segregation within the hazardous materials storage building is explained, and what each room's envelope and the materials of construction should be is also described. It also explains the requirements for storage arrangements for drums and their grounding, for storage arrangements of smaller containers over shelves, for doors and blast panels, and for ventilation of the rooms. Requirements for explosion-proof lighting and mechanical and electrical equipment within hazardous chemical storage rooms are also discussed.

This chapter also addresses most of the standards governing the storage of different gas cylinders. It deduces from these standards a set of guidelines to be followed when storing gas cylinders in enclosed rooms or open space. It categorizes gases according to their hazard levels and indicates the code governing the storage of each type. It explains how gas cylinders should be segregated in different rooms or spaces both within the main laboratory building and in the hazardous materials storage building and specifies what each room or space's envelope and materials of construction should be. It also explains storage arrangements for gas cylinders and their required grounding.

The eighth chapter, "Recommendations for the Architect/Engineer Interaction," aims to familiarize the laboratory owner and other professionals with the process of programming and designing a laboratory. It defines a laboratory as a complex building with many conflicting, complicated requirements and unique, often unprecedented problems. The relationship between architects and engineers programming and designing such buildings needs to be tailored to the requirements and problems to be solved. This relationship, therefore, differs significantly from that of architects and engineers designing most other types of buildings. This chapter investigates and contrasts both types of relationships. It also discusses why the project mechanical engineer should take the lead in programming and designing laboratories and how the different engineers and the project architect should integrate their activities. It shows how each of these disciplines leads initial program development and the project design in the following phases. In essence, the chapter covers most of the important activities involved in programming and designing and suggests who should do what, and when.

The ninth chapter, "Guidelines for Design of the Mechanical Systems for the Laboratory Wing," suggests principles of sustainable design as applied to the plumbing and mechanical systems of laboratory buildings. It describes devices and systems that promote the health, safety, and productivity of the occupants, as well as energy and resources conservation. This chapter recommends the use of modularity and flexibility in the mechanical systems providing services and utilities to the laboratory modules and explains how such modularity can be achieved. It also states laboratory mechanical requirements, which fall into three groups: plumbing, laboratory services, and heating, ventilation, and air conditioning (HVAC). The plumbing category includes the domestic and laboratory water supply and waste water, the laboratory's waste water sampling, and the acid neutralization system. This category also includes all piping systems, the emergency shower and eye washes, and the sprinkler system and steam service. The laboratory services category includes the distribution systems for natural gas and for inert, oxidizing, and flammable gases. This category also includes the compressed air and vacuum systems as well as the systems that deionize water. The HVAC category includes the supply, exhaust, and treatment of the laboratory wings' air. It addresses air purity in laboratory rooms and in clean rooms, the level of hazard and risk, and the requirements such levels dictate. This category also includes the temperature and humidity control and air circulation in the laboratory rooms and in the special-purpose rooms and explains the HVAC balancing requirements. It deals with the supply and exhaust ducts and their connections. It states the precautions to take in order to avoid cross-contamination. It defines "constant supply and exhaust airflow" and "variable air volume supply and exhaust airflow."

This chapter identifies and categorizes the different types of ventilation protective devices, including fume hoods, glove boxes, biological safety cabinets, and clean benches. Fume hoods are designed mostly for the protection of the operator and may be equipped with filters or scrubbers to protect the environment. The chapter details the functions of several types of fume hoods: the constant volume conventional fume hood, the constant volume bypass fume hood, the constant volume auxiliary air fume hood, the variable air volume fume hood, the perchloric acid fume hood, the radioisotope fume hood, the distillation fume hood, the California fume hood, the walk-in fume hood, the ductless fume hood, the ADA fume hood, the canopy hood, the slot hood, the downdraft hood, and the elephant trunk. Glove boxes are designed for the protection of the operator and are generally equipped with filters that also protect the environment. This chapter explains the functions of the two types of glove boxes: the controlled atmosphere glove box and the ventilated glove box. Biological safety cabinets are designed mostly for the protection of the operator. However, some types are equipped with filters that also protect the environment and/or the operation conducted in the cabinet. This chapter explains the functions of the types of biological safety cabinets: class I, class II A, class II B, class II B1, class II B2 , class II B3, and class III. Clean benches are designed (and equipped with filters) to protect the operation. This chapter explains the functions of clean benches.

This chapter also describes the requirements of fume hood exhausts, as well as their manifolding, heat recovery, treatment and/or scrubbing, and filtration either by HEPA filters or by gas-phase filtration. Finally, this chapter addresses the requirements for the materials of fume hood exhaust ducts.

The tenth chapter, "Guidelines for Design of Electrical Systems for the Laboratory Wing," sets the electrical design criteria for laboratory rooms and wings. This chapter, like the previous one, incorporates principles of sustainable design. It applies these principles to the electrical design criteria for systems in laboratory buildings. It describes devices and systems that promote the health, safety, and productivity of the occupants as well as energy and resources conservation. The three most important electrical design criteria are flexibility, reliability, and safety. The chapter explains how to use these criteria to determine the needed service capacity to respond to full design loads, including load needs for lighting and for the existing and projected mechanical and instrumentation. It also gives advice on laboratory instrumentation, equipment, and mechanical needs for electric power, particularly regarding switching, gearing, circuitry, tripping, and emergency power. It also provides guidelines for systems that protect laboratory wings from lightning, fire, explosion, and intrusion.

The eleventh chapter, "Guidelines for the Design of Special Laboratory," explains the design requirements for partial- and full-containment laboratories, biosafety laboratories, and clean rooms. It discusses when a partial-containment laboratory should be used and when a full-containment laboratory is necessary. It also provides information on support areas for the partial- and full-containment laboratories. This chapter explains the design requirements of biosafety laboratories. It categorizes them into three biosafety levels and gives criteria governing the design and use of laboratories of each level. It discusses when a biosafety laboratory room of a given level should be used and when it becomes necessary to use a laboratory room of a higher level. It also provides information on support areas of each of the three biosafety laboratory levels. This chapter also defines clean rooms and explains their design and use requirements. It categorizes them into different levels of cleanness and provides information on support areas for each of the levels.

The twelfth chapter, "Indoor Air Quality in Laboratory Buildings and Rooms," identifies the eleven major indoor pollutants, how these pollutants are generated, how they reach the human body, and what their effects on human beings are. It also specifies ways of controlling their introduction into the laboratory and other rooms and how to expel them if they are already present. It deals with means of providing quality indoor air in order to promote occupants' health, safety, and productivity.

The thirteenth chapter, "Strategic Master Planning for Laboratory Facilities," explains the procedures used by corporations or government agencies wanting to build or alter general chemistry or biochemical laboratories. It shows how to analyze the corporation or agency's stated mission, goals, and objectives and how to determine the tasks needed to achieve them. The chapter explains how to identify the resources needed to perform these tasks (human, facilities, and instrumentation and equipment), how to analyze them, and how to determine with the corporation management or the government agency the means to obtain them. Also addressed is how to develop both an organizational and a facility strategic master plan and how to derive long-range, intermediate-range, and short-range organizational and facility master plans. The chapter shows how to determine the space size and requirements for each projected task and, once these are identified, how to determine the support space needed and related requirements. It also explains the activities required for procurement or acquisition. The chapter discusses how to deduce future needs from projections and how to develop and use customized standards responsive to the needs of entity (i.e., corporation, government agency, or business). The chapter also proposes approaches for the institutionalization of the master planning process and of implementation mechanisms to monitor, evaluate, and upgrade facilities on an ongoing basis in order to keep them responsive to continuous changes in tasks.

Finally, the title of the fourteenth chapter, "List of Common Laboratory Instruments," sufficiently explains the chapter's content.

Chapter 1

Predesign Activities: Planning to Program the Requirements

THIS chapter attempts to guide the planning and designing of all laboratories covered in this book. Its intent is to introduce many of the factors that must be considered, including criteria that will make the laboratory a green building and incorporating principles of sustainable design for renovations. The actors who should be involved in such projects are also identified, as are the roles each actor must play.

Every construction project begins with an idea. Rarely, however, does this idea lead to real action until funds become available. The programming and the requirements developed at this early stage of the project are general in nature—they are oriented toward identifying and securing the funds required rather than used for the development of construction plans. Once a decision is made and requested funds are available, there is a temptation to have construction work start soon.

Having worked to obtain approval and funds for the project, the chief executive officer (CEO) and fund-raisers crave quick results: perhaps a groundbreaking ceremony or a similarly tangible activity. The CEO may be tempted to immediately hire architects and engineers and ask them to start preparing plans for the project. This often occurs prior to the determination of the project's real space needs and requirements.

Suppose you are this CEO. You hire an architect as soon as the project is approved and the funds are available, and you let him and his engineers determine your space needs and requirements. In so doing, you let them decide how your laboratory and the operations undertaken in it are going to be conducted. Knowing little about building design or about the needs of architects and engineers trying to design a laboratory, you assume that they are the most qualified people to collect and determine the information needed to design it. This may be true, but only partly true, as you will soon see.

Laboratory space and other requirements must be identified and determined by the laboratory users. This should be done through a process involving all the users. Once these requirements are identified and coordinated, they should be packaged into what is called a "program of requirements." After being approved by the CEO or administrator, this package is translated by the architect and engineers into plans and specifications. Later, the contractor translates these plans and specifications into a facility that responds to the requirements. Throughout these different phases, the program of requirements is used as a yardstick to measure the goals expected to be achieved.

If you were this CEO or administrator with secured funds for your project, you would appropriately decide to organize all activities related to this project again, perhaps from the very beginning. You would use the help of the technical staff who will be working in and operating the laboratory. They would first decide on the procedures of their operations and then determine what is needed to conduct them.

Should you not yet have a staff, you would hire one or more consultants. One of them should be specialized in laboratory operations and should be familiar with the kind of laboratory work you plan to undertake. You would then ask either your staff or the consultants to reanalyze all information and data, prepare an operation plan for the laboratory's expected activities, and determine all needs and requirements. Your role would then be to follow through and to keep the project on schedule. You would divide the activities related to the procurement of this laboratory into three phases:

1. *Planning and programming.* In this phase, which is the subject of this chapter, pertinent information is collected and analyzed, a plan for laboratory operations is prepared, the facility's needs and requirements are determined, and an analysis and allocation of funds is made.
2. *Design.* In this phase, plans and specifications for the facility are prepared and other related factors are dealt with.
3. *Construction.* This is the physical realization of the two previous phases.

Facility Planning Team

Once the project is approved and the funds become available, the administrator determines what needs to be done. He defines the required activities in terms of major objectives and forms a group of "consultants" selected from his staff or by contract to undertake the necessary tasks for planning and programming the project. He then either heads this group himself or appoints a chairperson from the group. This group constitutes the "planning team" and should include:

- A head for the laboratory section who will be in charge of the work undertaken in the projected laboratory.
- The administrator of the body for whom the laboratory is being designed or his representative. This person will chair the team. If the authority is delegat-

ed by the administrator to one of his subordinates, the subordinate should have sufficient authority to make the process fruitful.

- An industrial hygienist who is familiar with the type of operations contemplated for the planned facility.
- A building engineer.
- An architect who has extensive knowledge and expertise in planning laboratory facilities. The architect should have an engineering team highly familiar with the intricate mechanical, electrical, and plumbing characteristics of laboratories.
- Others, either staff members or experts (such as a financial consultant) may become temporary members of the team as needed.

The head of the laboratory section may designate one or more persons of his department to deal with specific aspects related to requirements. The administrator may direct the chief of staff or head of housekeeping or others to specify certain requirements pertaining to their fields. The architect or engineer may ask other architects or engineers to deal with the aspects pertaining to their field of expertise.

The first four members of the planning team (the head of the laboratory section, the administrator, the industrial hygienist, and the building engineer) may already be staff members if the new facility is to serve an already-existing venture. If the organization does not have full-time architects or engineers on its staff, the laboratory administrator must choose them from architectural and engineering firms. These architects and engineers then become part of the planning team.

The firms of the architects and engineers on the planning team are asked to provide several types of services prior to and independently from the design of a building but all geared towards the development of the projected building program of requirements. The services to be provided will be selected by the planning team from the ten tasks necessary for planning and programming the project. It is essential that the firm or firms selected have both architectural and engineering capabilities with regard to laboratories. Experience shows that engineering capability—mechanical as well as electrical—in laboratory planning and design is of paramount importance. The architectural/engineering (A/E) team acts on behalf of the administrator to undertake selected tasks from the tasks necessary for the planning and programming of the project. It provides pertinent facts such as the project budget, the approximate project size, a time schedule for the assignment, and a description of the tasks expected as part of the assignment.

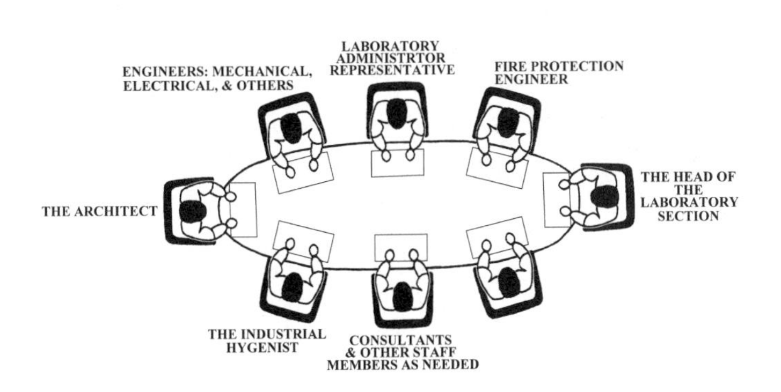

FIGURE 1-1
The Planning Team

Planning Team Tasks

1. To determine or confirm the need for a new laboratory or for the extension of an existing one.
2. To prepare a laboratory operation plan.
3. To participate in the selection of the consultants and the architect.
4. To prepare the program of requirements for the projected facility.
5. To consider the options and to prepare a cost estimate for each of them.
6. To undertake costs/benefits studies.
5. To review the consultant's and the architect's work.
6. To recommend the project type that best responds to the needs.
7. To coordinate the operation and staff requirements with the consultant's and the architect's outputs.
8. To assign space for different tasks and operations.

Planning Team Tasks

Figure 1-1 shows the members of the planning team and provides an outline of the tasks undertaken by this team. These tasks are:

1. *To determine or confirm the need for a new laboratory or for the extension of an existing one.* This requires the collection and analysis of new or additional data and information. This analysis should examine laboratory instrumentation and its effect on laboratory planning. It should also look at existing regional systems and consider the possibility of integrating into one of them.
2. *To prepare a laboratory operation plan.* Once the need for a laboratory project is recognized, the team determines the functions and tasks that will be performed and develop operational concepts for each task and function. The sum of these concepts, once coordinated, constitutes the operation plan. With assistance from planning team members, the users of the facility develop these concepts and plans. Once completed, the concepts and plans are provided to the planning team, which uses them to allocate space and determine space relationships.
3. *To participate in the selection of the consultants and the architect.* The planning team determines the tasks for which consultants are needed. These tasks should be categorized and explained, and the kind of expected outputs should be identified. The team also determines the criteria for selecting a consultant and the means to identify qualified firms. It analyzes consultant submissions, interviews considered firms, checks references and samples of work, discusses schedule and fees, and provides findings and recommendations to the selecting body.
4. *To prepare the program of requirements for the projected facility.* This program should include:
   - The needed net and gross space, subdivided in groups according to type of use and levels of fire hazard and health risk. This determination is made with the assistance of an industrial hygienist and a fire protection engineer familiar with the type of operations considered.
   - A description of the tests and operations to be undertaken in each space, laboratory, or room; of the frequency of these tests or operations; of the chemicals or other matter used; of their quantities and level of concentration; and of the requirements related to their use, storage, or handling.

- Subdivision of spaces into laboratories and rooms and description of the requirements pertinent to each room.
- A determination, by an industrial hygienist, of the levels of hazard and risk associated with operations and chemicals, and recommendations on how to respond to them.
- A statement and explanation of the criteria and the standards (architectural, mechanical, and electrical) to which the projected facility should respond.

5. *To consider the options and to prepare a cost estimate for each of them.* The amount of funds available and the type of project considered will point to possible options for the laboratory project. An accurate cost estimate of each option is essential to investigate, analyze, and prioritize. If the option considered is renovation, it is necessary first to investigate the existing conditions. This enables the identification of necessary modifications and allows for their cost estimate. In all cases, the estimate should include not only the construction costs but also all other costs pertaining to the considered options.
6. *To undertake costs/benefits studies.* These studies analyze the options to determine the costs and benefits of direct building as compared to:
   - The costs and benefits of leasing an existing facility, modified to respond to the stated needs.
   - The costs and benefits of leasing a facility built especially to respond to the stated needs and leased for a determined period of time.
   - The costs and benefits of leasing a facility—either existing and modified to suit requirements or built especially to respond to these requirements—for a given period of time during which the tenant may have a purchasing option at a predetermined price or at a predetermined method of pricing.
7. *To review the consultant's and the architect's work.* The coordinating role of the team and the nature of its activities make it the body that will generate issues requiring study by outside consultants. Review of the consultant's work by the team, therefore, assures that the information provided is in line with the contract intent and is in usable form. The architect, when hired either as a consultant or to develop preliminary drawings, construction drawings, specifications, and other contract documents, also has his work reviewed by the planning team. This assures that proposed arrangements and materials respond to program needs, entity policies, and requirements.
8. *To recommend the project type that best responds to the needs.* The planning team is the body that, by the nature of the tasks required, acquires a comprehensive view of the situation. Its awareness of the laboratory operations, requirements, options available, costs, and all types of advantages and disadvantages makes it the body most able to prioritize the options and back its recommendations by facts and data.
9. *To coordinate the operation and staff requirements with the consultant's and the architect's outputs.* Many of the planning team tasks are undertaken to identify and determine operation and staff requirements and to translate them into information useful to acquire a facility. The consultant's and architect's roles are to assist the team in their area of expertise. The planning team sets the scope of work and program of requirements and identifies the necessary outputs. It should, therefore, throughout the process review and coordinate as needed to assure that the right information is channeled to the consultant and that the consultant is addressing the right issues and providing the expected outputs.

10. *To assign space for different tasks and operations.* The laboratory mission and the operation plan should be used to determine the space requirements for the different tasks and operations. The planning team decides and proposes what group of tasks can or should be undertaken in one space or what number of or combination of spaces can or should be used for a given task. This decision and proposal results from the analysis of the tasks' requirements and of several other factors. It should be the role of the planning team to assign space and coordinate its use among its different users.

These ten tasks do not have to be undertaken separately. Rather, they can be intermingled as needed to obtain the required results. This can be done in various ways. The following procedures will help you to identify the factors that must be considered and develop a method to analyze them.

Procedures to Be Undertaken by the Planning Team

In performing the ten tasks just described, the planning team must follow certain procedures:

I. *Procedure for determination or confirmation of needs and requirements.*
II. *Procedure for analyzing the economic factors.*
III. *Procedure for selection of consultants or of an architect.*

A detailed description of these procedures follows.

I. PROCEDURE FOR DETERMINATION OR CONFIRMATION OF NEEDS AND REQUIREMENTS

Several different issues must be considered for this procedure:

1. *Identification of tasks and operations.* One of the first tasks undertaken by the planning team is to analyze all facts and data that confirm or deny the need for the facility as initially conceived. As part of this analysis, the team may propose alternate options and, through the administrator, present its findings and recommendations to the decision makers.

   The planning team should analyze the operations and tasks for which a new facility is sought. In order to make the right recommendations, team members must have extensive knowledge of the subject field, of the available resources, and of the needs of the population to be served. The team should analyze all relevant facts and identify all tests and programs to be conducted in the projected facility. From there, the team identifies the tasks to be performed in each specialized part of the laboratory and the volume of work these tasks represent.

   This may be a good time to evaluate the entire operation, including assessing or reassessing the population to be served and goals to be achieved and the effects of the operation on both of them. This process can also be used to identify and analyze changes in population numbers, characteristics, and needs. It should look at all pertinent facts and predict any changes that could affect them. Due regard should be given to technological advances or other developments in laboratory operations and processes.

New equipment and instrumentation should be studied according to their cost, functions, and the level of skill required for their use. The cost-effectiveness and effect of their use in the projected operation should also be considered. This type of analysis, if thoroughly undertaken, sometimes leads to a complete reorganization of the existing operations. It negates or confirms the need for new laboratory space and clearly defines the programs and activities for which the new facility is needed.

2. *Laboratory operation plan.* Once the need for a new facility is confirmed and the tasks and operations that will take place in it are determined, the team member in charge of the laboratory section decides how these tasks will be undertaken and in which laboratory space or room each task should be performed. He or she details the procedures, determines paths for each task, decides the approximate volume of tests, the methodology to be used, and the chemicals and instrumentation required. He or she also categorizes chemicals and operations according to hazard levels and determines paths for chemicals, gases, and personnel. This constitutes the laboratory operation plan, which is a detailed record of how the laboratory is operated. It describes each tasks and operation undertaken in the laboratory and the circulation, processing, handling, and manipulation of chemicals and other matters in the different spaces, from reception to dispatching. It identifies inputs and outputs, chemicals used and their level of hazard, individual laboratory space specifications, ventilation requirements, and other needs for each operation. This plan will help the architect determine the kind of space needed for each task and how this space will relate to the rest of the building. Figure 1-2 is a schematic showing the types of paths in an operation plan.

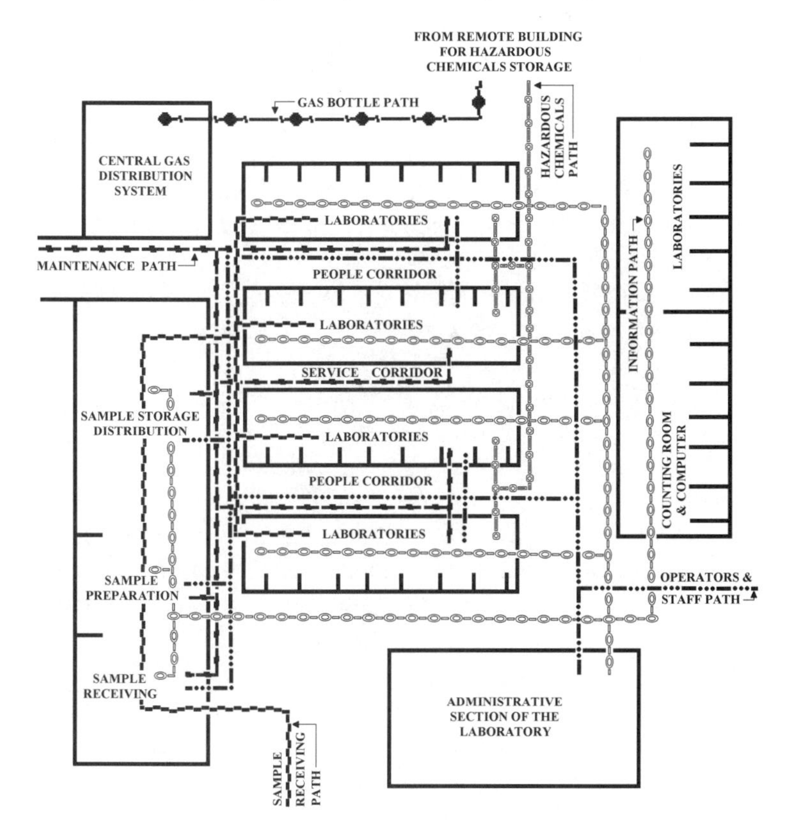

FIGURE 1-2
Schematic showing types of paths in an operation plan.

The planning team must also assess the technical and other staff in relation to the projected tasks. If the staff is already existing does it need to be supplemented for the projected tasks? Do they have the required skills? Is there a need to hire a new work force for the contemplated venture? In any of these cases, the team should look at the availability of staff and training in the area. The team can then determine the size and formation of the staff and the space they will need to conduct their activities.

3. *Laboratory instruments.* The importance of laboratory instruments must be singled out and dealt with separately in this context. In the last decade, laboratory instruments have undergone an unmatched revolution. They can now be used to undertake thousands of tasks with extraordinary accuracy in only a fraction of the time once required. As their efficiency increased, so did their cost, which jumped from a few hundred dollars to several hundred thousand dollars. This has not reduced the appetite for their use. Today, in many laboratories all over the country, space conceived for wet chemistry or other types of operations is being cluttered with instrumentation. Benchtop space and passages are crowded and, ironically, often laboratory operators often find themselves competing with the instruments for space.

    Since the use of instrumentation is expected to increase, it must be accounted for in any planning process or projection of activities. Thus, prior to deciding on the acquisition of new instruments and their inclusion in either new or existing facilities, the consequences of such acquisition should be completely evaluated. All facts related to instrument acquisition should be included in the evaluation and all costs and benefits should be considered. This should include the lead time prior to the delivery of the instruments, the installation time, the impact of the instruments on the facility (what modifications they will require), the extent and costs of the modifications and the time required to undertake them, and, finally, the effect of all these factors on the laboratory operations.

    Such evaluation usually shows that most instruments will require more space and electric power than first thought. Some instruments generate enough heat and humidity to require substantial and costly modifications of, or addition to, the facility's HVAC system. The intent of such an evaluation should not be to discourage the use of new instruments. It should be to help the laboratory decision makers consider all costs and benefits and efficiently plan and prepare for their use.

    The planning team must remember that as time passes, the instruments' uses will become more generalized; thus, flexibility in the design of new facilities is needed. During the planning process, the planning team should realize that many instruments work automatically and have the ability to perform large volumes of work and to undertake many simultaneous tasks in a very short period of time. Their productive capacity is immense and often commensurate with their cost. These instruments, handled efficiently and at capacity, can service entire communities. Often, however, they are underutilized and duplicated well beyond present needs.

    It should be noted that these instruments can only be efficiently handled by highly trained personnel whose services are expensive and in short supply. Training for their use is not only costly but it is also time-consuming. Generally, a well-trained individual tunes himself to the use of his or her instrument. Once assigned, he or she becomes responsible for it and almost literally teams with it in determining results from his or her readings. All these reasons—cost and requirements and effects on the facility—make the exclusive use of certain instruments justifiable only when they can be used efficiently, i.e., close to their capacity, or when the use of such instruments by more than one entity is not feasible or permitted by law.

4. *Laboratory regional systems.* The increase in cost, use, and complexity of laboratory instrumentation is matched by cost increases in laboratory construction, equipment, and organization. In order to cope with cost escalation and to increase efficiency, many laboratories located in the same region started to organize themselves and to subdivide the services rendered. Each laboratory specialized in tasks and tests, purchased the instruments it could use efficiently, and hired or trained personnel to use such instruments.

   This approach allowed a substantial increase in the quality of service and decrease in the cost of operations. When analyzing the laboratory's operations and determining the tasks to be performed, the planning team must seriously consider this type of cooperation between laboratories. This option often results in a better distribution of work and has favorable impact on the size and cost of new laboratory space, as well as on the quality of the services rendered.

   If the projected facility is governmental, the planning team should search for existing governmental systems, either regional or national, with which to integrate the new facility. They may be within the same agency or in other agencies. This should not preclude the possibility of integration with nongovernmental systems, perhaps by contract, if the nature and specifics of the tasks allow it, if it makes sense economically, and if it does not create a conflict of interest.

II. PROCEDURE FOR ANALYZING ECONOMIC FACTORS

Before the planning team starts its work, the laboratory decision makers should review and analyze all related economic factors. How would the procurement of this facility be financed? How much money is available? Are these funds sufficient? If the projected laboratory is a commercial venture, would a financial institution be willing to finance the venture? To what extent and for how long? What would their conditions be? How much could be procured with the total available funds?

If the projected laboratory is a governmental or semigovernmental venture, have all the proper authorities, legislative and administrative, approved the project and made funds available for it? How much could be procured with the available or suggested funds? These data, financial and other, are extremely important. They will be used by the planning team to determine whether leasing or owning is most convenient or efficient. The planning team will also have to know the span of time for which the new laboratory is needed: Is it to lodge a short-term (two- to five-year) program or is it for a permanent, long-term set of activities? Is the new laboratory space needed as an addition or an extension of an existing facility? Or as a whole new independent laboratory building? If it is an addition to an existing facility, is this facility owned or leased? What are the conditions of ownership or lease?

Once this information is obtained, the planning team should analyze it to determine the best response to the needs. A life-cycle cost/benefit analysis of needs and resources, as well as goals to be achieved, may be the most logical if not the best way to respond to the needs. The following chapter suggests how to approach such an analysis. The choices at the conclusion of this analysis can be summarized as follows:

1. Look for one or more facilities or spaces already available for lease or sale. Determine their suitability (regarding their location, orientation, and characteristics). Identify and estimate the cost of modification that may be required, including overhead and design costs, and project their operation and maintenance costs. Then, if the facility is for sale, add the sale price and associated costs; if it is for lease, calculate the rent and other abnormal costs, such as higher HVAC costs due to inefficient systems. Sum up the costs for each option: for the duration of the lease (if leased) or for the expected life of the building (if built or owned). With this information, determine total costs and benefits in present value for each option.
2. Look for one or more available pieces of land and determine their suitability (regarding zoning, location, and characteristics). Identify each site's purchase or lease costs and estimate its utilities-extension and improvement costs. Also estimate the engineering, architectural, legal, overhead, and other costs directly related to the facility acquisition, design, and construction. Add all these costs. Determine the total cost estimate for each site and identify all the advantages and disadvantages (economic and otherwise) of locating the facility in each of them. Translate monetary costs and benefits in present value. Note the noneconomic advantages and disadvantages that will be considered in the final determination. (A noneconomic advantage would be that locating the laboratory in a depressed area could help uplift the area by generating business and providing educational and employment opportunities to the communities. A noneconomic disadvantage would be the resistance of a community to having a laboratory built.) The cost analyses to determine present value are described in detail in the next chapter.
3. Invite interested parties to offer proposals to build and lease a new facility or to modify and lease an existing one. (This manner of acquiring the use of a facility for a long period of time, generally 10 to 30 years, is know as a build-lease contract.) The proposal should address the cost of leasing, operating, and maintaining the facility. The proposal may include two sets of costs, one for a determined period of time at the end of which the premises will be evacuated, and the other for the option of purchasing the premises during or at the end of the contract period. The determination of the best course of action will result from feasibility studies for each serious option. If those options are existing buildings, the planning team will establish baseline information about the buildings' physical conditions and other factors (such as facility of payment, willingness of seller or landlord to pay up front for improvements, or to provide services over and above those required) for each of the options. This information is then utilized to develop a life-cycle analysis for each option.

The comparative analysis of these life cycles will permit an economical prioritization and point out the best course of action. There are advantages and disadvantages to each of the options of buying or leasing.

- The purchasing of an existing facility or the purchasing of land and the building of a new facility requires a major up-front capital outlay. The buyer may have to use funds from its own capital or secure financing at a cost. In this case, the facility, land, and improvements remain the property of the entity that acquired it.
- If the entity decides to lease either an existing facility or a build-lease facility, the up-front capital outlay is the landlord's responsibility, although a sharing of responsibility may be negotiated. The tenant may pay part of the required up-front capital, thus reducing the amount it will have to pay in rent. In the case of leasing the land, the facility and the improvements remain the property of the landlord and, at the end of the contract, the tenant leaves the premises owning nothing. This arrangement may be unsuitable for certain entities but very advantageous to others (due to short-term contracts, a spend-as-you-earn program, tax advantages, etc.).
- The entity may be able to lease an existing facility or a build-lease facility with the option of purchasing it during or at the end of the contract period. This type of negotiated arrangement takes into account the needs and resources of the contracting parties and leads to a contract that is mutually convenient to the landlord and the tenant. The outcome of the feasibility study will influence the course of action taken in the acquisition of the facility. How best to meet the needs with the available resources will be evaluated. The procedures related to the feasibility studies and the life-cycle analysis are addressed in the next chapter.

III. PROCEDURE FOR SELECTION OF CONSULTANTS OR OF AN ARCHITECT

Investigation of the most qualified consulting and architectural firms interested in such projects should make it possible to select the right firm for the right task. Professional organizations keep information on their members' fields of expertise and provide lists of qualified firms. Among these organizations are the American Institute of Architects, the National Society of Professional Engineers, the American Chemical Association, and the American Institute of Chemists. It may also be helpful to ask these organizations to point out the names of firms known for their experience in the field appropriate to the laboratory. Once a list of qualified firms has been obtained and analyzed, the administrator appoints a selection panel. This panel should include planning team members and should meet with representatives from each firm, including the technical personnel who will be involved in the work. Visiting the offices of these firms is recommended, as it will give the panel first-hand knowledge of how the firm is organized and of the working atmosphere of its staff.

The panel should determine the level of knowledge, interest, and experience as well as the flexibility of each firm. This panel should also request references and contact the referees or see the referenced work and talk with the users. The panel members then analyze the findings, discuss them with the firm's representatives and technical personnel, and talk to the referees until they become fully satisfied with their findings. The panel may also at this time request estimates or fees (and their explanation) from each firm and subsequently discuss them.

Once the choices are narrowed down to the two or three firms that respond best to the stated criteria, each of the firms should receive a detailed description of what services would be expected from them. These firms should, in turn, submit proposals describing how they will render these services. Eventually, by a process of analysis and negotiation, one or more firms will be selected to perform the tasks required. If their number and nature allows it, the tasks can be performed independently and serially. This means they may not have to be performed by only one firm, which, in certain circumstances, is advantageous to the project.

The A/E firm or firms that will be involved in the different phases of the project are selected in the same manner. These firms may be involved in the preparation of the program of requirements and may be asked to undertake the feasibility studies, master plans, and the prioritization and selection of the type of facility to be procured. After the program of requirements is finalized and the laboratory site is selected, the major planning activities are completed, and the team focuses on the design and cost estimates of the facility.

The architect orchestrates this phase of the process. Several team members and others (e.g., consultants or members of the laboratory staff who have certain expertise or requirements) should support the architect by reacting to his or her recommendations until agreement is reached on the most efficient design for the laboratory. How the new facility is acquired will determine the architect and engineer's tasks and final product. If the decision is to build a facility or to add to an existing facility, the tasks will be to:

- Prepare designs and estimate costs for the facility, for site improvement, and for utilities extensions.
- Prepare contract and construction documents and orchestrate the bidding process.
- Observe the construction process and administer the technical aspects of the contract.

As stated earlier, other options may be chosen. One option is to lease an existing facility and have it modified by the lessor. Another option is to acquire a build-lease facility, built to respond to required specifications or to performance standards. The architect's and engineer's tasks in these cases are similar to those described above, except that they are performed for and at the request of the lessor, who interprets and responds to the tenant's requests. In this case, the tenant's architect and engineer must describe the required specifications and performance standards in such a manner as to assure that the resulting design, space, equipment, and systems will efficiently respond to their requirements. They will also have to assure that all costs—initial, operational, as well as those of maintenance—are considered technically and contractually.

The process that should be followed to design a laboratory is different from those used to design conventional buildings. It requires a complex relationship and interaction between the architects and the engineers. Recommendations for and description of the interaction are discussed in chapter 8. Ultimately, the architect and the engineers will have to consider and incorporate many complicated and sometimes conflicting architectural, mechanical, and electrical factors that result from programmatic needs. Because of the nature of these factors and the need to continuously define and clarify requirements, frequent reviews of the architect's work by the planning and design team are needed.

Chapter 2

Laboratory Space: Renovated Versus New, Owned Versus Leased

LABORATORY owners in need of new space are often faced with the problem of determining what is the best way to acquire such space. Should they build a new building or renovate an existing building? Should they purchase a new space or lease an existing one?

In the last two decades, the renovation began competing with the "wreck and build anew" concept, which had been predominant. Several factors contributed to this change of approach. The most important one is that in many cases it became more economical to renovate than to wreck and build anew. A substantial rise in the cost of energy increased the cost of building materials far more than the cost of labor and substantially increased the cost of transportation. This, plus the fact that renovation often is less time-consuming and that work can be performed in adjacent spaces or building's without major disruption of ongoing activities, rendered renovation very attractive.

Can renovation be an option for acquisition of laboratory space? Yes. Is it always the best option? This chapter looks at different options and proposes a method of analysis and prioritization to help make this kind of determination.

Narrowing the Options

The following options are open to laboratory owners or management in need of new or additional space:

1. Lease an existing building and use it as is.
2. Lease an existing building and modify it to suit needs.
3. Lease a new building built to suit needs.

4. Purchase a piece of land and build a new building on it to suit needs.
5. Build a new building on a piece of land already owned.
6. Add to or modify a building already owned.
7. Purchase a building and then add to it or modify it to suit needs.

These options may need to be combined, as they generally aren't as straightforward as the list suggests. A lease option may be offered with the possibility of purchase or with very convenient financial arrangements. A building or a lot for sale may offer advantages not necessarily related to its condition or location.

Although all of these options may be available, certain factors may eliminate the viability of some and narrow the choice. If the laboratory owner owns a piece of land in a convenient location, he or she may not want to consider the options of leasing or adding to a purchased or owned building in a less convenient area. If, on the other hand, the laboratory owner does not own a site, does not want to face a large initial capital outlay, or needs only short-term laboratory space, the option to lease may be quite attractive.

With this in mind, the laboratory owner or management's first task should be to reduce these options to those that are indeed feasible. In order to make this decision, the laboratory management should answer the following questions:

- Why is new space required?
- How soon and for how long is the new space needed?
- How much funding is available? How much money is already lined up?
- What is the projection of income for the project's life cycle?

These and many other similar questions must be answered by the members of the board of directors, the laboratory management, and the planning team in order to decide which options to consider and further analyze. These members should analyze the laboratory's present situation and projections; they should consult with their accountants, attorneys, and bankers; and then they should determine their expansion direction.

Let's assume, for example, that the laboratory's operations need to be lodged in a type of space not readily available, that they do not own a piece of land, and that they will need the space for about thirty years. Let's also assume that their income shows that they have, or will be able to get, sufficient funds for the down payment and financing of the required building, should they choose to purchase or build, and a steady income should they choose to lease. The options then would be narrowed to numbers 2, 3, 4, and 7, stated at the beginning of this chapter. These four options can be summarized and restated as follows:

- To lease a building (new or modified to suit the laboratory's needs).
- To purchase an existing building or build a new one on a purchased piece of land.

These two questions must be carefully evaluated and answered. The members of the board of directors and the laboratory management should carefully look at their needs, circumstances, and resources and, based on their findings and judgment, decide to:

- Use a new laboratory building (which will have to be built); or

- Use an existing laboratory building owned by them and modify or extend it to meet the new needs.

In order to simplify the responses to these sets of choices, the site evaluation that appears later in this chapter will first address the first two choices. It will then use the same data to address the second two choices.

Preliminary Gathering of Data

While the board of directors and the top management are considering the issues of options, direction, and projection, the laboratory technical team should be determining particular space and building needs and the requirements for each type of space. Derived from the tasks, functions, and operations that are or will be conducted in the facility, these needs and requirements are generally described in the laboratory operation plan. Because of its importance, a laboratory operation plan for the operations needing a new facility should be developed. If the new building needed is an extension to an existing facility (which will be built to accommodate new or extended functions), the laboratory operation plan should encompass all activities—those presently undertaken as well as those projected. The laboratory operation plan was described in detail in chapter 1. If the facility already has an operation plan and this plan is efficient and satisfactory, it should be used to identify space, equipment, and system needs for the projected facility. If, on the other hand, the facility is operating without a recorded operation plan, it is essential to prepare one.

Laboratory space requirements should be derived from the operations and tasks described in the operation plan. Each operation or task may require hood space and floor and bench space for equipment and experimentation. The nature, quantities, and concentration of the chemicals used will affect whether operations and tasks are undertaken in separate rooms or take place in one laboratory room. The hazard and risk levels associated with the chemicals will determine the fume hood requirements and ventilation needs of each room.

Using the analysis described in chapter 1, the laboratory planning team determines the instruments' needs in bench or floor space and in utilities and power. This team also determines how much heat and humidity is produced by these instruments and how it will affect the ventilation of the room and possibly of the entire facility. The laboratory planning team also identifies, among other factors, the chemicals involved, their physical circulation from reception to dispatching, and their processing from concentrate to dilution, and then determines the grouping and location of the laboratory rooms in which the tasks are conducted.

The following information should be derived or extracted from the laboratory operation plan, as it will be needed in the analysis:

- Space requirements for the floor and fume hood
- Total linear feet (meters) of casework required
- The amount and types of chemicals that will be used in the operations conducted in each of the laboratory rooms

- The hazard and risk levels of the operations that will be conducted in each laboratory room
- The instruments and equipment needed in each laboratory room
- The utility and service requirements for the instruments and equipment needed
- The effect of these instruments and equipment on the HVAC system

This and other pertinent information should be expressed in the program of requirements, the development of which was described in chapter 1. For this evaluation, the program of requirements must state information on the laboratory and its operations in the following manner:

- Space requirements (in square feet or meters), including:
  Laboratory space, identified by class
  Laboratory support space (shops, electronic storage, etc.)
  Office space and office support space
  Laboratory related offices
  Space used for storage of hazardous chemicals
  Other types of required spaces
- General architectural, structural, mechanical, and electrical requirements for each type of space. These requirements should adhere to the principles of sustainable design, which will render the building as close as possible to a green building.
- Specific types of space and other specific requirements for given areas, rooms, or group of rooms
- Flow charts summarizing operations, people, and processes that dictate circulation and require grouping of rooms

This information should be explained and detailed as needed.

Once the decision is made to go ahead with the new space—after the procurement options have been defined and the program of requirements completed—attention should be directed to the identification of sites and buildings available for sale and to the hiring of an A/E firm that will undertake all required investigations and evaluate the proposed facilities. The laboratory management should contact a real estate company and ask it to find available lots and/or buildings that may be used for the new laboratory. The laboratory management should also inform the real estate company of the facility's requirements and conditions.

The A/E firm should be selected in the manner described in chapter 1. It must have extensive experience in laboratory design and strong engineering and architectural capabilities. It should be capable of performing building and building-component analyses as well as life-cycle cost analyses. The conception of a laboratory can be described as the result of intermeshed efforts between the project architect and the project mechanical engineer. The interaction between the architect and the engineer is discussed in chapter 8. The location choice, plan, volumes, and exterior facade of the building derive from the dual design efforts responding to the program of requirements.

With an existing building, the process is somewhat reversed—the building is already located on a site, its plan is outlined, and its volumes, masses, and exterior facade are pretty much established. The A/E in this case determines how the design can be accommodated or adapted, in a cost-effective manner, to the existing situation. The A/E conceptually supplements or modifies the existing facility, as necessary, in its attempt to satisfy the project's space and other requirements.

The Six-Phase Site Evaluation Process

Once available sites and buildings are identified and the services of an A/E firm are secured, the main evaluation process begins. This process consists of six phases and can be summarized as:

Phase I. Initial site selection. The laboratory management and board of directors, perhaps assisted by their attorney, banker, and architect, select properties from the options provided by the real estate agent.

Phase II. A/E preliminary evaluation. The A/E is then asked by the laboratory management and board of directors to undertake its own preliminary screening of the sites and buildings selected in phase I, sort them, and identify the most promising options, which are then further analyzed.

Phase III. A/E feasibility study. In this phase, the A/E evaluates each promising option, determines the extent of alterations and modifications needed in each of them, estimates their costs, and determines to what degree the remodeled building will respond to the project requirements. The A/E should then sum up all costs by categories, extend them for the building's life cycle, and convert them to present value,* thereby enabling a comparison of the different options.

Phase IV. New facility concept-feasibility study. Simultaneously with the evaluation of the selected options, the A/E should develop a concept for a new facility that accommodates all the project requirements. This facility should be conceptualized within each of the selected sites. For each situation, the A/E should estimate all costs—initial construction costs as well as operation and maintenance costs—sum them up by categories, extend them for the building's life cycle, and translate them into present value. The A/E can then use this information to compare the new construction options to those requiring alterations.

Phase V. Life-cycle comparative analysis for owned options. In this phase, the A/E uses the data developed in phases III and IV to develop a life-cycle comparative analysis.

Phase VI. Owned versus leased analysis. In this phase, the advantages and disadvantages of ownership versus lease are compared. Also addressed are the advantages and disadvantages of each method of acquisition/procurement.

* Present value or present worth refers to the value of a future payment or series of payments discounted to the current value or to "time zero" of a given project. Undertake a comparative present-value analysis of all the options considered. These options are limited to the new or remodeled buildings that are owned by the laboratory.

The following is a detailed description of the tasks required by this six part or phase process.

PHASE I. INITIAL SITE SELECTION

The laboratory management and board of directors should undertake the first evaluation of all proposed properties. They should determine who owns each property, in what jurisdiction it is located, what the jurisdiction's land and improvement tax base is (whether it is large or small), what percentage of market value is used for appraisal; what the legal requirements related to the property are, whether the zoning and building-code regulations allow the projected use for the property, and whether there is a restrictive covenant.

The laboratory management and board of directors should also look at how the acquisition and alteration work will be financed, and what the cost of financing will be. All this information, once collected, should be given to the A/E, since it will use most of it in its analysis.

The initial prioritization of the sites under consideration should be subjectively performed by the laboratory management and by selected staff members, as needed, by the laboratory's accountants, attorneys, bankers, and even by the A/E. The management and staff should familiarize themselves with each option's facts and to identify their preferences by ranking each options according to its feasibility. Each site available or offered will therefore undergo a management and midmanagement screening that considers its convenience of location, the adequacy of its size, and the availability of main utilities. Existing buildings offered for sale have to be screened to determine whether sufficient space is available, whether they offer the possibility of expansion, and whether the present use would negate projected use as a laboratory.

PHASE II. THE A/E PRELIMINARY EVALUATION

The A/E then conducts an overall preliminary evaluation of each of the properties proposed: land alone and land with building(s). It reviews the management's findings on all properties to confirm the availability of main utilities and services and that the size of the properties will accommodate the laboratory requirements.

For properties encompassing already-built buildings, the buildings should be surveyed to determine if they could efficiently be altered and/or be incorporated into the required laboratory. The A/E should obtain and study a set of "as-built drawings" for these structures. It should also look at the age and general conditions of the building and identify the zoning and building-code restrictions that apply to each structure.

Once the A/E becomes familiar with the as-built drawings and the code requirements and restrictions, it should conduct a preliminary field survey to investigate the existing conditions and test the HVAC, plumbing, and electrical systems and their suitability, condition, and existing capabilities. The as-built drawings are of vital importance for the serious study of an option. Thus, the A/E should request them for each building under consideration and update them when they survey the site. If the as-built drawings are unavailable and the

A/E feels that having them is essential, the A/E should let the laboratory management know that it is willing to undertake such work for a cost.

The overall preliminary survey of each structure must be undertaken systematically. This will ensure that the information on existing conditions is available and accurate. This approach also quickly exposes any major factors that might disqualify an option as grossly inadequate. The intent of this survey, therefore, is to verify the suitability of the option and the accuracy of the information included in the as-built drawings. Through this survey the A/E pinpoints any information that may be needed but is not included. The A/E then identifies ways to acquire that information and to check the actual condition of features that may be of key importance.

In undertaking this survey, the A/E may have to dig alongside foundations to determine their size and depth, remove finishes in some areas to check the size and condition of structural members, and look at pipes, wiring, and other hidden materials and systems. The A/E may have to cut steel pipes to determine the level of rust and need for replacement. The A/E must inspect the size of and loads carried by the electric wiring and circuitry, the size of the air systems' components, and the loads, location, and size of the duct dampers. Once all this information is collected and recorded, the systems should be tested to determine their working condition and real capability.

The plumbing system should be overpressurized in order to determine its suitability for carrying the design loads. The electrical circuitry should be checked to determine its load capacity. The HVAC system should be activated, tested, and studied to determine if it is capable of meeting the loads required by the new laboratory.

The A/E must also investigate the age of the building and who was in charge of its maintenance throughout the years. The age of the building indicates the kind and quality of its materials and the structural methods used to build it. This information, as well as that regarding the quality of maintenance the building received, will help establish whether the building has special concerns, such as the presence of asbestos or aluminum wiring, or is structurally sound and able to be renovated. It will also determine its life expectancy. The building materials, and the methods used to put them together, should be examined to determine their life expectancy, their structural qualities, their present condition, and what it will take to correct any damage that may have occurred. The structural methods used in constructing the building also need to be evaluated to determine the present structure condition, the building's structural qualities, and the structural capability to respond to changes in loads or spans as may be required.

The A/E firm must verify with the zoning authorities that the building being considered is in a zone authorizing its use as a laboratory and that the modifications contemplated could be implemented without infringing zoning restrictions. If the zoning classification does not permit the proposed laboratory, the A/E should find out the procedure for zoning reclassification or for obtaining a zoning variance, which is a one-case waiver of the existing zoning classification. The A/E should also investigate with the zoning officials the possibility of obtaining a reclassification or a variance and assess—from its previous experiences—the chances of obtaining such reclassification or variance.

Existing buildings may have code violations not detected because of code changes after construction. To bring such buildings into total conformance with codes, when renovating them, may be either physically impossible or very costly. When this is the case, the A/E needs to discuss the existing code violations with the building officials. The A/E should identify ways to conform with the code intent, to upgrade deficiencies, and to make the building safe and secure. It should then attempt to get approval for the proposed solutions and identify what procedures must be followed to obtain a waiver or to appeal decisions.

Construction work can be very disruptive to unrelated adjacent activities. If the space considered for renovation is an extension of an operating laboratory, the construction work would definitely disrupt the laboratory routine and experiments. In order to plan for reduction and control of this disruption, the A/E firm must investigate the nature of these experiments and what their requirements are. The A/E must be told which experiments are long-term and which are routine or daily tasks, as well as which experiments and tasks would be affected by noise, dust, vibration, and temporary or extended breaks in utilities and services.

There is yet another form of disruption that may take place. Often renovation calls for changing the room arrangements of the existing laboratory space and the relocation of casework, equipment, and instruments. These renovations may require extensive interruptions of services and utilities, not only in areas affected but also in other areas using the same lines of services and utilities. Naturally, when all these activities are taking place, the personnel working in the areas affected must interrupt their work and vacate the premises until they are again ready for occupancy. The A/E must understand and discuss the above issues and their implications and evaluate their impact with the laboratory management. Also, when necessary, the A/E should develop a tentative construction schedule for certain sites, as well as procedures for planning and coordinating moves, interruptions, and other precautions if these factors become important in the evaluation of an option.

The A/E should summarize all the findings for each option being considered and give a short list of the most promising options to the laboratory management. These would undergo the feasibility study described in the next part. This list preferably should have at least one of each type of option (existing and new structure).

PHASE III. A/E FEASIBILITY STUDY

In the feasibility study of each of the short-list options the A/E evaluates five major areas. It determines and describes the existing conditions as well as the extent of work required to modify the condition to respond to needs. It then estimates the costs for such modifications.

The five areas of evaluation are:

1. Site and location evaluation
2. Utilities and services evaluation
3. Building condition evaluation and assessment of possibilities
4. Mechanical and HVAC conditions evaluation and assessment of possibilities
5. Electrical system inspection, evaluation, and assessment

The evaluation of these areas is conducted in the following manner.

1. Site and location evaluation. The site on which a laboratory is or will be located is important and its impact on the performance of functions should be evaluated. The A/E should analyze each site's characteristics and location, the lot size and accesses, the utilities and services available in or around it, and the immediate surrounding uses. The A/E should also evaluate any other factors that would be an asset or a liability and that may affect the laboratory's efficient functioning or its initial and operational costs.

A laboratory also generates traffic, fumes, and other wastes and includes operations that, even if not hazardous, may be bothersome or threatening to environments not tolerant of this type of laboratory byproduct. The A/E should evaluate the nuisances and threats that will be produced by the laboratory and determine their immediate and long-term effects on the laboratory. The A/E then subdivides the nuisances and threats into two groups: One group includes those that cannot be changed and must be accepted (with their favorable or unfavorable impact), assuming the option is not rejected on the grounds of these nuisances and threats. The other group includes the nuisances and threats that, if unfavorable to the feasibility of the project, could be changed at a cost. For this group, the extent of the cost will be a factor in the feasibility of the option.

The A/E then determines if the fumes and noises generated by the neighbors' activities would affect the laboratory operations or experiments. Will the location of the lot, with respect to its neighbors and the predominant winds, allow or prevent cross-contamination between the laboratory and its neighbors?

In evaluating the characteristics and location of a site, the A/E should consider all the factors that may affect the conduct of the laboratory operation. Is the lot big enough to accommodate the new extension as well as the additional required driveways and loading areas? Is there enough space for parking lots to provide the required number of parking spaces for personnel, suppliers, and visitors? Would it be necessary to allow for further future expansion and if so, is there enough space available for that? If the lot size is not sufficient to accommodate all requirements, the option may be rejected. The A/E should also look at how neighbors might affect the laboratory accessibility.

Each of the factors mentioned in the above paragraphs will affect the design and operation of the laboratory and the initial and operational costs. Once all the factors have been evaluated, the A/E conceptualizes the site arrangement with driveways, parking areas, landscaping, and other requirements and estimates the related costs. If the location, although acceptable, has certain characteristics that would make the laboratory operation more costly, these costs should be estimated for the life cycle of the building, translated into present value, and accounted for in the evaluation.

The cost of some of the impacting factors may not always be estimated with a dollar value for the life of the project. When that is the case, the A/E must identify the advantages and disadvantages of these factors and provide this information, with an evaluation, to the laboratory management and board members. These factors should be taken into account in studying the different options.

It should be noted that if the option being evaluated is an extension of an existing laboratory that is being used in the same way the proposed laboratory

FIGURE 2-1

1. Main entrance road connecting the laboratory to the main highway
2. Office wings
3. Laboratory wings
4. Parking lot
5. Direction of predominant winds
6. Day-care center for children of this and other laboratory employees
7. Location of other laboratories (the closest is at 600 feet [180m])
8. Line along which water supply and sewage lines are located
9. Location of the electric power connection

will be used, all of the above factors may not need to be reevaluated. This, however, presumes that a previous evaluation has already established the suitability of the factors.

Figure 2-1 shows an existing laboratory building and site. The building is surrounded by a wooded area quite distant from other structures. In this location, the building does not affect and is not affected by its immediate environment. The laboratory is neither a threat nor a nuisance, as any noise or fumes dissipate prior to reaching any neighboring building. It is also appropriately oriented with respect to the predominant winds. The wooded area is in a light industrial park and is connected to a major highway. Additionally, all the utilities and services needed by such a building are readily available.

2. Utilities and services evaluation. The site also must be evaluated according to whether the utilities and services necessary to the laboratory's operation are available. How is the existing building presently disposing of its wastes? Are the wastes domestic or are they industrial, chemical, or biological? Are there any municipal sewage lines in proximity to the lot? How far away are they? How is their capacity related to the laboratory needs? Can they treat laboratory waste or would the waste have to be treated prior to being discharged? If so, what type of treatment is available? The answers to these questions will determine how the laboratory waste can be managed and disposed of as well as the required work and cost for waste disposal. Once the A/E knows the answers to these questions, it can select the waste disposal system most appropriate for the option, estimate its installation and operation costs, project these costs for the life cycle of the building, translate them into present value, and include them in the life-cycle cost analysis of the option considered.

Is the municipality supplying water to the existing building? If so what use is made of this water? Is it for domestic use only or could it be used for tasks similar to those contemplated for the new laboratory? If the building under consideration is not used as a laboratory, the A/E must determine the type of water that would be supplied. Is it hard or soft? What is its level of purity? What is its mineral and ferrous content? What is its level of conductivity? What must be done to this water to render it usable for laboratory experimentation? Does it need deionization, softening, or ferrous removal? Does it need to have its conductivity lowered? Once these questions are answered, the A/E can determine if the existing supply lines are sufficient or need to be supplemented or replaced. The A/E will also know what new system must be included in order to adjust the water's conductivity, to soften or deionize it, or to remove ferrous content.

The A/E then estimates the costs of installation and operation of all the systems needed, projects them for the building's life cycle, translates them into present value, and includes them in the life-cycle analysis of the option under study.

What source of energy is presently accommodating the existing building's energy needs? Would the same source be efficient for the new laboratory? Is there another type of energy available that may be more efficient for the life cycle of the building and that may provide energy for heating, ventilating, and air conditioning, as well as for operating the laboratory tasks? Is municipal gas available? If so, and it is not presently used, how close to the laboratory site are its lines? What would it take and cost to extend a line to the laboratory?

The answers to these questions will provide the A/E with the information needed to estimate the costs of providing the laboratory with energy as well as the cost of energy consumption for the life cycle of the building. The A/E projects the costs of energy consumption for the building's life cycle, translates all costs into present value, and uses them in the life-cycle analysis of the option under consideration.

Is the site located in an area close enough to its suppliers and to the provider of services it routinely needs? Are there bottled gas and chemicals suppliers whose location and volume of work make them capable of routine delivery within an acceptable lapse of time? Are there mechanical, electrical, and other contractors located nearby who can undertake routine maintenance and repair tasks as may be required by the laboratory operations?

These and other similar questions must be answered, as they may have long-range impact on the operations and costs of an option. If they do, all costs must be projected for the life cycle of the project, accounted for, and included in the life-cycle cost analysis. If there are inconveniences affecting laboratory efficiency that cannot be costed, they should be stated and considered as subjective factors in the comparative analysis of the options.

3. Building condition evaluation and assessment of possibilities. Each building as a structure must be evaluated according to how efficiently, safely, and economically it can respond to requirements set by the laboratory functions and the program of requirements. The A/E firm evaluates its configuration, in its three dimensions, surveys the building with the as-built drawings in hand, and records the findings. The building's method of construction is studied first to determine the flexibility of the structure. Is it a wall-bearing structure with fixed interior dividing walls used to bear loads or is it a framed building with regularly spaced bearing columns? If it is a wall-bearing structure, are the bearing walls spaced in a manner that determines the layout of rooms and corridors?

How would that affect utilities distribution and location and arrangement of equipment?

When evaluating building configuration and its factors, the A/E must have a good grasp of the most important and determining requirements, of the laboratory functions (as stated in the program of requirements) and of the flow, and their impact on use of space.

Are the room sizes and arrangement already acceptable? Does the fact that they cannot be rearranged without altering the structural system—at high cost and with substantial disruption—disqualify the structure? If the structure being evaluated is a framed building, how would the location of columns and windows affect the use of modular spacing to subdivide space into laboratory rooms?

Do factors such as the width of the structure, the distribution of space, the number of floors, the number of square feet (or meters) per floor, the configuration of each floor plan, and the location of service cores and rooms indicate that the structure is suitable for the tasks and operations that will be performed in the laboratory? Can these tasks and operations be performed in a manner responding to the Program of Requirements? Would all these factors allow the designer to adequately distribute space for different functions and let the functions and the requirements determine the interrelation of rooms? It is possible that a building configuration factor may force the rejection of the building.

Are any of the building components not in conformance with the applicable building code? How much labor and cost would be necessary to upgrade it to meet code requirements? If a building cannot meet code requirements to the letter, can it meet the spirit or intent of the code? Can such a solution be approved by the enforcing authority?

Can the building be made barrier-free accessible to the handicapped without excessive costs and efforts? If the building has more than one floor, can an elevator be added to provide vertical transportation for the handicapped? If so, what would this involve in work and cost?

Can the floor slabs and the structural bearing members (beams and columns) bear concentrated loads resulting from the relocation of partitions and other building components or from the installation of required casework and equipment? If some do and some do not, how would that affect the design and arrangement of the space? Are the building components' structural bearing members adequately fireproofed? Is the fire resistance adequate for separating partitions required by the new design? If not, can they be economically upgraded without having to be demolished and rebuilt?

Another important building factor that must be evaluated is the floor-to-floor height. Does this height allow sufficient overhead space for additional mechanical ductwork, piping, and other distribution systems that may be required? Would the existing ductwork and piping that must remain interfere with the new requirements? Once combined, would these systems and equipment, existing and new, allow sufficient floor-to-ceiling height? Would ceiling height or systems and equipment interfere with the installation of hoods and hood ductwork and bonnets?

Figure 2-2 shows a service catwalk over a set of laboratory rooms. This figure clearly illustrates how crowded the area containing the mechanical ductwork, piping, and other distribution systems is. These systems extend from the

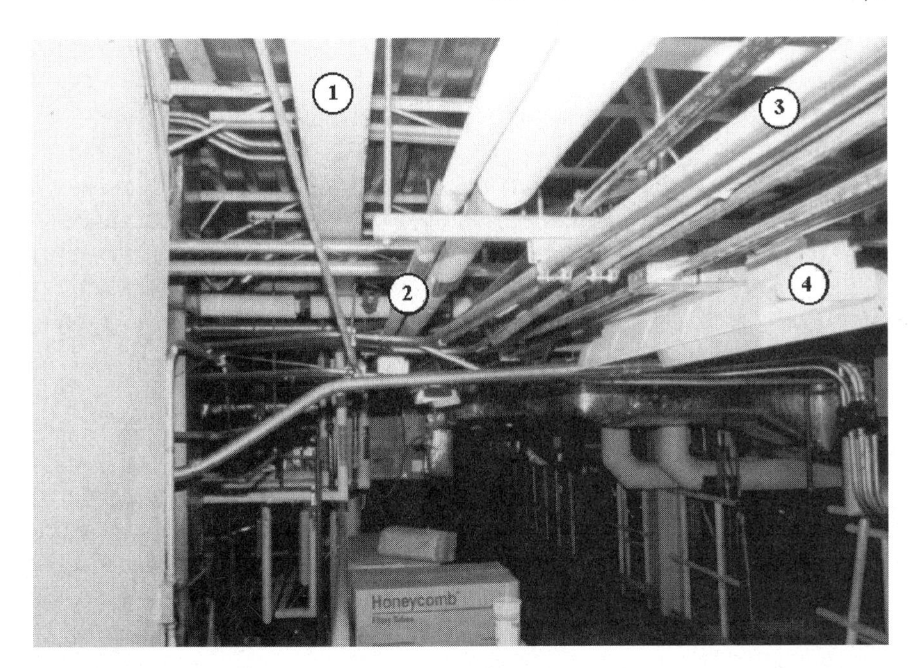

FIGURE 2-2
1. Air duct
2. Chilled water pipes
3. Conduits
4. Light fixture

mechanical room to all the rooms being serviced. They are intermingled, they require quite a bit of space, and they should be located in such a way that servicing them is possible.

Last, but not least, of the building factors that must be evaluated are the building materials and finishes. The existing floors, walls, and ceilings may be made of or finished with materials that are not acceptable for the projected uses and are difficult or impossible to substitute or cover in a reasonable manner. Certain rooms in a laboratory require special conditions and treatment that are difficult to attain in a new building and even more difficult in an existing one. Some rooms may need to be shielded for sound, radiation, vibration, or electromagnetic interference. Others may need to be "clean" rooms with special types of floors, walls, and ceilings. Some rooms need finishing that can cope with overexposure to heat, cold, corrosive materials, or moisture and humidity. One of the most demanding tasks is evaluating how the existing building materials and finishes may be used or altered to respond to the requirements of the projected uses.

A balance room lodging very sensitive balances, for example, may need to be vibration-free. This room's floor would have to be isolated from the rest of the structure and have separate foundations resting on a sand base. The mechanical and other system components such as ducts and pipes entering the room also would have to be vibration-free and isolated by vibration pads at points of contact with the walls and ceilings. The volume and velocity of air induced in this room must not force the ductwork to vibrate, and the airflow must not create turbulence as it leaves the duct. The same is true for water supply pipes. Usually a ground floor space with no basement under it, should be selected for this room. The floor slab of this room can be broken at the room perimeter and the proper flooring and subflooring installed. If existing ductwork and piping cross the room, they may have to be rerouted a certain distance from the room or have their vibration reduced to an acceptable level. Duct and piping supplying the room with air humidity and water need to be designed to respond to the required criteria. Certain floor, wall, and ceiling finishing also must be selected to avoid any possible vibrations. If the existing finishings are not compatible or easily removable, this may cause a substantial increase in cost.

This room illustrates how building configuration factors can conflict with the program of requirements. If this room must be at ground level in order to have independent foundations and needs to service laboratories that, because of the nature of their work, mandate that no floor be placed over them, it might disqualify multifloor options, as the two requirements could not be satisfied. If these requirements cannot be modified or accommodated, the building would have to be rejected.

The evaluation of the previously mentioned factors should be undertaken first to determine if the building must be rejected. Once the building passes the viability test, the A/E conceptualizes a design for floor plan arrangements and utilizes the available space to respond to the requirements. The A/E then identifies the modifications and additions that need to be undertaken to alter the building for the new use and estimates their costs. This estimate should cover the demolition and removal costs as well as the construction and modification costs.

The mechanical and electrical system evaluation should be undertaken separately. The structural and architectural impact on modifications of these systems should be considered and costed as part of the building conditions evaluation. On the other hand, if in order to retrofit a space some ductwork, wiring, or other mechanical or electrical component needs to be relocated, this relocation should be part of the mechanical or electrical evaluation.

Once the A/E estimates all initial and construction costs, it identifies what would be required in order to operate and maintain the altered building. The A/E estimates the yearly costs for operation and maintenance and projects these costs for the life cycle of the building. Once this is done, it translates into present value first the initial and construction costs and then the operation and maintenance costs. It then accounts for them in the evaluation of the different options.

4. Mechanical and HVAC condition evaluation and assessment of possibilities. The mechanical systems of a building, including all components, are very important factors in determining the viability of a building. Their capabilities and conditions and the modifications that may be needed to make them respond to the new requirements must be known in order to estimate the cost of upgrading them. It may be that the existing systems are not usable but are easily and economically replaceable. They also may be modified at a reasonable cost or at a cost that makes the option unacceptable. The mechanical systems required in a laboratory are quite special and different from those acceptable in most other buildings. The general, brief description that follows will help evaluators focus on the factors which are important to this feasibility study, as they analyze the mechanical conditions and each of the four other areas of evaluation.

The mechanical systems in a laboratory fall in one of three groups. The first group deals with heating, ventilating, and air conditioning; the second group includes plumbing, supply and wastes of water and other fluids; and the third group encompasses all other mechanical devices and special systems. Most often in laboratories, the heating systems are integrated with ventilating and air-conditioning systems. That, of course, is where the abbreviation HVAC comes from.

The ventilation of a laboratory is often the most important factor in laboratory design. As a general rule, any laboratory in which hazardous or risky tasks or chemicals exist should be equipped with fume hoods. Further, these laboratories should not have recirculated air; they should be ventilated with what is called "one-pass air." The volume of air needed to ventilate laboratories with many hoods may become too large. When that is the case, for reasons of efficiency—mechanical as well as economical—auxiliary air fume hoods are sometimes used. As described in chapter 9, an auxiliary air fume hood takes air partly from the room where it is located and partly from a bonnet above it. The air exhausted from the room is called primary air and the air supplied through the bonnet above the hood is called auxiliary air.

The room air, or primary air, is supplied by the building's main air handlers and is generally heated or air conditioned. The auxiliary air, introduced through the hood bonnet, is generally only heated or dehumidified when conditions are extreme, although cooling may be required under certain outside conditions, depending on the operations in the hood. Under these conditions, a laboratory should have two sets of supply air ducts. One is for distribution of the primary air to the different rooms of the building and the other is for introduction of the auxiliary air into the bonnets of the auxiliary air hoods. These ducts should extend from the location of the air handlers supplying them to the room and hoods they service. They should be sized to handle a very large volume of air in order to respond to the number of air changes required by the serviced rooms (in the case of primary air ducts) and the required face velocity in hoods (in the case of the auxiliary air ducts).

In either of these two cases, the air should enter the room or the hood bonnets without causing any significant turbulence. This means that the air should flow, when inside the ducts, at a lower velocity than that usually used to service other types of rooms. This means that the cross section of the duct and the supply air diffuser must be larger than if the same volume of air was to be supplied without turbulence control at the points of delivery. The air is exhausted from the laboratory room, where one or more hoods are located and left in operation, through the hood exhaust ducts. These ducts extend from the hood to vertical shafts to and well above the building roof. The air, combined with the fumes generated in the fume hoods, is mechanically exhausted from the hoods to these ducts and dispersed in the environment.

If a laboratory room does not contain a hood, or if the hood it contains is not to remain operational at all times, the air should be exhausted through an exhaust grille, along with air similarly exhausted from other rooms, and ducted to the outside in the same manner the hood air is exhausted. Air is recirculated in sections where laboratory work is not performed or likely to be performed. Thus, a third set of ducts is needed from the rooms where recirculation is allowed to the location of the air handlers supplying them.

When the A/E analyzes the HVAC systems of a building, it must look not only at the existing system conditions and potentials, but also at the space required for added or new air handlers, furnaces, and cooling units in the mechanical equipment room and elsewhere. It also needs to estimate the size of the supply and exhaust ducts and project their paths. The previous description of how laboratory air is supplied and exhausted should be used as a guide.

In the preliminary evaluation, the HVAC system was tested. Its condition and capability and the extent of its adequacy and flexibility were identified. With these data and the knowledge of the requirements, it becomes possible to determine if the existing system can be used "as is" or needs to be modified, supplemented, or replaced. Once this is decided, the projected system should be conceptualized and the equipment and its components—existing and new—should be identified. The A/E may then survey the building and equipment again to determine how much of the existing system components will remain and where and how the new equipment and its components could be installed.

If an obstacle or impediment exists where a piece of equipment should be located or alongside the path of a duct, the necessity of the obstacle or impediment should be evaluated. If it is needed, its relocation should be determined and costed. If it is not needed, its removal should be costed and considered.

The A/E then estimates all the costs involved in providing the building with a new or modified HVAC system. Included in these costs should be all the costs of removal and relocation of the obstacles and impediments. The A/E also estimates the operation and maintenance costs of the proposed system, projects them for the life cycle of the building, and calculates the present value of the systems.

5. Electrical system inspection, evaluation, and assessment. In the preliminary survey the A/E inspects each segment of the electrical distribution system in a building, including the service entrance, all derived systems, transformers, panelboards, lighting fixtures, wiring, and devices. The physical and electrical condition of all equipment is identified, along with the electrical load capacity of service equipment, feeders, and branch circuits. In this phase, the A/E determines the building's base power and lighting requirements and compares them with the build-out for laboratory areas housing special equipment and instrumentation, special environmental rooms, and administrative space as dictated by the program of requirements. Consideration of mechanical system requirements and of implementation of energy conservation features completes the laboratory facility assessment. Specifically, the A/E then looks at and records each room's and function's needs in electrical power. It identifies these needs in regular, emergency, and uninterrupted types of power and estimates the number of outlets, connections, and circuits required to provide each room and piece of equipment with the appropriate type of power.

In laboratory design, all circuits servicing a laboratory room should come out of one distribution panel. This panel should be dedicated to the circuits of this room and located in the corridor next to its entrance door. This helps the A/E to accurately estimate the number of circuits serviced by a panel and the total loads per panel. With this information it can estimate the total power required, the emergency equipment needed, the number and kind of batteries needed for the uninterrupted power, and all the other required design factors.

The A/E then determines if additional loads are needed as part of the renovation. If the existing power panel boards do not have space for the required new circuit breakers, then one of two options must be considered: to upgrade one or more of the existing panel boards so they can receive the additional power

FIGURE 2-3

1. Transformer
2. Emergency power distribution panel board
3. Transfer switch
4. Power distribution board

FIGURE 2-4

1. Main circuit breaker with a conduit and wiring entrance compartment in its lower part
2. Circuit breaker group
3. Individual circuit breaker
4. Empty spare cubicle

breakers, or to install one or more subpanels in a location adjacent to the existing panel boards. It may also be possible that the new power needed requires one or more distribution panel boards. If this is the case, the A/E must determine whether the existing transformers have enough capacity to receive additional loads or if new transformers are needed.

If part of the renovation requires the use of additional motors, the A/E must determine if space in the existing motor control center can be used. If there is no space available in the motor control center, then one or more new sections must be added to service the new equipment.

Finally, the A/E determines if the existing electrical room size is sufficiently large to accommodate all services, including the new devices. If it is not, the A/E should explore ways to either increase the size of the room or to build a new room nearby.

Figures 2-3 and 2-4 are views of two walls of an electrical room. Item 2 in figure 2-3 is an emergency power distribution panel board. It may also need to be upgraded if more load is required from the generator.

The A/E uses the information gleaned from the electrical condition evaluation to conceptualize a system to respond to the needs and then decides if all or part of the existing electrical installation can be incorporated in the new system. It finalizes its conceptualization in a manner allowing it to estimate all costs.

It then converts these estimates into present value and adds them to the other costs related to the option being considered.

PHASE IV. NEW FACILITY CONCEPT FEASIBILITY STUDY

In the previous two phases, the A/E evaluated all sites and determined their acceptability as to location, size, and proximity to services and utilities. It also analyzed the sites with existing buildings to determine if these buildings could be altered or expanded and estimated the cost of alterations and additions. In each evaluation of these two phases, all requirements were translated into a conceptualized design that was juxtaposed over the existing building's design in order to derive the extent and cost of needed alterations or additions.

This phase of the evaluation is much simpler. The same translation of requirements into a conceptualized design must be undertaken, but this time the design requirements determine the shape and content of the building. Thus, all of the architectural, mechanical, and electrical requirements are translated into a conceptualized design that responds to needs and criteria and is not subject to existing building restraints. The result will be a new building concept with the most efficient and adequate mechanical and electrical systems.

The A/E estimates the construction costs of this building with all its structural and architectural components. The same is done for the mechanical and electrical systems and equipment. The A/E also conceptualizes the mode of operation and maintenance of each of the buildings and of the mechanical and electrical systems and estimates the yearly costs for each of them. All construction costs will be summed up by components and translated into present value. All operation and maintenance costs are projected for the building's life cycle and translated into present value. These values are included in the life-cycle cost comparison analysis table.

Once a conceptualized building—complete with its systems—has been designed, located on the lot, and had its construction and operation costs estimated, the A/E examines how the site surrounding this building will be developed and how the building will be provided with all required utilities and services. First, it locates the utilities and service lines and conceptualizes a utilities and service design that connects the building lines with the municipal lines. It estimates all costs—construction and others—related to these connections and translates them into present value. It also estimates the yearly costs of utilities and services (water, sewage, gas, etc.), projects them for the building's life cycle, and translates them into present value.

Once this is done, it conceptualizes a site plan sensitive to all that is needed to make the laboratory operational on this site. The A/E conceptually grades the site as needed for drainage and use, develops an evacuation system for storm water, and locates driveways and parking spaces. It also includes landscaping in its conceptualization, estimates all related construction and erection costs, and translates these costs into present value. It should also estimate the yearly maintenance and operational costs of the grounds (if these costs are not included elsewhere), project them for the life cycle of the building, and translate them into present value. These values will be included in the life-cycle cost comparison analysis table.

The evaluations covered so far have been for options that consisted either of sites with existing buildings that could be used as a laboratory or of vacant sites that could be developed to accept the projected laboratory. If the options available do not include a site without a building, the A/E could only estimate the cost of upgrading existing properties and compare the cost of the different options. The A/E would not be able to determine if upgrading costs would be lower or higher than those of building an entirely new structure on a given site. It could, however, evaluate one (or more) of the options both with the existing structure located on it and without it. This will determine whether it would be more economical to build a new building rather than to add to or to renovate an existing one.

If it appears from the evaluations that the cost of refitting is higher than new construction, it may be worthwhile to estimate the cost of demolishing the existing structure and replacing it with a new one. If this cost appears reasonable, this alternative can become a new option.

PHASE V. LIFE-CYCLE COMPARATIVE ANALYSIS FOR OWNED OPTIONS

In each of the previous phases the options were evaluated; their construction, operation, and maintenance costs were estimated; and their advantages and disadvantages identified. These evaluations were performed in order to identify data needed to compare them and to determine the best, most efficient, and most advantageous option for the life cycle of the project. Would it be a new building on a purchased piece of land? Or a building purchased, altered, or added to in order to respond to the requirements? The life-cycle comparative analyses that follow rank these options from an economical point of view and indicate the advantages and disadvantages of factors that cannot be objectively measured in monetary value but may have an impact on the laboratory operation or cost.

The life-cycle cost comparison analysis consists of a spreadsheet similar to the one for purchased properties illustrated in Exhibit 2-1. Items in the table should be input by the person undertaking the analysis, who, after computing the numbers, will be able to compare the real value of the properties.

The table includes a column for each option, plus a first column (in which the cost item being compared is identified) and a last column (which lists any attachments or footnotes for additional information). In this spreadsheet there are twenty-five rows listing the following:

1. *Purchase cost.* The cost entered in this row for each option should include the total cost of purchasing the land and the present value of any structure or building erected on it. If one of the options being considered is to build on a piece of property already owned, the estimated value of the property, in present value, should be entered anyway for purposes of analyses. A note in the Remarks column should state this fact, since it may have an impact on the final analysis.
2. *General expenses.* These will include attorney, accountant, and real estate fees and any other general expenses related to a given option and not included in any other cost item.

EXHIBIT 2-1 LIFE-CYCLE COST COMPARISON ANALYSIS FOR PURCHASED PROPERTIES

COST ITEM	OPTION 1	OPTION 2	OPTION 3	OPTION 4	REMARKS
1. Purchase cost					
2. General expenses					
3. Interest cost					
4. Taxes and insurance					
5. A/E fees					
6. Site improvement costs					
7. Site maintenance					
8. Utilities and services					
9. Building construction					
10. Building operation and maintenance					
11. Mechanical construction					
12. Electrical construction					
13. Mechanical operation and maintenance					
14. Electrical operation and maintenance					
15. Other construction					
16. Utilities, energy, and other					
17. Minor repairs and alteration					
18. Major repairs and alteration					
19. Interim lease					
20. Interim move cost					
21. Operational and personnel interruption costs					
SUBTOTALS					
22. Residual value (to be deducted)					
23. Total present value					
24. Adjustment (if applicable)					
25. Present value adjusted					

3. *Interest cost.* All costs related to financing a given option and to procuring funds (such as interest and points) should be translated into present value and entered in the row under the option's column.
4. *Taxes and insurance.* The funds required each year for taxes and insurance should be estimated independently for each option (for land alone, or land and structure). These estimates should then be projected for the building's life cycle and translated into present value. The resulting number should be entered in this row under the option's column.
5. *A/E fees.* Include all fees. The consultant fees for the feasibility study may be subdivided equally between all options. The A/E should estimate its fees for

each option, either as a percentage of construction cost or in some other manner. These estimates should include all fees—those to prepare the design and construction documents as well as those for construction management. Since these fees will be disbursed in more than a year, for the purpose of this analysis they must be translated into present value.

6. *Site improvement costs.* All site improvement costs for each option, as determined by the feasibility study, should be inserted, in present value, in this row.
7. *Site maintenance.* The site maintenance costs indicated by each option's feasibility study should be inserted in present value in this row.
8. *Utilities and services.* The cost of installation, extension, and connection to the utilities and services lines for each option should be translated into present value and inserted in this row. The costs of utilities and service use should not be included here.
9. *Building construction.* All construction costs estimated in each option's feasibility study should be inserted in present value in this row.
10. *Building operation and maintenance.* Any cost determined by the feasibility study as necessary to operate and/or maintain an existing or projected building or any of its systems or components should be inserted in present value in this row.
11. *Mechanical construction.* All mechanical construction costs indicated by each option's feasibility study should be inserted in present value in this row.
12. *Electrical construction.* All electrical construction costs indicated by each option's feasibility study should be inserted in present value in this row.
13. *Mechanical operation and maintenance,* and
14. *Electrical operation and maintenance.* The mechanical and electrical operation and maintenance costs as estimated by the feasibility study should be projected for the life cycle of each option, translated into present value, and then inserted in the relevant row. The cost of utilities and the energy consumption necessary to run or maintain the equipment should not be included under this heading, although it was necessary for comparison purposes to have these costs computed as per equipment needs (mechanical and electrical) in order to differentiate between options and compare the efficiency of their systems. The volume of water and gas and the electricity used in a laboratory are more easily and perhaps more accurately estimated on a total basis. Their costs should be added to the utilities and energy costs.
15. *Other construction.* Any construction or erection costs that were found necessary by the feasibility study and that do not fall under any named category should be inserted in present value in this row.
16. *Utilities, energy, and other.* Each laboratory operation generates needs for water, gas, electricity, and other utilities or sources of energy to energize, cool, or operate instruments or equipment. When the cost of these is not a factor of the building or of its equipment efficiency, these costs could be assumed to be equal regardless of the building or site where they occur. The utilities and annual energy needs for the mechanical and electrical equipment and systems—existing or proposed—may vary substantially from option to option. They were estimated when the costs of operation and maintenance of mechanical and electrical systems were evaluated. These

estimates should be added to the utilities and energy costs needed to operate the laboratory along with any other operation or maintenance that may be necessary but does not fall under any named category. The sum of these should be for the life cycle of the option, translated into present value, and inserted in this row.

17. *Minor repairs and alterations*, and
18. *Major repairs and alterations.* Any building, old or new, will need minor repairs and alterations at some point in its life cycle. These will vary from building to building depending on the age of the building, the condition of its systems and equipment, the change in functions, and other identifiable (and not-so-identifiable) factors. In undertaking the feasibility study of an option, the A/E may also discover that certain components or systems—perhaps a roof, a boiler, or an air handler—that are still in good working condition are nevertheless approaching the end of their life expectancy and may need to be replaced or reconditioned at a substantial cost before the end of the building's projected life cycle. When undertaking the feasibility study of a given option, the A/E must therefore decide how many funds should be allocated and when these funds should be available. Once this is done, the A/E projects these costs, as required, for the building's life cycle, translates them into present value, and inserts them in the relevant row.
19. *Interim lease*,
20. *Interim move cost*, and
21. *Operational and personnel interruption costs.* An existing structure housing a laboratory that needs to expand may have some additional space that could be altered to become laboratory space. It may also have adjacent land that could be used to expand the building. In either case, or in other similar cases, undertaking the expansion or alteration to interrupt the entire laboratory's operation for a short period of time during certain construction phases and for long spans of time in some areas of the building. This may call for a temporary reorganization of the laboratory work. The laboratory management may decide to completely interrupt the operations during these spans of time, to diminish the workload and relocate certain operations within its own space, or to relocate these operations elsewhere. In any of these cases, there will be a loss of income and increase in expenses, which should be accounted for in considering the option costs. The A/E must therefore estimate and account for the cost of space leased to relocate operations and personnel and for all moving costs, including those of personnel and equipment. To these costs must be added the costs resulting from the interruption of operations (loss of income and loss of man-hour productivity) as estimated and provided by the laboratory management. These costs are then categorized, translated into present value, and entered in their respective rows.
22. *Residual value.* If a building's "economic life span" is assumed to be thirty years, for amortization reasons, this building, theoretically, will be worth about half its value in present dollars at the end of a fifteen-year period and will not have any residual value at the end of the thirty-year period. The land where the building is erected would still have value, which would have either increased or diminished during the thirty years. In real life, however, a building that is well built, operated, and maintained or that is located in a

sought-after area will still have a remaining commercial resale value. This value, added to the land value, constitutes the commercial residual value of the property. Either the amortization or the commercial residual values may be used for the purpose of this analysis. The selected value should be translated into present value dollars and entered under the option's column. Note that this value should be deducted from, rather than added to, the costs summed up in this analysis.

23. *Total present value,*
24. *Adjustment (if applicable),* and
25. *Present value adjusted.* Each column's value should be totaled and entered in the Total Present Value row. Sometimes in real estate transactions, an offeror may find it convenient (for tax purposes or to attract a given industry, for example) to offer a major advantage such as a free piece of land or long-term free use of the land or of the building erected on it. When such a situation occurs, the A/E should undertake feasibility studies and evaluations independently of this advantage. This will permit it to have, without distortion, the real comparative values between the options. The present value of the advantage should be entered in row 24 and deducted from the total value of this option to reflect the proposed advantage.

    The figures appearing in these three last rows summarize the cost of each option in present value and help rank the options from an economic point of view. The laboratory management and board of directors looks at these figures, evaluates as objectively as possible the advantages and disadvantages of each option, and decides which option to select.

PHASE VI. OWNED VERSUS LEASED ANALYSIS

If we use the same options and the same assumptions as used throughout this analysis, there is no doubt that it will cost more to lease than to own a laboratory. This is true if for no other reason than the fact that a certain percentage—the lessor's profit on the investment—must be added to all the costs considered. Thus, generally speaking, it is more economical for an entity to own the laboratory rather than to lease it for long spans of time.

Often, however, there are circumstances that make it more convenient to lease laboratory space rather than to own it. These factors are generally known to the board of directors and top management and are considered when the initial decisions are being made. A listing of some of the advantages and disadvantages of owning and leasing follows, as they will help clarify the decision-making process. Advantages of owning a laboratory are:

- It usually requires the least amount of capital outlay in the long term.
- It allows for direct control in alteration, modification, and repairs.
- Building administration is simpler and geared to user's needs.
- All improvements remain the entity's own property.
- Investment offers long-range tax advantages.

Disadvantages of owning a laboratory are:

- Large initial capital or capitalization is required.
- Total staff involvement may be required during construction phase.

Advantages of leasing a laboratory are:

- Compared to owning, almost no initial capital is required. Rent is paid as income is earned and rental obligation is incurred.
- At the end of the lease, the entity can walk away from the building with relative ease.
- The building can be used for as short a period as the lease allows and could be extended by agreement.

Disadvantages of leasing a laboratory are:

- If the lease is extended for long periods of time, the total capital outlay may be much higher than if the facility were owned.
- Staff or contract is needed to operate and maintain facilities through the lessor.
- At the end of the lease agreement, the entity walks out of the facility owning nothing.
- The laboratory management has no direct control over the construction, the operation, or the maintenance of the building.
- All funds spent by laboratory management to improve or alter the facility are lost.
- The building administration is much more complex.

The laboratory board and top management then examine their time and space needs, as well as their resources (i.e., cash flow, tax rate, and other factors) and review the advantages and disadvantages of owning and leasing options offered. Even if they choose to lease, their decision will have been based on what the feasibility studies and life-cycle cost comparison analyses indicated was the most efficient choice.

Chapter 3

Selecting a Site for the Laboratory

THE evaluation used to select a site for a laboratory incorporates several groups of criteria. Of these groups, the health, safety, and environmental criteria must be addressed prior to the engineering, operational, and public perception criteria.

Health, Safety, and Environmental Evaluation

The site that houses the laboratory inevitably has an impact on the laboratory's operations. It is also true that the laboratory may have an impact on the surrounding environment. First, it is important to identify the operations that will take place in the laboratory and determine their level of hazard and risks. If these operations involve highly toxic chemicals and must be conducted in a partial- or full-containment laboratory, the site must respond to different criteria than those used for locating a laboratory without or with low toxic operations. In any case, the health and safety criteria selected must be such as to minimize the risk to nonlaboratory personnel and to the surrounding environment. These criteria should be based on the assumption of a worst chemical release and of mechanical and human failures.

The health and safety criteria functions as a yardstick measuring the potential impact of the laboratory operations on people, property, and natural resources and of any accident that may result from such operations. The most disastrous accident should normally be considered in the assessment of the potential impact of a laboratory on a given site. This may mean that a potentially harmful chemical is released out of the laboratory, is dispersed by some means, and affects people, property, or natural resources. Such accidents may fall into one of three categories: mechanical failures, human failures, or critical events. Each of these categories can be further subdivided into more specific types of failures. When and if such failures occur, the released chemicals may be ingested by mouth, be inhaled into the lungs, or permeate the body through the skin. Once absorbed by the body they are distributed to the organs and tissues.

Existing laboratory control technologies minimize the risk of chemical contamination, rendering these risks so low that in effect these technologies may eliminate the environmental and safety concerns in the selection of a laboratory site. However, the occurrence of some event that, despite existing controls, can contaminate the laboratory or its surroundings is always a possibility. These events, their probability of occurrence, and the resulting harm must be identified and evaluated.

If the proposed laboratory incorporates a full- or partial-containment facility, the best location would be a single-occupancy building in a rural environment with a buffer zone around it and a low population density. It may also be in a single-occupancy building in an industrial park, also with a buffer zone around it. It would also be acceptable to have the laboratory building on a moderately developed site with a medium population density but with a buffer zone around the laboratory building. In any of these cases, access to the laboratory must be restricted.

It is preferable that the site be remote from sensitive natural resources such as streams, rivers, and lakes. The site may, however, be at a limited distance from such resources, as long as the distance prevents any contamination. A full- or partial-containment laboratory facility should not be located in a multiple-occupancy building and/or in a multiple-use, developed site. It is also preferable and highly recommended that conventional laboratories with some level of hazard and risk not be located in a multiple-occupancy building and/or in a multiple-use, developed site.

Engineering and Operational Evaluation

Site and Location Evaluation

The site's impact on the laboratory's performance of functions must be evaluated. This evaluation should analyze the site characteristics and location, lot size, and accessibility, as well as the availability of utilities and services. It should also consider the uses of the immediate surroundings and other site factors regulated by covenants, zoning, or building codes that may affect the efficient functioning of the laboratory and its initial and operational costs.

Another criterion that should be considered during the screening process is the site's response to the principle of sustainable design. Are there naturally occurring resources such as solar and wind energy, natural shading, native plant materials, and good topography and drainage on the site? When selecting a site, the evaluator should consider the possibility of the laboratory's using these resources.

General Characteristics and Location of the Site

The first screening of sites looks at the zoning and covenants governing the site's permitted uses. Do the zoning laws allow the site to be used for a laboratory? Is there any covenant prohibiting the use of the land for the type of activity to be conducted in the laboratory?

The environment surrounding the site and the kind of activities permitted and presently taking place there should also be critically examined. Is the proposed site in a rural area surrounded by fields and woods? Is it in an industrial park where similar or conflicting activities are presently conducted? Is it in an area where mixed uses are permitted? In each of these cases, how would the laboratory affect or be affected by the surrounding environment and by its neighbors' activities?

The site's accessibility to both public and private transportation also should be evaluated. Is its location central enough to the place of residence of the majority of its prospective employees to allow them to commute without undue hardship? How important is it for the projected laboratory to be accessible to the population or to the geographical area it serves? Would the laboratory's geographic location affect the functions performed in it?

The site's proximity to its suppliers and to the providers of services it routinely needs also must be examined. Are there bottled gas and chemical suppliers whose location and volume of work give them the capability of routine delivery to this location within a day? Are there nearby mechanical, electrical, and other contractors with the capability to undertake routine maintenance and repair tasks? These and similar considerations should be evaluated, as they will affect the operation and efficiency of the laboratory.

Lot Size and Accessibility

The site size should be evaluated according to its capacity to meet the laboratory's present and expansion needs. Is the lot size large enough to accommodate the projected laboratory and its future expansion as well as the required driveway, loading areas, and parking spaces for personnel, suppliers, and visitors and their future expansion?

If the available lot is marginal in size, several possible footprints of the projected laboratory structures, with required square footage, should be assumed and placed on the site. These footprints should include the driveway, loading dock, and parking space needs for both the proposed building and its future expansion.

The site's accessibility is another factor. Is it important for the projected laboratory site to be easily accessible? If yes, how and to whom? Should it be accessible by public as well as by private transportation? Who other than the staff and the suppliers should have easy access to the laboratory?

Once the specific accessibility needs are defined, the site should be evaluated according to its response to these needs. Is the site surrounded by roads or is it alongside one road? Is the traffic on the surrounding roads heavy or light? Are the site configuration and topography impacting on its accessibility? How do the site and surrounding roads relate to each other in terms of accesses? Is the site located in an industrial park where other tenants with other uses generate heavy and perhaps incompatible traffic? What would the effect of this traffic be on the projected laboratory functions and accessibility? Do the different site arrangements, with footprints of buildings, parking spaces, and internal roads, fit well with the surrounding roads and the normal, logical accesses to and from them?

Utilities and Services

The availability of the utilities and services necessary to the laboratory's operation also must be considered. It is more economical to develop a site where basic utilities (storm and sanitary sewers, water, gas, telephone and power lines) are available on-site or nearby, as these are very high-cost items in laboratory construction.

If the lot under consideration is an industrial park, how are other buildings presently disposing of their wastes? Are their wastes domestic or are they industrial, chemical, or biological? Are there any municipal sewage lines near the lot? How far away are they? How is their capacity related to the new laboratory's needs? Can they treat laboratory waste, or would the waste have to be treated prior to being discharged? If so, what type of treatment would be needed? If the lot under consideration is in a rural area, how close are the basic utilities it would need? What would it take in cost and time to extend them and, if needed, increase them?

Planning the provision of utilities to a laboratory is as important as planning its building units. A preliminary identification of requirements for each of the new laboratory buildings should be made at the conceptual stage, followed by an assessment of the entire facilities combined needs. Once the requirements are identified, they should become important factors in the site selection, as utilities and service provisions are very high-cost items.

Prior to selecting a site, a determination should be made about how the laboratory waste will be managed and disposed of. The required work for and cost of waste disposal (by hooking up to an available waste line or by other means) should be assessed. Are there any municipal sewage lines near the lot? How far away are they? How is their capacity related to the laboratory needs? Can they treat laboratory waste or would the waste have to be treated prior to being discharged? If they can treat the waste, what type of treatment is available? Once the answers to these questions are known, estimate the waste disposal system's installation and operation costs, project these costs for the life cycle of the building, and compare them to the other options considered.

Is the municipality supplying water to the site and other nearby sites where functions similar to those of the projected laboratory are taking place? If so, what use is made of this water? Is it for domestic use only or could it be used for the proposed laboratory's tasks? What type of water would be supplied? Is it hard or soft? What is its level of purity and its mineral and ferrous content? What is its level of conductivity?

What must be done to this water to render it usable for the projected laboratory experimentation? Does it need deionization, softening, or ferrous removal? Does it need to have its conductivity lowered? Once these questions are answered, determine if the existing supply lines are sufficient and if the quality of the water is acceptable or needs to be corrected. The installation and operation costs of any system that needs to be included (to adjust the level of conductivity, to soften the water, to remove ferrous content, or to deionize the water) should be projected to the building's life cycle, translated to present value, and added to the cost of the contemplated site.

What source of energy presently provides the structures surrounding the contemplated site with their energy needs? Would the same source be available and efficient enough for the new laboratory's energy needs? Is there another type of energy that might be more efficient for the life cycle of the building and could provide energy for heating, ventilating, and air conditioning, as well as for operation of laboratory tasks?

Is municipal gas available? If so, and it is not presently used, how close to the laboratory site are its lines? And what would it take and cost to extend a line to the laboratory? The answers to these questions will provide the information needed to estimate the cost of providing the laboratory with energy, as well as the cost of energy consumption for the life cycle of the building. These costs should be translated to present value and added to the cost of the contemplated site.

Public Perception

When the health and safety factors of a given site are weighed and found safe and acceptable, some other critical factors may override some of the engineering and operative factors in the selection process. One of these is the public perception of the risk.

If the perception of the risk is based on solid facts, such factors should be addressed frankly and directly. Precautions and corrective actions should be designed and in place and should be clearly explained and discussed with the public up front. An ongoing communication should be established between members of the laboratory's scientific staff and interested members of the community. This group should meet regularly, in a public forum, and openly address all concerns.

If the perception of risk is not based on facts, measures should be taken to educate and improve relations with the public. Those responsible for the laboratory conception, design, and construction should detect at a very early stage what the public perception of the projected laboratory may be and establish an ongoing communication with the concerned public in a public forum. Responsible parties should provide information on the laboratory's future operations, explain them in easy language, and communicate possible risks, planned precautions, and the actions that would be taken should an accident occur.

Using The Criteria: Three Examples

The following site evaluations illustrate the use of the criteria discussed above. Sites A, B, and C were offered for a new general chemistry analytical laboratory near Washington, D. C. Here is how this evaluation took place.

SITE A

The site plan and a photograph of this site are shown in Figures 3-1 and Figure 3-2.

LOCATION AND SIZE

The site is located in the southeast corner of a wooded industrial park, west of a major highway. This area is designated as a transitional zone. The site size is

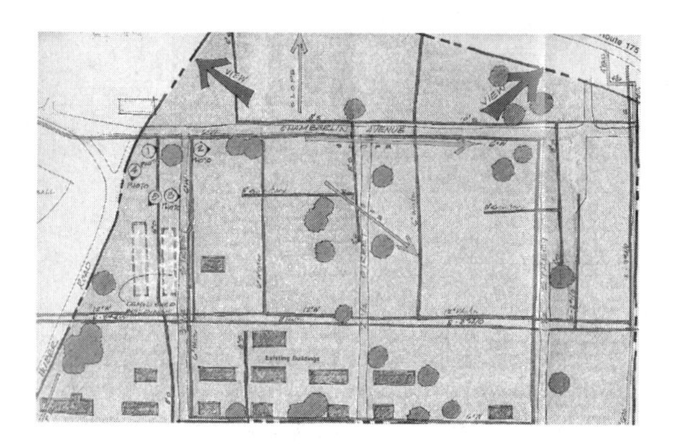

FIGURE 3-1 (left)
Site A Plan

FIGURE 3-2 (right)
Site A showing the trees lining the existing road network

sufficient for the laboratory structure and allows for future expansion. North of the site are softball fields. Southwest of the site are many residential facilities. Mature trees line the existing road network on the site.

GREEN BUILDING FACTORS

Solar and wind energy: Possible.

Natural shading: Not too many trees. Mature trees line the existing road network.

Native plants: Not too many trees.

Topography and drainage: Flat terrain, good topography and drainage.

ZONING AND COVENANTS

The structure's appearance and location on the site must follow a covenant dictated by the landowner. However, this covenant, restrictive as it is, will not conflict too much with the laboratory requirements.

HOW WOULD THE LABORATORY AFFECT AND BE AFFECTED BY THE ENVIRONMENT SURROUNDING THE SITE?

Southwest of the site are many residential facilities used by people who work in and around the industrial park. However, the prevailing winds will not direct laboratory fumes toward the residential facilities.

PROXIMITY TO SUPPLIERS AND PROVIDER OF SERVICES

The site is very close to the suppliers and the provider of services.

ACCESSIBILITY TO MAIN THOROUGHFARE

The site is nearby and accessible to the main thoroughfare.

CONFIGURATION AND TOPOGRAPHY OF SITE AND IMPACT ON ACCESSIBILITY

Substantial road improvement will be needed to connect the site to the main thoroughfare.

AVAILABILITY OF SERVICES AND UTILITIES

Availability of on-site or nearby storm and sanitary sewer lines: Yes, but existing lines will need to be repaired and extended.

Availability (and quality) of on-site or nearby water supply lines: Yes, but existing lines will need to be replaced as their condition is questionable because of their age.

Availability of nearby burning gas lines: Yes.

Availability of on-site or nearby telephone and power lines: Telephone and power lines will need to be extended from quite a distance.

MEANS FOR WASTE MANAGEMENT AND DISPOSAL

Yes.

SOURCES OF ENERGY (GAS, OIL, ELECTRICITY) TO FEED THE LABORATORY WITH SUFFICIENT ENERGY

Gas and power lines are available but will need to be extended to the site from quite a distance.

SUMMARY OF FINDINGS AND PRELIMINARY RECOMMENDATION

Although acceptable, this location was rejected because a much better location was available.

SITE B

This site is shown in plan (figure 3-3) and a photograph (figure 3-4).

LOCATION AND SIZE

The site is located in the northeast section of a wooded industrial park in a somewhat remote and isolated area. A portion of the site is paved with several small structures presently used for the storage of vehicles. The majority of the site is an undeveloped parcel of land heavy in vegetation.

GREEN BUILDING FACTORS

Solar and wind energy: Possible.

Natural shading: The wooded area is composed of mature trees that could provide shading if the building is placed in cleared area central to this wooded area. This is possible, but quite a few trees may have to be cut.

Native plants: All trees are native.

Topography and drainage: The paved area is slightly sloping. The wooded area slope is much steeper.

FIGURE 3-3 (left)
Site B Plan

FIGURE 3-4 (right)
Site B showing the wooded area with the cleared area central to this site

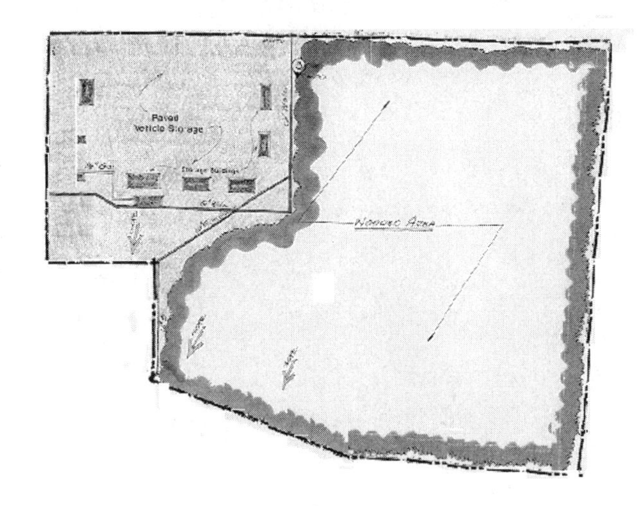

ZONING AND COVENANTS

The structure's appearance and location on the site must follow a covenant dictated by the landowner. However, this covenant, restrictive as it is, will not conflict too much with the laboratory requirements.

HOW WOULD THE LABORATORY AFFECT AND BE AFFECTED BY THE ENVIRONMENT SURROUNDING THE SITE?

From an environmental and economical standpoint, the main drawback of this option is the site cleaning it would require. A large number of existing mature trees will have to be cut down to make room for the facility. The site around the vehicle storage area may be contaminated due to the previous use of oil and of other petroleum products. The prevailing winds will not direct laboratory fumes toward any facilities around this site, nor would these winds direct these facilities' fumes toward this site.

PROXIMITY TO SUPPLIERS AND PROVIDER OF SERVICES

The site is very close to the location of the suppliers and the provider of services.

ACCESSIBILITY TO MAIN THOROUGHFARE

The site is nearby and accessible to the main thoroughfare.

CONFIGURATION AND TOPOGRAPHY OF SITE AND IMPACT ON ACCESSIBILITY

Substantial road improvement will be needed to connect the site to the main thoroughfare.

AVAILABILITY OF SERVICES AND UTILITIES

Availability (and quality) of on-site or nearby water supply lines: Yes, but existing lines will need to be replaced as their condition is questionable because of their age.

Availability of nearby burning gas lines: Yes.

Availability of on-site or nearby telephone and power lines: Telephone and power lines will need to be extended from quite a distance.

MEANS FOR WASTE MANAGEMENT AND DISPOSAL

Yes.

SOURCES OF ENERGY (GAS, OIL, ELECTRICITY) TO FEED THE LABORATORY WITH SUFFICIENT ENERGY

Gas and power lines are available but will need to be extended to the site from quite a distance.

SUMMARY OF FINDINGS AND PRELIMINARY RECOMMENDATION

Although acceptable, this location was rejected because a much better location was available.

SITE C

This site is shown in plan (figure 3-5) and photograph (figure 3-6).

LOCATION AND SIZE

The site is located right off a main highway at the edge of an industrial area for light industries. This site is larger than sites A and B. It is wooded with mature

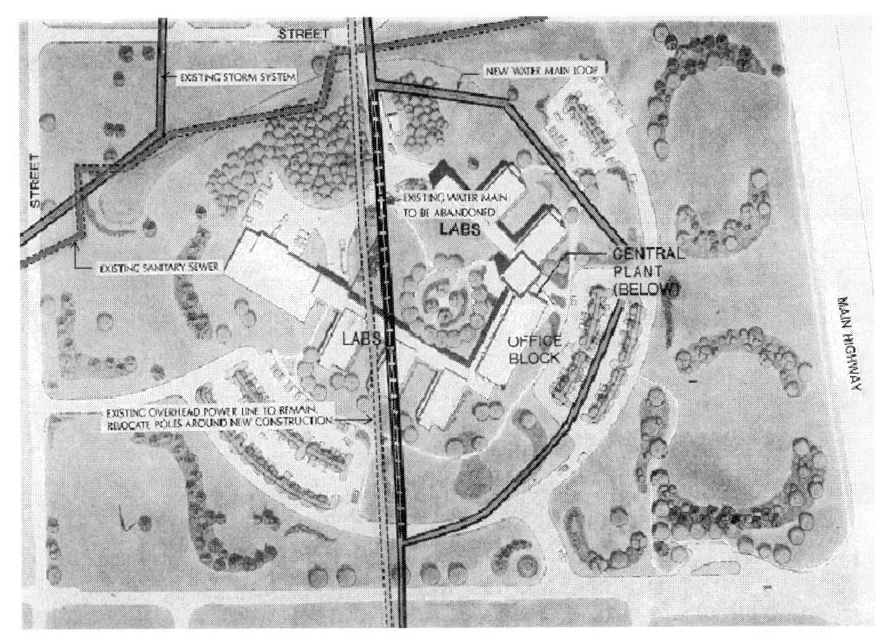

FIGURE 3-5
Plan of Site C showing the building location and the service and utility lines as modified to service the laboratory facility

FIGURE 3-6
Site C showing the cleared area central to this site

trees and has areas heavy in vegetation. It also has large cleared areas at its center, offering an ideal site to locate the building without cutting too many trees. The site is also within walking distance of many amenities: restaurants, a day-care center, and shopping and recreational facilities. These amenities were a factor in the selection of this site.

GREEN BUILDING FACTORS

Solar and wind energy: Possible.

Natural shading: The wooded area is composed of mature trees that could provide shading if the building is placed in the cleared area central to the wooded area.

Native plants: All trees are native.

Topography and drainage: The entire site is almost flat. Some places within the wooded area and at the site's periphery are slightly sloping.

ZONING AND COVENANTS

The structure's appearance and location on the site must follow a covenant dictated by the landowner. However, this covenant, restrictive as it is, will not conflict too much with the laboratory requirements.

HOW WOULD THE LABORATORY AFFECT AND BE AFFECTED BY THE ENVIRONMENT SURROUNDING THE SITE?

The development of this site will have almost no impact on its surroundings, nor would it be impacted by the surrounding uses. A small number of mature trees will have to be cut down to make room for the facility. The prevailing winds will not direct laboratory fumes toward any facilities around this site, nor would these winds direct these facilities' fumes toward this site.

PROXIMITY TO SUPPLIERS AND PROVIDER OF SERVICES

The site is very close to the location of the suppliers and the provider of services.

ACCESSIBILITY TO MAIN THOROUGHFARE

The site is accessible to the main thoroughfare and not too far from it. Secondary roads to the main highway may require minimal upgrading. No additional local roads need to be added.

CONFIGURATION AND TOPOGRAPHY OF SITE AND IMPACT ON ACCESSIBILITY

Most of the site is on flat land and it is already connected to the main thoroughfare.

AVAILABILITY OF SERVICES AND UTILITIES

Availability of on-site or nearby storm and sanitary sewer lines: Yes, existing lines are on site and will not need to be repaired or extended.

Availability (and quality) of on-site or nearby water supply lines: Yes, existing lines are on site and will not need to be repaired or extended.

Availability of nearby burning gas lines: Yes.

Availability of on-site or nearby telephone and power lines: Telephone and power lines will need to be extended to the site.

MEANS FOR WASTE MANAGEMENT AND DISPOSAL

Yes.

SOURCES OF ENERGY (GAS, OIL, ELECTRICITY) TO FEED THE LABORATORY WITH SUFFICIENT ENERGY

Gas lines are available but will need to be extended to the site.

SUMMARY OF FINDINGS AND PRELIMINARY RECOMMENDATION

Of the three locations evaluated, this location's response to the evaluation criteria was the most positive. It was therefore found the most adequate site for this laboratory facility.

Chapter 4

Designing the Laboratory Room

THE design of a laboratory room starts with the preparation of a room data sheet (RDS) and a preferred room arrangement. The preparer(s) of the RDS and of the preferred room arrangement should have an understanding of what the modular system is and of how this system is used in designing each laboratory room. The laboratory room user, guided by management, should prepare for each room an RDS and a sketch showing the preferred room arrangement, using the typical layouts described in this chapter. The RDS, which is explained later in this chapter, contains all the requirements pertaining to the operations that will be conducted in the room.

Use of Modules for Laboratory Rooms

The "modular" concept provides a system of wall location, lighting arrangement, HVAC distribution, and utility distribution that allows maximum flexibility in changing laboratory use and size without requiring major physical changes to building systems.

A laboratory room is composed of one, two, or several modules. The definition of a module and the way in which a module's minimum size is determined is described in chapter 6. Once the minimum size of the module is determined, the sizing of each laboratory room should be determined in terms of modules. The operations and tasks that will take place in the room, the equipment required to undertake those operations and tasks, and the code requirements help determine the size of each laboratory room.

A description of typical layouts for different sizes of laboratory rooms follows. The final design of the room should determined by the requirements of the functions that will take place in it. Consequently, final layout of a laboratory room may not necessarily look like the layouts proposed in this chapter.

Typical Layouts for a One-Module Laboratory Room

In a general chemistry or biochemical laboratory, a one-module laboratory room has casework along the length of the room. Sometimes, however, floor space alongside the length of the room is needed to provide for large equipment or instrumentation.

If a fume hood is required by the operations and tasks that will take place in the room, it should be located opposite the primary exit door. The fume hood may be located on the end wall of the laboratory if there is not a window there, or on the side wall if the room has a window in the back wall. When this is the case, the fume hood should be located a minimum of 12 to 16 inches (0.30 to 0.41 m)—preferably 16 inches (0.41 m)—away from the back wall (which will be on its side). Such an arrangement is shown in figure 4-1.

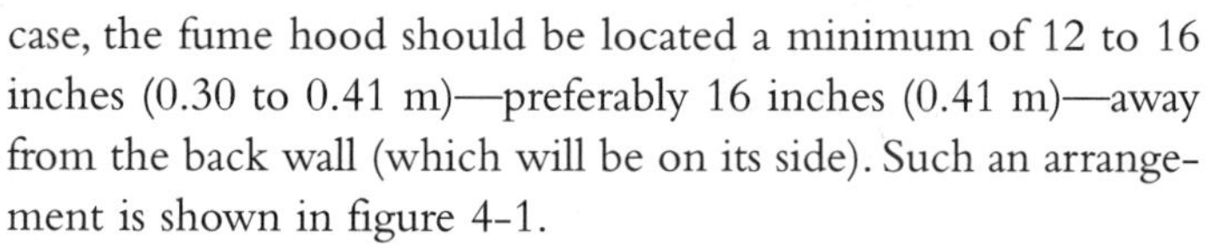

FIGURE 4-1

These locations for fume hoods, on or next to the end wall of the laboratory room, are considered the best because the hood operator is essentially the only one who enters the hood's zone of influence. Additionally, the fume hood should be away from all doors, should the room have more than one door, or from any other source that would create air movement that would interfere with the planned air flow from the sources of supply to the hood face.

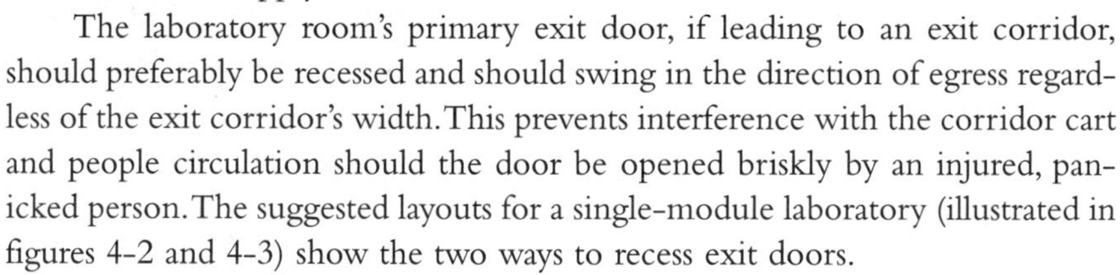

The laboratory room's primary exit door, if leading to an exit corridor, should preferably be recessed and should swing in the direction of egress regardless of the exit corridor's width. This prevents interference with the corridor cart and people circulation should the door be opened briskly by an injured, panicked person. The suggested layouts for a single-module laboratory (illustrated in figures 4-2 and 4-3) show the two ways to recess exit doors.

Often it is necessary to move large equipment in and out of the room. This could be achieved by using either one 4′-0″ (1.22 m) door or, preferably, a double door with a 3-foot (0.91 m) active leaf and a 1-foot (0.30 m) inactive leaf. The inactive leaf is kept closed most of the time, with one bolt at the top and one at the bottom. It is good practice to always have a 4′-0″ (1.22 m) opening for each laboratory room.

If operations with explosion hazards are to be conducted in a one-module laboratory and if the location of those operations is such that they would block the path to the exit door if an accident were to occur, a second exit is required. The location of the second exit should allow the person in the laboratory to escape from the location of the accident. Figures 4-2 and 4-3 do not show a second exit door because the fume hoods (which are where any hazardous work would take place) are located on the back walls, opposite the exit doors.

Each laboratory room where chemicals are used should have an eye wash device as well as an emergency shower. The location and other requirements of these devices is provided at the end of the description for layout of laboratory rooms.

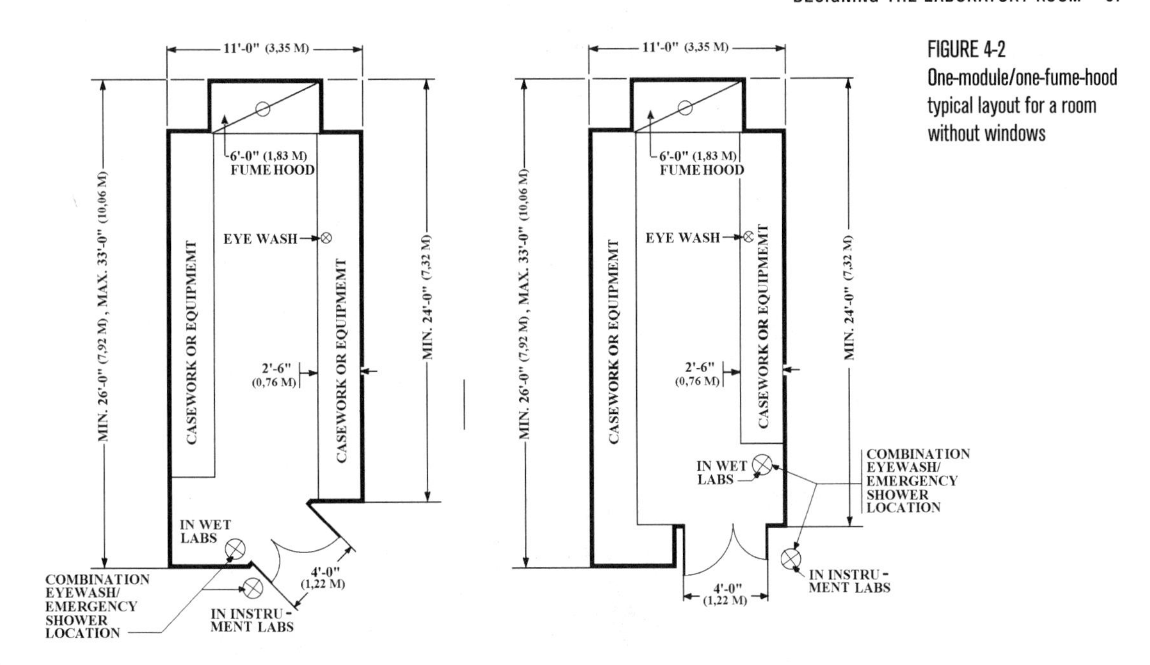

FIGURE 4-2
One-module/one-fume-hood typical layout for a room without windows

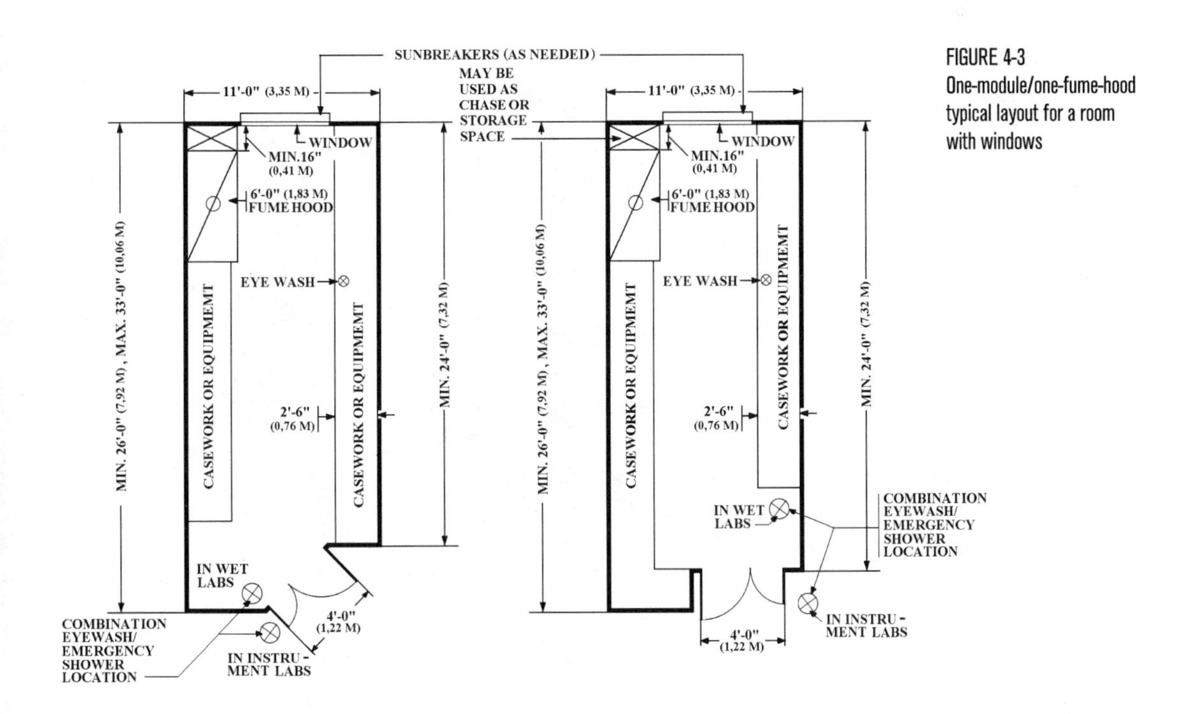

FIGURE 4-3
One-module/one-fume-hood typical layout for a room with windows

Typical Layouts for a Two-Module Laboratory Room

In most laboratories, and particularly in general chemistry and biochemical laboratories, a two-module laboratory room has casework along the length of the two side walls. A peninsula-type work area is centered on the imaginary line dividing the room into its two modules. If needed, floor space should be provided for large equipment or instrumentation. This space should be located along the length of the room, at the end of the peninsula, or in the wall along the corridor where the exit door is located. Figures 4-4 through 4-11 fully illustrate these arrangements. If a two-module room classifies as a class A laboratory, or if operations with explosion hazards are to be conducted in the room, a second exit door is required. The location of this second door should allow for escape if the access to the first door is blocked by the hazard.

One or two fume hoods can be lodged in a two-module laboratory. These fume hoods should be located on the end wall of the laboratory, opposite the primary exit door. These locations are considered the best because the hood operator is essentially the only one who would enter the hood's zone of influence. These fume hoods may be located against the back wall, if the laboratory room does not have a window on the back wall (figures 4-4 and 4-5). In rooms with windows, the fume hoods can be located either back to back at the end of the peninsula or on the two side walls. These four fume hood arrangements are illustrated in figures 4-6 through 4-11. In these six room arrangements the fume hoods should be located a minimum of 12 to 16 inches (0.30 to 0.41 m) away from the back wall. When the fume hoods are located opposite to each other, a space of at least 4'-6" (1.37 m) should be left between them to provide each hood with sufficient space to accommodate the hood operator. Although acceptable, this arrangement is not recommended because, being face to face, the two hoods compete for air. This, added to the traffic of one hood operator, creates eddies at the face of the hoods. Figures 4-8 and 4-9 show two hoods next to each other. This arrangement is also acceptable but not recommended, as it generates conflicting airflows. (This arrangement, however, does not pose a problem in organic chemistry laboratories.)

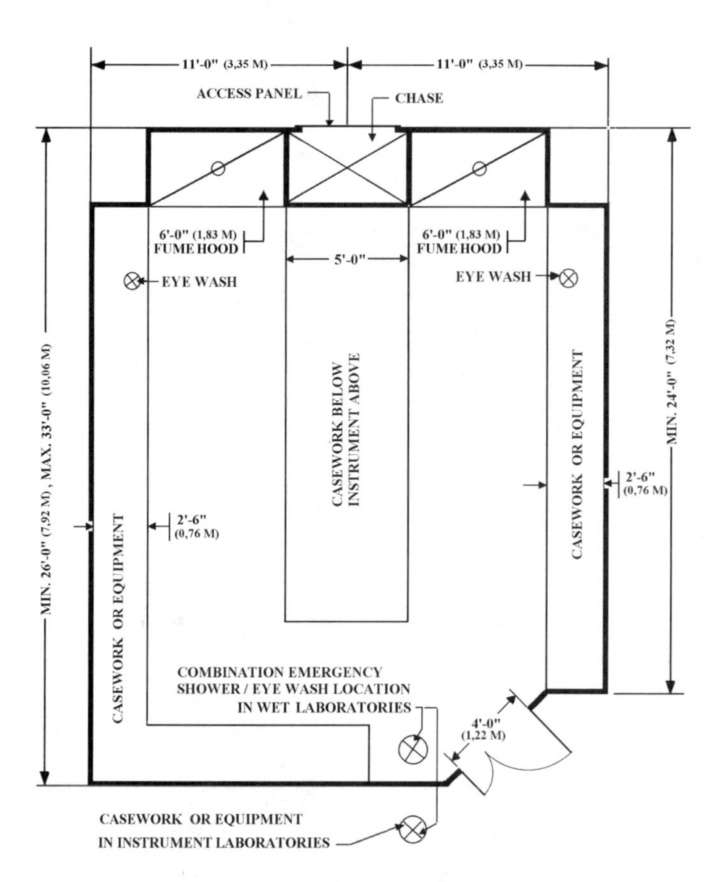

FIGURE 4-4
Two-module/two-fume-hood typical layout for a room without windows

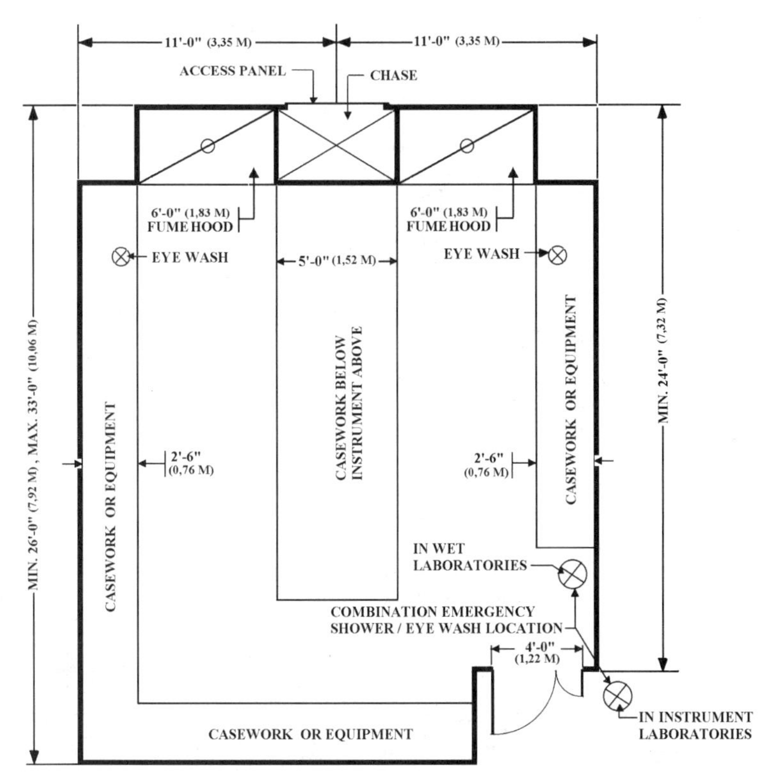

FIGURE 4-5
Two-module/two-fume-hood typical layout for a room without windows

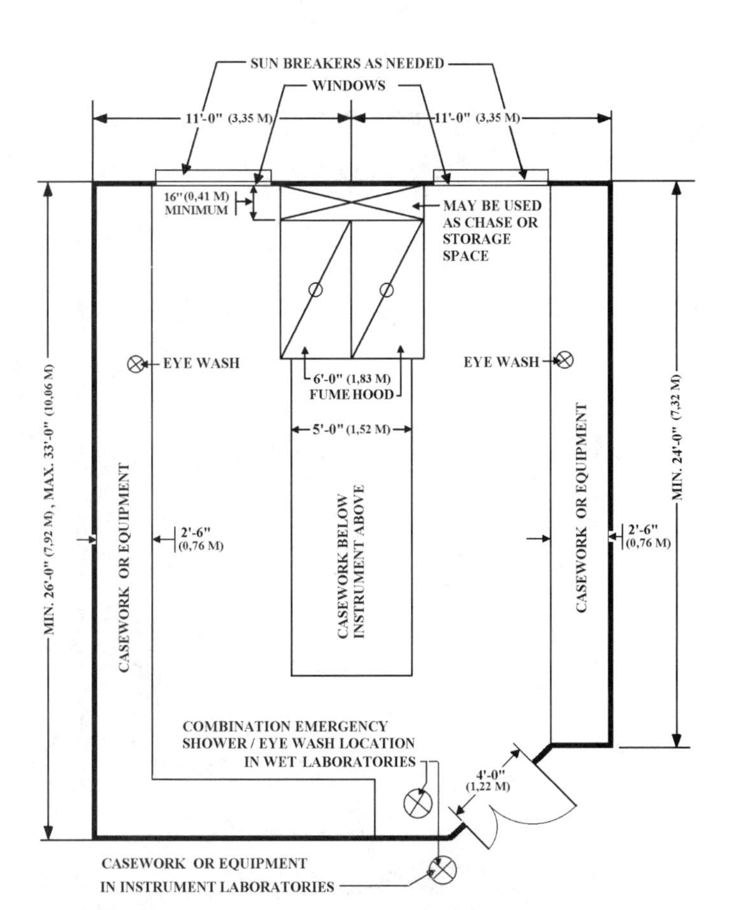

FIGURE 4-6
Two-module/two-fume-hood typical layout for a room with windows

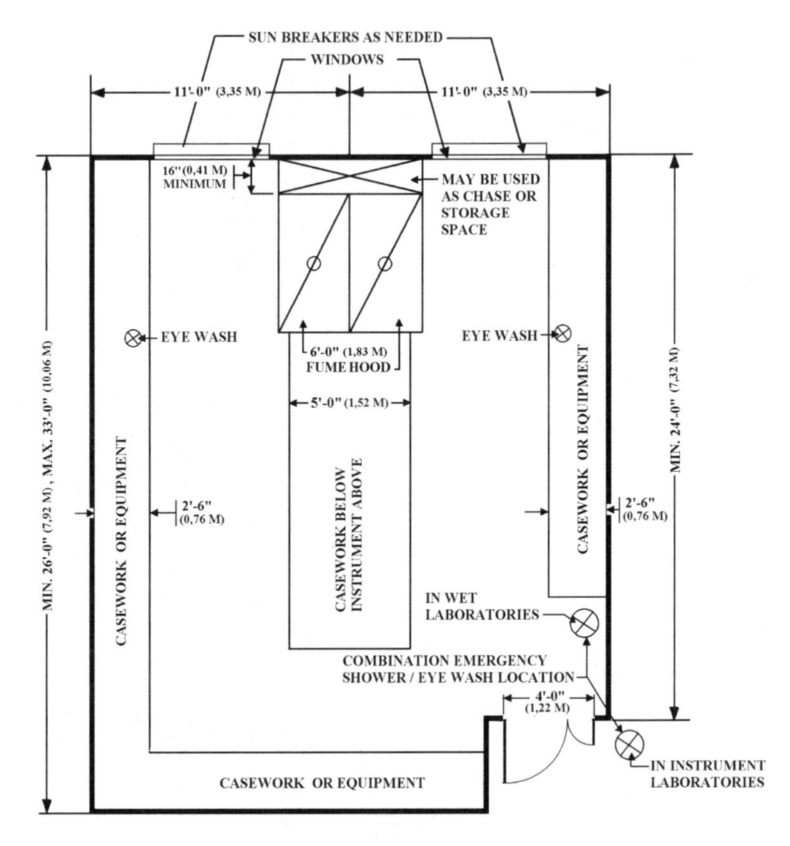

FIGURE 4-7
Two-module/two-fume-hood typical layout for a room with windows

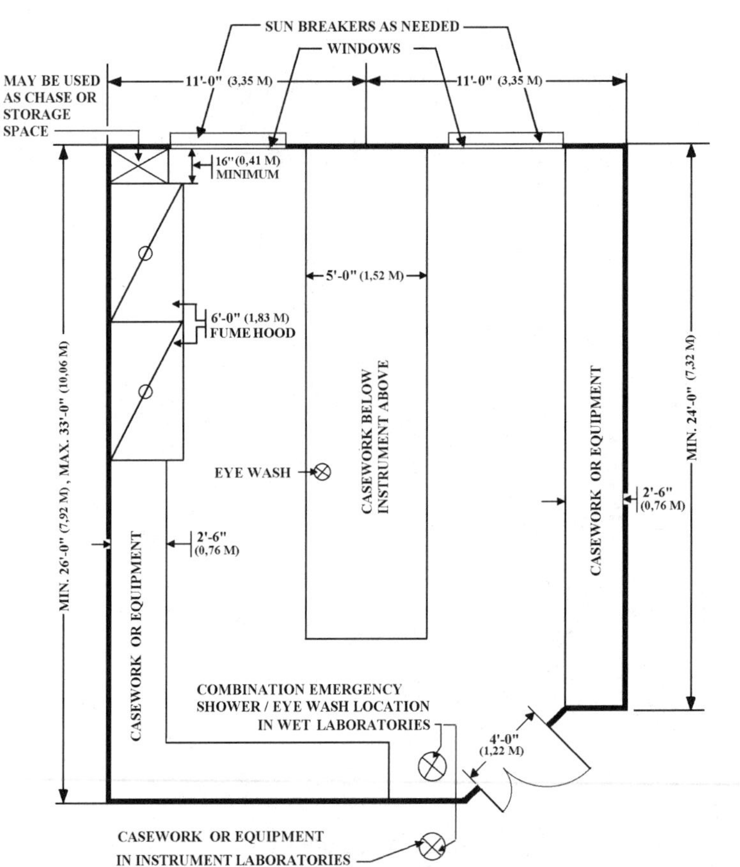

FIGURE 4-8
Two-module/two-fume-hood typical layout for a room with windows

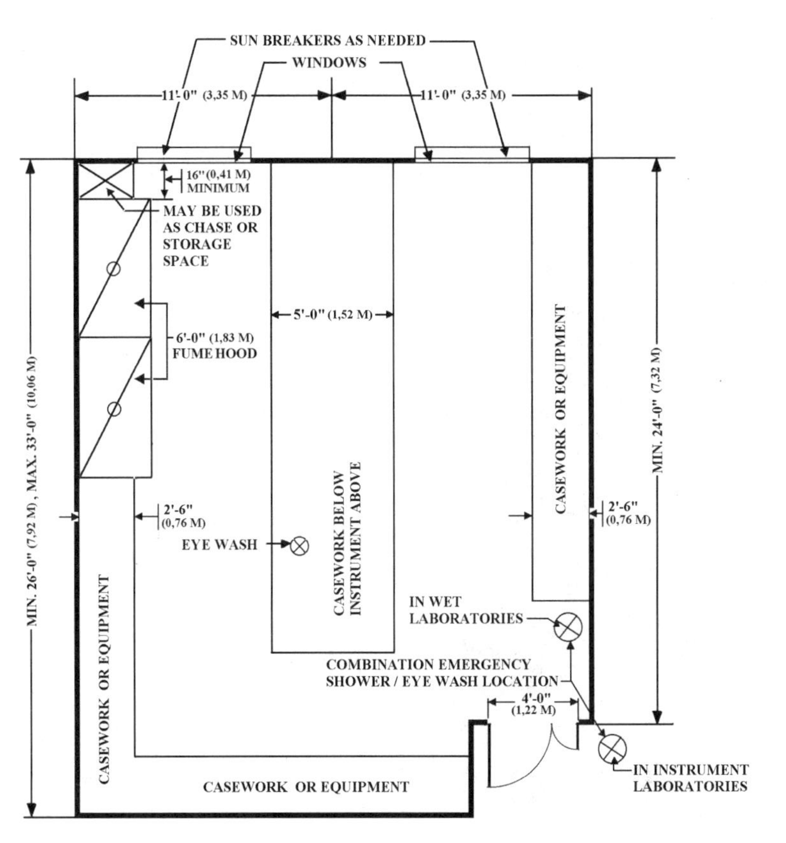

FIGURE 4-9
Two-module/two-fume-hood typical layout for a room with windows

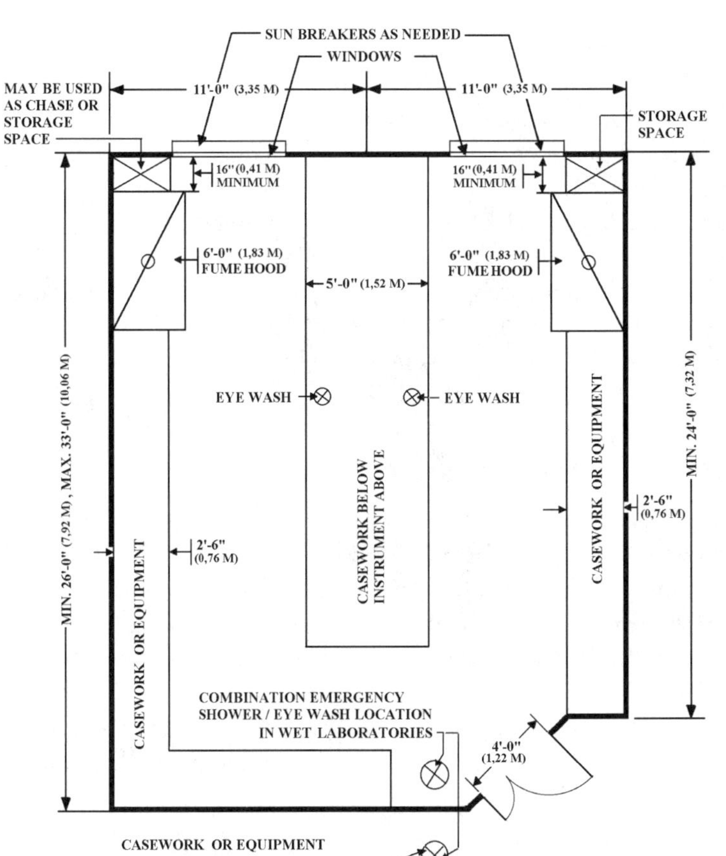

FIGURE 4-10
Two-module/two-fume-hood typical layout for a room with windows

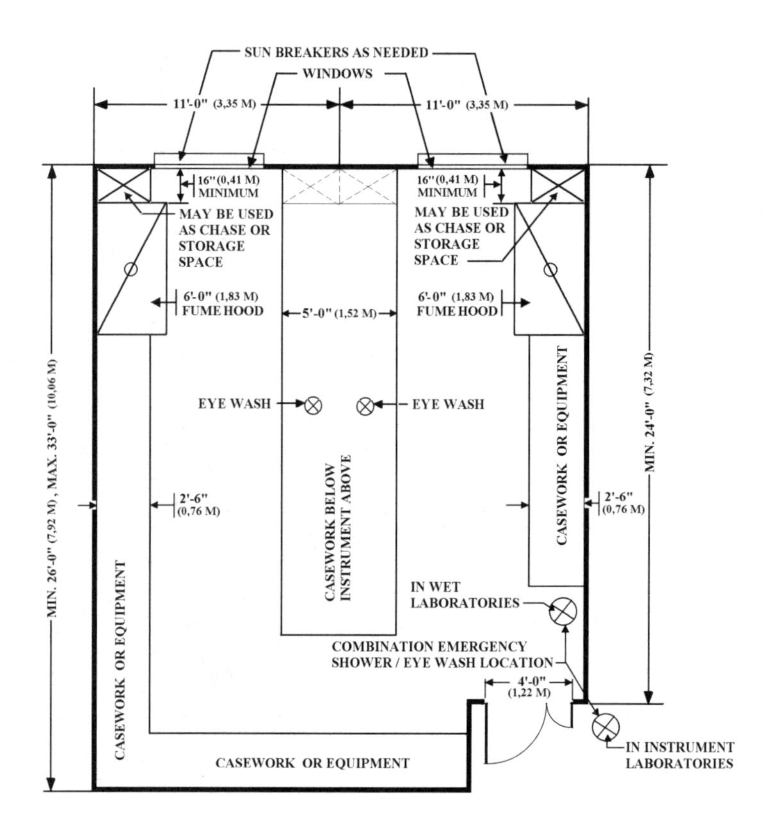

FIGURE 4-11
Two-module/two-fume-hood typical layout for a room with windows

As is the case with single-door rooms, doors in rooms with more than one door should be placed away from the fume hoods. The designer also should avoid anything else that would create air movement that would interfere with the planned airflow from the sources of supply to the hood face. Most codes allow a two-module research laboratory room to have only one primary exit door, as long as no hazardous work (as defined by NFPA-45) is undertaken in the room. This door leads to a corridor and preferably is recessed, regardless of the corridor width. As explained before, this avoids interference with the corridor cart and circulation of people should the door be opened briskly by an injured, panicked person.

It is good to have one eye wash device per hood, as well as one emergency shower per room in two-module laboratory rooms where chemicals are used. (The location of these devices is shown in the figures. Other requirements related to these devices are provided in chapter 9.)

Typical Layouts for Laboratory Rooms With Three or More Modules

Laboratory rooms with three or more modules also have casework along the length of the two side walls of the room. In a three-module laboratory room, peninsula-type work areas are centered on the imaginary lines dividing the modules. Thus, a three-module laboratory room has two peninsulas, and a four-module laboratory room has three peninsulas. Often, the nature of the functions may require empty space for special equipment in the location of one or more peninsulas. When this is the case, the designer naturally must oblige but should be careful to provide convenient services and utilities to the equipment that will be placed there.

Three fume hoods can be lodged in a three-module laboratory. Four fume hoods can be lodged in a four-module laboratory, and so on. The fume hoods' location requirements and relationships should be the same as those described for two-module laboratory rooms. When a laboratory room has three or more modules, it must have a minimum of two exit doors. These doors should always be placed away from any fume hood. If leading to an exit corridor, these doors should be recessed. No source of air disturbance that would create air movement that would interfere with the airflow from the sources of supply to the hood faces should be placed in the room. Suggested layouts for a three-module laboratory room are shown in figures 4-12 through 4-15.

It is good to have one eye wash device per hood in laboratory rooms with three or more modules where chemicals are used. It is also good to have a minimum of one emergency shower in such a laboratory. The location and number of the eye wash devices depend on the location of the hoods with respect to each other and of the arrangement of the room.

The preferred location of these devices is shown in the figures. Other requirements related to these devices are provided in chapter nine.

Eye Wash Devices and Emergency Showers

Eye Wash Devices

There are several different types of eye washes, which are described and regulated by the American National Standard Institute (ANSI). The advantages and disadvantages of each type are discussed in the industrial hygienist manuals. The eye

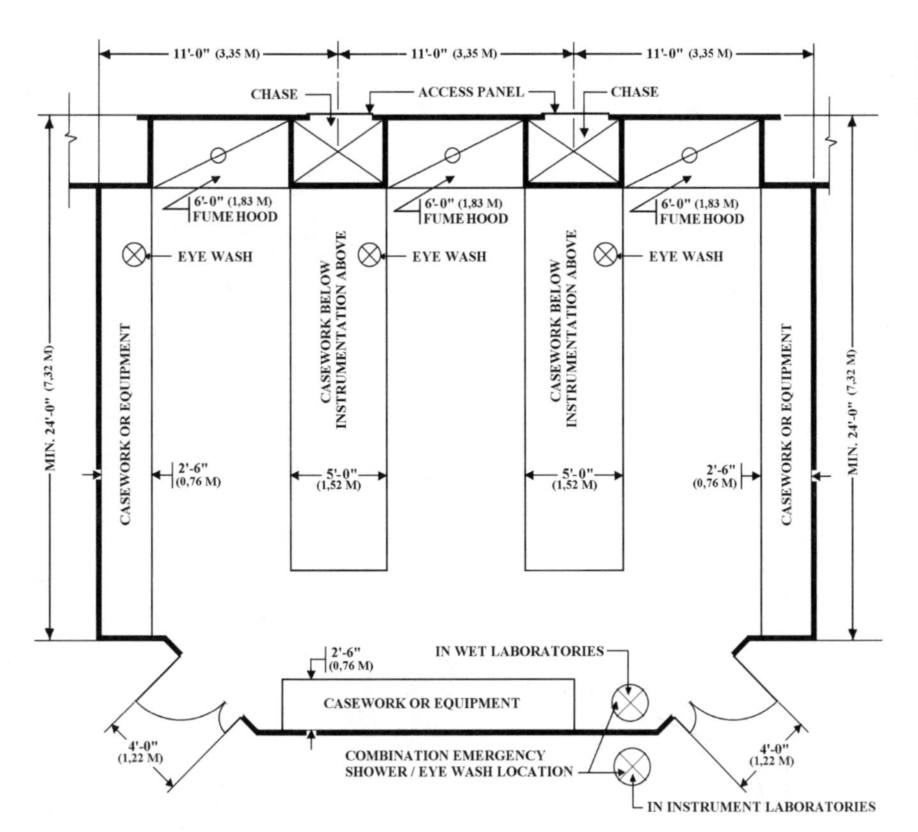

FIGURE 4-12
Three-module/three-fume-hood typical layout for a room without windows

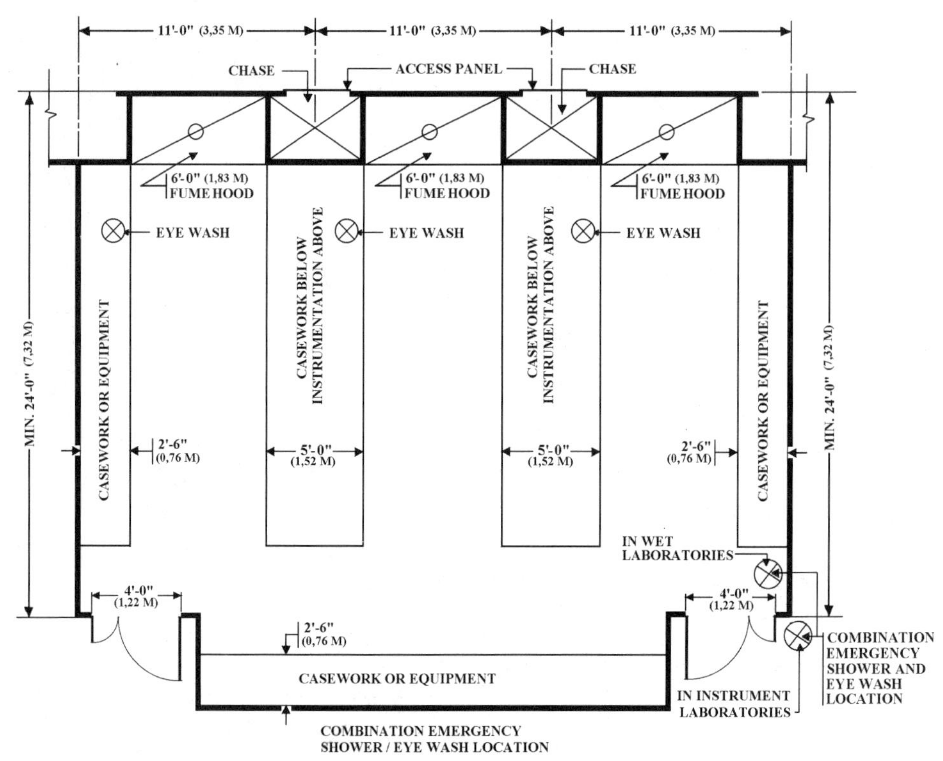

FIGURE 4-13 Three-module/three-fume-hood typical layout for a room without windows

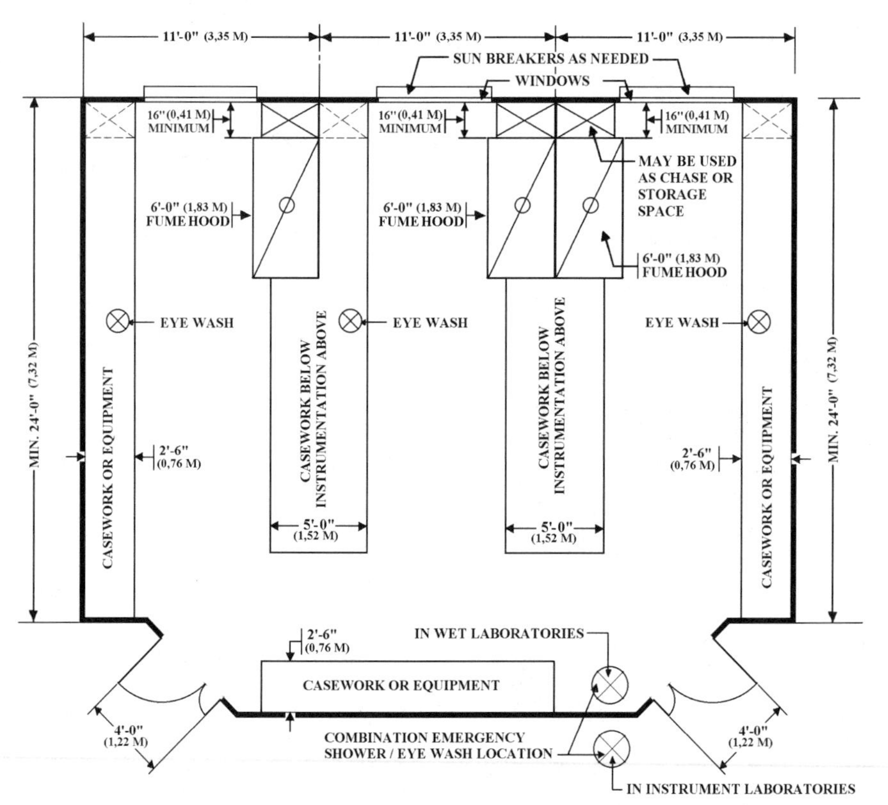

FIGURE 4-14 Three-module/three-fume-hood typical layout for a room with windows

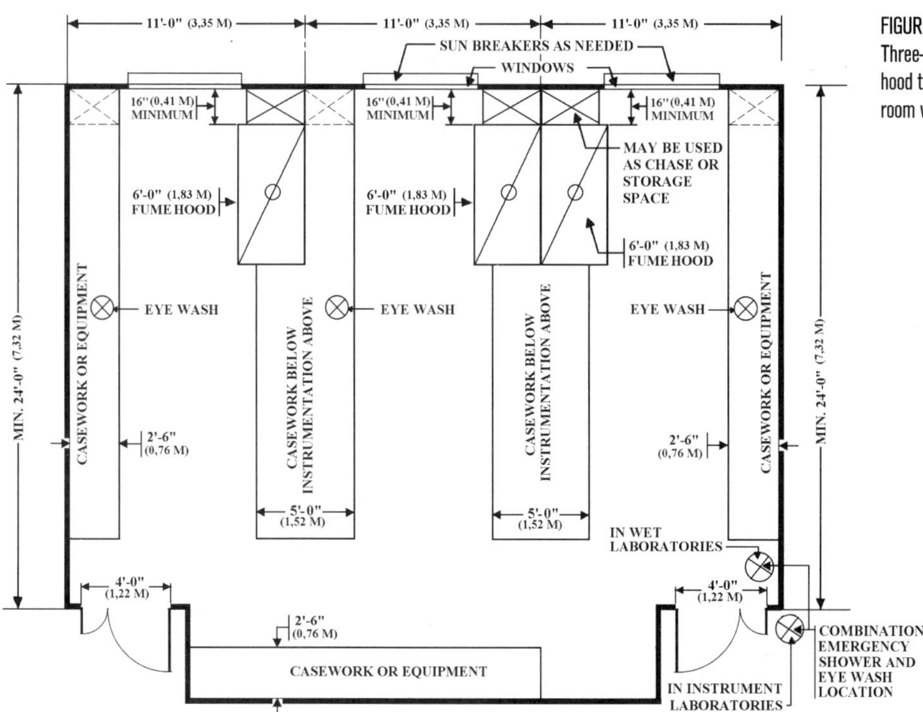

FIGURE 4-15
Three-module/three-fume-hood typical layout for a room with windows

wash device used in a general chemistry or biochemical laboratory must be permanent and located near the potential source of risk. This location is almost always near a fume hood, as shown in the layout figures. With very few exceptions, if any, the location of the eye wash device must be consistent in all the laboratory rooms. This means that it would always, for example, be two or three steps to the left (or right) of the potential source of risk. Eye wash devices may be freestanding or located next to a sink. This allows an injured person to find the eye wash.

As stated previously, it is good to have an eye wash device for every fume hood. There are exceptions to this rule. For example, when two fume hoods are side by side or next to each other, they can share one eye wash device. In such cases, industrial hygienists and engineers should discuss and analyze the situation and arrive at a sensible solution.

Emergency Showers

The emergency shower must be located next to the exit doors, either inside or outside the laboratory room. In a wet laboratory without electrical instruments, the emergency showers are generally located inside the laboratory room. In an instrument room (today most of them may fall under this category) with electrical instruments scattered over counters and elsewhere, the emergency showers preferably should be located outside the laboratory room. These locations must also be consistent in all the laboratory rooms. This means that the location of the emergency shower would always, for example, be two or three steps to the left (or right) of the exit door. This allows an injured person to go blindly from the eye wash to the emergency shower. Figure 4-16 shows a good design for an emergency shower installed out of the room and next to the entrance door.

FIGURE 4-16
An emergency shower installed out of the room and next to the door

In laboratory rooms with more than one exit door, it may be acceptable to have one emergency shower for every two or even three doors. As indicated earlier, however, the number of emergency showers depends upon the number of fume hoods, the potency and quantity of the chemicals used, the level of risk of the operations undertaken in the room, and other legal and operational factors and requirements. In any case, the emergency shower should always be located near the same door of each room (the door on the extreme right or extreme left side of the room) to allow an injured person to find it blindly.

It is the designer's responsibility to give a certain rhythm to his design for safety reasons. Once the individual rooms are designed and arranged into the laboratory wing, the designer should take a second look at the design to assure that there is a consistent relationship between the locations of the fume hoods, the eye washes, and the doors next to which the emergency showers are located.

Room Data Sheets

As noted at the beginning of this chapter, the design of a laboratory room starts with the preparation of an RDS. Each room must have its own RDS. The RDS is an information gathering tool as well as an information provider and, therefore, should contain all information and requirements pertaining to the laboratory room.

Exhibit 4-1 shows a blank RDS. There are two types of information included on the RDS: information provided by the users and their management, and information determined by the planning team (based on the information provided by the users). The planning team may provide an abbreviated version of the RDS for the users once the two sets of information are incorporated into the document. Then, if additional revisions are needed, the planning team reviews and adds the new information to the RDS to complete the document.

Guided by their management, the laboratory room users should identify the information related to the room: the room's I.D. and size, the activities that will take place in the room, and the number of people who normally work in the room. This information must be inserted in each room's RDS. The users should also describe the operations and tasks that will take place in the room, detailing the chemicals used and the flow in and out of the room and within the room. They may also use a flow chart to illustrate the way the operations are conducted within the room. They should list the chemicals that will be used in these operations, their concentration, the quantities used during a certain period, and how they will be stored.

The users should also list the instruments and equipment required to perform the tasks and operations. They should provide the sizes, footprints, and utilities and services requirements of the instruments and equipment, as well as the manufacturer information on how much heat and humidity the instruments and equipment may generate. They should also provide the location of the utilities and services outlets needed for each instrument in their proposed room layout. This information is then used by the health and safety specialists, industrial hygienists, fire protection engineers, and others to determine the hazard classification (laboratory class A, B, or C). Each of these three classifications has specific construction and safety requirements, which are discussed in the section on standard requirements.

EXHIBIT 4-1

ROOM DATA SHEET

ROOM NAME: ....

ROOM'S IDENTIFICATION (I.D.): ....
....

ACTIVITY/PROGRAM NAME: ....

ROOM SIZE: .... MODULE(S)

LOCATION OF ROOM IN BUILDING
OR BUILDING SECTION:
....

NUMBER OF PEOPLE WORKING
IN THIS ROOM: .... PERSONS

DESCRIPTION OF OPERATIONS UNDERTAKEN IN THIS LABORATORY ROOM:
....
....
....
....
....
....
....
....
....
....

IS THIS A WET LABORATORY .... IS THIS A DRY LABORATORY ....
(SEE DEFINITION OF WET AND DRY LABORATORIES IN STANDARD REQUIREMENTS)

SPECIAL REQUIREMENTS FOR THIS ROOM LOCATION:
....
....
....
....
....
....
....

LIST OF THE CHEMICALS USED IN THIS ROOM:

CHEMICAL NAME	CONCENT-RATION	QUANTITY USED IN GALLON/DAY	REMARKS

LABORATORY CLASSIFICATION:

CLASS: A ...., B ......, C ......

BIOLOGICAL CLASSIFICATION: ....

REMARKS: ....

LIST OF SPECIAL EQUIPMENT USED IN THIS ROOM:

EQUIPMENT NAME	MANUFACTURER	MODEL ID:	VOLTS/AMPS REQUIRED	REMARK SEE NOTE N*
			/	
			/	
			/	
			/	
			/	
			/	
			/	
			/	

EQUIPMENT NAME	MANUFACTURER	MODEL ID:	VOLTS/AMPS REQUIRED	REMARK SEE NOTE N*
....	....	....	.... /	....
....	....	....	.... /	....
....	....	....	.... /	....
....	....	....	.... /	....
....	....	....	.... /	....
....	....	....	.... /	....

ARCHITECTURE REQUIREMENTS:

ROOM FINISHES:

FLOOR: STANDARD OR

....

....

BASE: STANDARD OR

....

....

WALL: STANDARD OR

....

....

CEILING: STANDARD OR

....

....

SOUND CONTROL: STANDARD OR

....

....

DOORS:

DOOR NO.1 TYPE: STANDARD N* ....OR **DOOR MATERIAL:** ....**OPENING SIZE:** ....

DOOR HARDWARE: ....

REMARKS ....

....

DOOR NO.2 TYPE: STANDARD N* ....OR **DOOR MATERIAL:** ....**OPENING SIZE:** ....

DOOR HARDWARE: ....

REMARKS ....

....

DOOR NO.3 TYPE: STANDARD N* ....OR **DOOR MATERIAL:** ....**OPENING SIZE:** ....

DOOR HARDWARE: ....

REMARKS ....

....

CABINETS

STANDARD TYPE .... **MODULAR TYPE** ....

BASE CABINETS MATERIAL: STANDARD OR ....

DRAWER UNITS LENGTH: .... LINEAR FEET OR METERS

N* OF DRAWERS PER UNIT ....

DESCRIBE TYPE OF DRAWERS REQUESTED

....

CUPBOARD DRAWER CABINETS: .... LINEAR FEET OR METERS

SINK CABINETS: .... LINEAR FEET **CORNER CABINETS** .... LINEAR FEET OR METERS

FUME HOOD BASE CABINETS: .... LINEAR FEET OR METERS

FLAMMABLE STORAGE BASE CABINETS: .... LINEAR FEET OR METERS

SINGLE PLATE GLASS DOORS: .... LINEAR FEET OR METERS

SWINGING GLASS DOORS: .... LINEAR FEET OR METERS

SLIDING DOOR CABINETS: .... LINEAR FEET OR METERS

SWINGING PANEL DOORS: .... LINEAR FEET OR METERS

CORNER CABINET-SWINGING GLAZED DOOR: .... UNITS

METAL FULL HEIGHT STORAGE CABINETS: .... UNITS

CHEMICAL STORAGE CABINET .... UNITS **SIZE .....X .....X** FEET OR METERS

VENTED: YES ...... **NO** .....

OTHER CABINETS (SPECIFY):

....

....

....

....

WORK TOPS:

WORK TOP MATERIAL .... **WORK TOP LENGTH** .... LINEAR FEET OR METERS
WORK TOP WIDTH: .... LINEAR FEET OR METERS
WORK TOP SPECIAL REQUIREMENTS: ....
....
....

OTHER ARCHITECTURAL REQUIREMENTS:
....
....
....
....

MECHANICAL REQUIREMENTS

HVAC

HVAC SYSTEM: STANDARD .... OR DESCRIBE: ....
....
....

AIR SUPPLY/EXHAUST CONSTANT VOLUME .... **VAV** ....

AMBIENT CONDITION REQUIRED: STANDARD **...** OR
DESCRIBE: ....
....
....
....

EXHAUST SYSTEMS:

FUME HOODS:

NUMBER OF FUME HOODS: .... **SIZE OF FUME HOODS:** .... FEET OR METERS
FOR EACH FUME HOOD INDICATE:

1- TYPE OF FUME HOOD: STANDARD .... OR
DESCRIBE: ....
....
....

DESCRIBE SPECIAL REQUIREMENTS RELATED TO FUME HOOD:
....
....

2- TYPE OF FUME HOOD: STANDARD .... OR
DESCRIBE: ....
....
....

DESCRIBE SPECIAL REQUIREMENTS RELATED TO FUME HOOD:
....
....

3- TYPE OF FUME HOOD: STANDARD .... OR
DESCRIBE: ....
....
....

DESCRIBE SPECIAL REQUIREMENTS RELATED TO FUME HOOD:
....
....
....

EXHAUST VENT(S):

FOR EACH EXHAUST VENT INDICATE SIZES IN FEET OR METERS:

1- LOCALIZED EXHAUST VENT(S): ..... **SIZE OF VENT CROSS SECTION:** ....
EXHAUST VENT(S) MATERIAL: ....
TYPE: ....
EXHAUST VENT(S) LOCATION: ....

2- LOCALIZED EXHAUST VENT(S): ..... **SIZE OF VENT CROSS SECTION:** ....
EXHAUST VENT(S) MATERIAL: ....
TYPE: ....
EXHAUST VENT(S) LOCATION: ....

3- LOCALIZED EXHAUST VENT(S): ..... **SIZE OF VENT CROSS SECTION:** ....
EXHAUST VENT(S) MATERIAL: ....
TYPE: ....
EXHAUST VENT(S) LOCATION: ....

DESCRIBE SPECIAL REQUIREMENTS RELATED TO EXHAUST VENTS: ....
....
....
....
....

VENTED CHEMICAL STORAGE CABINETS:

FOR EACH VENTED CHEMICAL STORAGE CABINET INDICATE SIZES OF VENTS IN INCHES OR CENTIMETERS:

1-VENTED CHEMICAL STORAGE CABINET: .... TYPE: SOLVENTS .... ACIDS ......
SIZE........ SIZE OF VENT .... LOCATION: ....

2-VENTED CHEMICAL STORAGE CABINET: .... TYPE: SOLVENTS .... ACIDS ......
SIZE........ SIZE OF VENT .... LOCATION: ....

3-VENTED CHEMICAL STORAGE CABINET: .... TYPE: SOLVENTS .... ACIDS ......
SIZE........ SIZE OF VENT .... LOCATION: ....

DESCRIBE SPECIAL REQUIREMENTS RELATED TO VENTED CHEMICAL STORAGE CABINETS:

....

....

....

OPERATING HOURS HVAC & EXHAUST SYSTEMS: ....

....

ADDITIONAL INFORMATION ON HVAC AND EXHAUST SYSTEMS: ....

....

....

....

PLUMBING:

PLUMBING FIXTURES REQUIRED (SPECIFY SIZES IN INCHES OR CENTIMETERS):

FIXTURE	SIZE	# OF UNITS	LOCATION
....	.....x.... x.....	....	....
-................................	.....x.... x.....	....	....
....	.....x.... x.....	....	....
....	.....x.... x.....	....	....
....	.....x.... x.....	....	....
....	.....x.... x.....	....	....
....	.....x.... x.....	....	....
....	.....x.... x.....	....	....
....	.....x.... x.....	....	....
....	.....x.... x.....	....	....
....	.....x.... x.....	....	....

UTILITIES REQUIRED:

OUTLET FOR:	# OF UNITS	LOCATION
BURNING GAS:	....	....
HOT WATER:	....	....
VACUUM:	....	....
COMP. AIR:	....	....
DISTILLED WATER:	....	....
NITROGEN:	....	....
OTHER SPECIFY:	....	....
....	....	....
....	....	....
....	....	....
....	....	....
....	....	....
....	....	....
....	....	....
....	....	....
....	....	....

FIRE PROTECTION REQUIRED:

STANDARD .... OR DESCRIBE: ....

ELECTRICAL REQUIREMENTS

LIGHTING:

REGULAR ROOM LIGHTING: STANDARD .... OR
DESCRIBE: ....

DESCRIBE ADDITIONAL LIGHTING REQUIRED: ....
....
....
....
....

ADDITIONAL LIGHTING OUTLETS (HOW MANY AND WHERE): ....
....
....
....
....

POWER OUTLETS

CONVENIENCE ELECTRIC POWER OUTLETS: STANDARD .... OR DESCRIBE: ....
....
....

ELECTRIC POWER OUTLETS (IN ADDITION TO STANDARD REQUIREMENTS AND TO THE OUTLETS NEEDED FOR THE EQUIPMENT DESCRIBED IN THIS ROOM DATA SHEET):

OUTLET SPECIFICATION: .... **OUTLET LOCATION:** ....

OUTLET SPECIFICATION: .... **OUTLET LOCATION:** ....

OUTLET SPECIFICATION: .... **OUTLET LOCATION:** ....

TELEPHONE AND COMPUTER OUTLETS:

THIS ROOM SHOULD HAVE: .... **TELEPHONE OUTLETS LOCATION(S):**
PHONE
No1: ....
PHONE
No2: ....
PHONE
No3: ....

THIS ROOM SHOULD HAVE: .... **CENTRAL COMPUTER OUTLET LOCATION(S):**
OUTLET
No1: ....
OUTLET
No2: ....
OUTLET
No3: ....
OUTLET
No4: ....

ROOM LAYOUT REQUIREMENTS:

PREFERRED ROOM ARRANGEMENT

SHOW SCHEMATIC DRAWING WITH PREFERRED LOCATION OF HOOD(S), CASEWORK, INSTRUMENTS AND EQUIPMENT, AS WELL AS PREFERRED LOCATION OF EACH OF THE SERVICES AND UTILITIES REQUESTED. YOU MAY USE THE DIFFERENT ROOM LAYOUTS SHOWN IN THIS CHAPTER AS A BASE. USE SYMBOLS TO IDENTIFY EACH SERVICE OR UTILITY REQUESTED.

SPACE FOR THE ROOM LAYOUT SHOULD BE PROVIDED IN THE ROOM DATA SHEET

STANDARD REQUIREMENTS AS USED IN ROOM DATA SHEETS

Once a laboratory room is classified, the architect and engineers involved in the programming of the laboratory requirements will, with the laboratory users' involvement, determine and describe the architectural, mechanical, and electrical requirements for each laboratory room. These requirements are derived from codes pertaining to health and safety and from the users' suggestions regarding the facilitating of operations. They are later used as specifications for construction; for the finishing, type, and size of the doors; for ventilation, temperature, and humidity controls; for electric outlets and lighting; and for other requirements.

Many of these requirements may be common to most rooms. Therefore, instead of repeating them each time reference is made to a different laboratory room, it is more practical to group them under a single heading, Standard Requirements, and to refer to them when they apply to the room being considered. It makes more sense to give complete required specifications for a floor system, a ceiling type and construction, or a common door size once. Every time this floor, ceiling, or door is referenced for a particular room, the standard name and/or N* is given in place of the full set of specifications. The list of standard requirements should be developed by the architect and engineers involved in the programming of the laboratory requirements. With the laboratory users involvement, they determine and describe the specifications of each architectural, mechanical, and electrical requirement. The final list of specifications constitutes what is called the "Standard Requirements as Used in Room Data Sheets."

Exhibit 4-2 is a made-up set of requirements put together to illustrate how the Standard Requirements as Used in Room Data Sheets should look. The requirements were extracted from those developed for a general chemistry laboratory and modified as needed for the purposes of this book. Many of the items on it are preceded or followed by an explanation or "how-to" description, and the reader is reminded of some definitions and standards that must be respected when the RDS is being completed. Note also that many of the explanations provided may or may not be necessary in the final document.

•

EXHIBIT 4-2
STANDARD REQUIREMENTS AS USED IN ROOM DATA SHEETS

It is always good to have the following statement at the beginning of the Standard Requirements as Used in the Room Data Sheets: "The narrative description of requirements in the Program of Requirements shall take precedence over drawings. If an item described in the narrative is not shown in a drawing, that should not be taken as a waiver of the requirement."

LABORATORY CLASSIFICATION

The preparer of the Standard Requirements As Used in Room Data Sheets should provide a definition of laboratory rooms' classification to assist the RDS preparer in determining how to classify each room. These definitions should be used as a guide:

- A dry laboratory (also called an instrumentation room) is a laboratory where instruments are used to do most of the work, such as to analyze samples. Once set, these instruments do most of the work for the operator.
- A wet laboratory is used for extraction, digestion, distillation, and similar sample preparation procedures. Wet laboratories require active operator involvement. A preparation room is an example of a wet laboratory.

The preparer of the RDS should know that laboratories are classified in NFPA-45 according to the level of hazard. Class A laboratories are high-hazard. Class B laboratories are intermediate-hazard. Class C laboratories are low-hazard. This classification is based on the quantity of flammable and combustible liquids used in the laboratory room. It takes into account whether the liquids are stored in safety storage cabinets or safety cans and whether the laboratory room is sprinklered or not.

Other NFPA classifications should also be looked at when preparing the RDS. Among these classifications are NFPA-101 (based on use of the building or room), NFPA-13 (based on the fire hazard of the use of building or area), and NFPA-231 (based on the type of materials stored and their burning characteristics).

The industrial hygienist and fire protection engineer should work with the laboratory users and management in determining each laboratory classification, as well as the applicable building codes. The classification of a laboratory determines its architectural, mechanical, and other requirements.

ARCHITECTURAL STANDARD REQUIREMENTS

Floors

The architect orchestrating the laboratory programming and the user should determine together what is the most adequate type of standard floor for each room. Several types of floors, each of which have advantages and disadvantages, are discussed in the general architectural requirements in the program of requirements (see chapter 1, fourth task). Common standard floors for laboratory rooms without specific requirements dictated by the tasks taking place in them are:

Vinyl tile floor, 12 x 12 x 1/8 (300 x 300 x 3.2 mm) inches thick, as manufactured by (give name of manufacturer) or approved equal; 35% to 40% reflectance, high density, meeting requirements of (give specific standards or code that the specifications (specs) of the floor should meet) specifications.

Chemical-resistant floor, 1/8 inch (3.2 mm) thick, as manufactured by ______ or approved equal, meeting requirements of ______ specifications.

Vinyl seamless floor, as manufactured by ______ or approved equal; 35–40 percent reflectance, high density, meeting requirements of ______ specifications.

Other types of floors, such as epoxy finished floors, quarry tile floors, wood floors, or concrete floors, may be required. If so, they should be specified here.

Bases

In most cases, bases could be specified as follows:

For vinyl tile floors, use a vinyl or rubber base.

For seamless floors, the base shall also be seamless at its joint with the floor and between its sections. A continuous seamless coved base 4 or 6 inches (0.1 or 0.15 m) high may also be used.

For epoxy floors, an epoxy base of at least 4 inches (0.1 m) shall be added.

For quarry tile floors, the base shall also be of quarry tiles.

For wood floors, the base may be wood, vinyl, or rubber.

Walls

In most instances, the walls and partitions are either insulated masonry or insulated gypsum board. In some cases, when containment is required, it is good practice to extend the walls from the floor to the underside of the upper floor or roof slab and seal the joints between the two. This diminishes the possibility of cross-contamination between two adjacent laboratories and adds to the sound insulation between the two rooms. In certain situations, the extension of the walls from the floor to the underside of the upper floor or roof slab is a code requirement.

Wall Finishing

Wall finishings could be specified as follows:

The walls' and partitions' surfaces shall be painted with glossy or semigloss latex-based paint.

In instrumentation rooms, walls shall be covered with sound-absorbing and -dampening panels. The panels must have a high sound-absorption coefficient (as close to 1.0 as possible) relative to the kind of sound emitted by the instruments in the room. Reverberation in walls shall be reduced to a minimum by the sound-absorbent panels.

Flame spread and smoke development specifications are specified in the general architectural requirements of the program of requirements. Walls must be reinforced to hold wall shelving and casework.

Ceilings

Usually laboratories' ceilings are suspended acoustical tiles. When that is the case, tiles should be of a nonflaking material. Ceilings in extraction, preparation, glassware washing, microbiology, and similar wet laboratories are generally of solid, water-resistant materials—either green board or cement plaster. When suspended acoustical tiles are used, ceiling grid suspension strips throughout the laboratory wing may be made of metal coated with PVC or other corrosion-resistant materials. The ceiling grid suspension strips must be made of metal or other nonflammable materials if the ceiling needs to be fire rated.

In all laboratory spaces that contain or may contain fume hoods, the ceiling height should be a minimum of 9′-8″ (3 m). It is good practice to require that these ceilings must respond to the same sound-absorption requirements previously described for the section on walls.

Doors

Most laboratory doors are a standard height of 7′-0″ (2.17 m). In a laboratory wing, doors opening from laboratory rooms should not protrude more than 6 inches (0.15 m) into corridors regardless of the corridor's width. This is to avoid hitting pushcarts loaded with chemicals when opening the door. Further, each laboratory room should have one 4′-0″ (1.22 m) opening to allow large equipment to be carried in and out of the room. This is best achieved with a 3′-0″ (0.91 m) active door (which will be regularly used) and a 1′-0″ (0.31 m) inactive door with top and bottom bolts.

The following examples show how six standard doors could be specified.

Door type A: Hallway access doors; pair doors; 3′-0″ (0.91 m) (active) with 1′-0″ (0.31 m) inactive door (with top and bottom bolts); wire glass (4 feet x 25 inches or 5 feet x 20 inches) (1.22 x 0.64 m or 1.52 x 0.51 m) vision panel; no threshold; push plate and pull bar; automatic closure.

Door type B: Interconnecting door (between laboratories); 3′-0″ (0.91 m); push plate; vision panel (4 feet by 25 inches or 5 feet x 20 inches) (1.22 x 0.64 m or 1.52 x 0.51 m); dual swing.

Door type C: Interconnecting doors (between wings); 4′-0″ (1.22 m) with panic bar hardware; wire glass on vision panel (4 feet x 6 inches) (1.22 x 0.15 m); automatic closures.

Door type D: Exterior fire doors; 4′-0″ (1.22 m) with panic bar hardware; automatic closure.

Door type E: Hallway access door; 3′-0″ (0.91 m) with glass vision panel; no threshold; push plate and pull bar; automatic closure.

Door type F: Interconnecting sliding door (between laboratories); 3′-0″ (0.91 m) with large vision panel; electric eye and mechanism for automatic opening and closing.

Door specifications are provided in the general architectural requirements in the program of requirements. In this laboratory facility, all laboratory doors shall be 7′-0″ (2.13 m) high.

Casework

There are two basic types of casework or cabinetry: fixed and flexible. These types are fully described in chapter 6. The following is one way to specify required casework in all laboratory rooms:

All standard casework shall be of a modular, flexible furniture system. Casework shall be of metal construction, unless wood, plastic laminate, polypropylene, or other materials are indicated on the RDS. Casework shall have components, configuration, materials, finish, and performance comparable to Company #1, Company #2, and Company #3, as manufactured by (___) Industries, (___) Corp., and (___) Company. Equipment manufactured by others may be acceptable, based on products of equal performance and similar appearance and construction.

Unless otherwise noted in the RDS, peninsulas shall not have reagent shelves. Six-inch (0.15 m) drawers are standard in the base drawer units. All units shall include label holders on all drawers and doors. With above-counter wall-mounted casework units, the wall above the unit shall be brought flush with casework front. Unless otherwise noted, the height of base cabinets shall be 36 inches (0.91 m), except where knee spaces are required. At knee spaces the countertop height shall be 29 inches (0.74 m).

Vented acid/base storage cabinets shall be x′ y″-wide (2m) metal cabinets. The inner surfaces of the cabinet shall be factory coated to resist acid/base fumes and spills. Two adjustable shelves shall be provided. Venting shall be as for vented chemical storage cabinets. Spill trays must be installed at the bottom.

Countertops

There are several types of countertops that are fully described in chapter 6. The following is one way to specify required countertops:

Standard countertops shall be man-made stone impregnated with chemical (e.g., acids, bases, solvents) -resistant epoxies as manufactured by Company #1, Company #2 and Company #3, or approved equal. A total width of approximately 60" (1.52 m) shall be standard for peninsulas. The standard depth for casework along walls shall be 30" (0.76 m). Countertops adjacent to sinks shall have grooved drain boards. All countertops shall be 1 1/4 inch (0.03 m) thick.

Countertops at Peninsulas

Countertops at peninsulas could be specified as follows:

In laboratory rooms with peninsulas where instruments need to rest on the countertops, the countertop of each peninsula shall be of the same height and be level across the entire width of the peninsula, except if otherwise specified.

Knee Space

In most cases, knee spaces should be 3 feet (0.91 m) in width and 29 inches (0.74 m) in height.

Hallway Closets

When called for, hallway closets usually are approximately 3 x 3 feet (0.91 x 0.91 m) and are placed in strategic locations in the laboratory wing. The following could be the specifications for hallway closets:

Hallway closets shall house laboratory supplies for spill cleanup and should include __ shelves on two walls. The closets should be located with equal travel distance between them.

MECHANICAL STANDARD REQUIREMENTS

HVAC Systems

The HVAC system for the laboratory wing of the building may be one of the two systems described hereafter, based on the need of the operations conducted in this wing and on a life-cycle cost analysis. The system with the lower life-cycle cost should normally be selected once needs of operations are satisfied. The two systems are:

System 1 is a two-position, constant volume terminal reheat system that offers the possibility of setback when hoods are not used or at night, weekends, and holidays. This system will provide the required volume of air in full during daytime operation hours when the fume hood sash is open. When the fume hood sash is closing, this system will, by controlling the speed of the supply and exhaust fans, reduce the airflow and thus the volume of air exhausted by the hood.

System 2 is a variable air volume system (VAV). It is designed to utilize variable volume hoods with variable speed exhaust fans, variable volume air supply terminals, and variable volume general area exhausts. The design modulates these three system components in such a way as to have them respond to a digital controller sensing hood position and differential pressure in the laboratory area.

Regardless of the system used, in most laboratories with one-pass air the fume hoods' exhaust systems are mostly used to exhaust air from the laboratory wing. When this is the case, all exhaust systems will be operated 24 hours a day. The setback capability of each of these two systems would generally provide a minimum exhaust volume for a 6-foot-long (1.83 m) fume hood of 250 cfm (0.12 m^3/s) with the sash fully closed. A minimum of four to six air changes per hour should be maintained in each laboratory room. This minimum is still required during the systems setback, regardless of whether the room has a fume hood in it. This minimum should also maintain the required room temperatures in winter and in summer.

HVAC Requirements

This is an example of the most often used HVAC requirements:

The standard heating, ventilating, and air conditioning of each laboratory room shall be one-pass air with maximum exhaust through hoods where hoods are used. The HVAC systems shall be continuously operational—24 hours a day, 7 days a week, summer and winter. The design temperatures shall be as follows:

Every laboratory room shall be controlled individually in accordance with the following:

summer, 72° FDB (22.2° C (DB) +/- 2° F (1.1° C) and 40% RH +/- 5%

winter, 72° FDB (22.2° C (DB) +/- 2° F (1.1° C) and 30% RH +/- 5%

For laboratories that are primarily instrumentation rooms, the standard shall be 70° FDB (21.1° C (DB) +/- 2° F (1.1° C).

Plumbing

Sinks

All the sinks in this example are made of epoxy resin. In some instances, stainless steel sinks may be a better choice. When this is the case the room user should inform the RDS preparer. This is an example of how sink standards could be specified:

Unless otherwise specified in the RDS, sinks shall conform to the standards described below and shall be as manufactured by Company #1, Company #2, or Company #3, or approved equivalent. Each sink shall be identified in the RDS as shown here:

TYPE	MATERIAL	SIZE (length x width x depth)
A	Peninsula single tub, epoxy resin	24″ x 13″ x 14″ (0.61 x 0.33 x 0.36 m)
B	Side-wall single tub, epoxy resin	25″ x 10″ x 15″ (0.64 x 0.25 x 0.38 m)
C	Side-wall glassware washing double tub, epoxy resin	37″ x 11″ x 15″ (0.94 x 0.28 x 0.38 m)

Eye Wash Device

One eye wash device should be required for every source of risk. This eye washer should be located near the source(s) of risk in all preparation rooms. The frequency of eye washers in instrument rooms should be one per small room, or two in larger rooms. See the preferred room arrangement of the RDS for location.

Emergency Showers

The standard for emergency showers in the laboratory wing could be as follows:

One emergency shower unit shall be located outside the corridor door of each instrument laboratory when the laboratory room has only one door. If the laboratory room has more than one door, the emergency shower unit shall always be placed outside the right or left door.

One emergency shower unit shall also be provided either inside each preparation rooms (close to its entrance door and in the path from the risk area to this door) or outside the preparation room door (if the laboratory has only one door). If the preparation room has more than one door, the emergency shower always shall be placed outside the right (or left) door. See the preferred room arrangement of the RDS for location.

Emergency showers preferably should have floor drains that empty into a holding tank before the water is wasted. See paragraph ____ of the program of requirements for complete specifications.

The design of floor drains shall be such that the floor around the drain is slightly pitched toward the drain. The pitch shall direct the shower water to the drain without causing a safety hazard. The floor drains for the emergency showers shall also be located in a depressed area that will collect the shower water and shall be covered with a grid that is level with the surrounding floor.

Central Deionized (DI) System

When a central DI system is required it could be specified as follows:

The required central DI system shall be a loop system designed to provide water with a resistivity greater than 10 megaohms at the tap. See paragraph __ of the program of requirements for complete specifications.

Nonflammable Gas Distribution System

The following type of information should be provided for the specification of nonflammable gas systems. Note that the gases selected would be different for each laboratory:

Outlets for the required gases shall be provided every (__) feet, or as specified in each RDS. Pressure regulators for the gases shall be placed at the gas bottles (or at the end of the manifolding system, if several bottles are manifolded together) and at a location inside the laboratory room where the gas is used. From there, the gas is distributed to each outlet as required by the RDS. The exact location of each outlet, by measurement, shall be provided by the client during the early design phase. The number of gas outlets required in each room shall be indicated in each RDS.

Unless otherwise noted in the RDS, the following gases shall be provided in each room at the stated purity. The number of tanks used per month (TPM) is given in the appropriate RDS. (This is an example for gas requirements.)

GAS	REQUIRED PURITY	MAXIMUM PRESSURE
Air	Breathing quality: 19–23% oxygen, 77–81% nitrogen	60 psi (pounds per square inch) (4.14 bar)
Compressed Air	Instrument-grade air; Oil- and water-free	60 psi (4.14 bar)
Nitrogen (N_2)	Ultrapure; 99.999% or better	60 psi (4.14 bar)
Helium (He)	Ultrapure; 99.999% or better	60 psi (4.14 bar)
Argon (Ar)	Ultrapure; 99.999% or better	60 psi (4.14 bar)
Argon/Methane (Ar/CH_2) (95%/5%)	High purity; <1 ppm oxygen, water, carbon dioxide	60 psi (4.14 bar)
Liquid Nitrogen (Liq. N_2)	Cryogenic grade	____
Liquid Argon (Liq. Ar)	Cryogenic grade; water-free	____
Carbon Dioxide (CO_2)	High purity; 99.5%	60 psi (4.14 bar)

Fixtures Materials

Fixtures materials could be specified as follows if work undertaken in the laboratory would be affected by metals:

Fixtures for utilities (e.g., water faucets and spigots, gas jets or nozzles, etc.) used in all laboratory rooms shall be made of PVC or equivalent corrosion-resistant materials; metal fixtures shall not be used.

Fire Protection

The whole structure (each room) should be sprinklered. Instrumentation laboratories should have a dry pipe sprinkler system. Class ABC dry chemical fire extinguishers should be provided in instrumentation laboratories; CO_2 fire extinguishers should be provided in preparation rooms at a frequency of one per module. Extinguishers should be no less than 10 lbs. (4.55 kg) each.

Fume Hoods

This is an example of the specifications for fume hoods:

Except as noted on specific RDSs, all fume hoods used in laboratory rooms shall be 6'-0" (1.83 m), welded stainless steel. Blowers shall be lined with acid-resistant material and shall be certified acid-resistant by the manufacturer. All fume hoods shall have two 120 V duplex outlets and one 208 V outlet on the face of the hood.

The following services shall be standard in all fume hoods: hot and cold water, vacuum, chill water, deionized water. Hoods shall be equipped with audible alarms to indicate malfunction.

Noise Control

The noise control standard for a given laboratory room should be:

The noise level at the face of the hood shall not exceed 70 dB(A) with the system operating; nor should it exceed 55 dB(A) at benchtop level elsewhere in the laboratory room.

Electrical Standard Requirements

A laboratory room should be provided with several types of electrical outlets, which fall into two categories: standard electrical convenience outlets and specific equipment outlets.

Standard Electrical Convenience Outlets

This standard for all laboratory rooms in the laboratory facility can be used as an example:

Outlets shall be duplex convenience 120 V/20 A outlets as defined in the general electrical requirements of the program of requirements. These outlets shall be provided in addition to the specific outlets called for, or shown in respective RDSs to feed the equipment and instruments used in each room.

These outlets shall be located either in the reagent shelf, or, if no reagent shelf is required, 8 inches (0.20 m) above countertop level when base cabinets are used and 18 inches (0.45m) above floor level in other locations. When required in peninsulas with level countertops, these outlets shall be quadruplex outlets at 3'-0" (0.91 m) on center line or center line to center line and placed in small posts raised at the center of the peninsulas. When specifically required, units shall have waterproof flush-mounted covers.

All outlets shall be no more than 3'-0" (0.91 m) apart in any of the above described situations. Face plates shall be steel. See also the general electrical requirements of the program of requirements for more specifications.

Specific Equipment Outlets

To guide the RDS preparer in identifying the need for special equipment outlets, the architect and engineer preparing the standard requirements document should provide a statement like the following:

Specific equipment outlets are outlets required to feed the equipment used in each room. Once a piece of equipment is identified under the listing of "Special Equipment Used in This Room," the RDS preparer does not need to indicate the required electrical outlets for this piece of equipment. However, the equipment location must be indicated on the room layout. Electrical outlet location should be near the equipment to be powered; the exact location of equipment and outlets should be determined by the user representative during an early design stage.

Instruments and Outlets Connected to Uninterruptible Power Supply and/or Emergency Generator

The abbreviations "UPS" and "EG" denote that instruments or outlets are connected to the uninterruptible power supply or to the emergency generator. Color-coded outlets should be used. Use white for UPS, red for EG, and black or gray for standard.

Additional Requirements

The following are examples of what the electrical additional requirements could be.

Special Electrical Outlet Requirements for Local Area Networks (LAN) and Laboratory Information Management System (LIMS)

In addition to the above requirements, six (6) electrical duplex 120 V/15 A outlets shall be provided for each LAN outlet and two (2) electrical duplex 120 V/15 A outlets shall be provided for each LIMS outlet. (These numbers may vary depending on the needs of the equipment used.) See LAN and LIMS outlet requirements stated later in this chapter.

The required outlets shall be located within 2 feet (0.62 m) of the associated computer outlet. The outlets must be marked "125 V, 15 A, UPS protected, for computer equipment only." These receptacles shall be red in color.

Lighting

Laboratory standard lighting shall be fluorescent, uniform lighting with two levels of lighting at bench top level (+36 inches from floor level) (+0.91 m). The high level shall be 100 foot-candles maintained at 36 inches (0.91 m) from floor level and the low level shall be 50 foot-candles. Light fixtures shall be activated with two switches: one to activate the low-level lighting and the other to activate the high level lighting. See also the general electrical requirements of the program of requirements for more specifications.

Emergency Lighting

Emergency lighting intensity shall be 5 foot-candles inside each laboratory room and 3 foot-candles (or higher) throughout the exit path. See the specific architectural requirements of the program of requirements for more specifications.

Switches

One switch shall be provided at each door that leads to a hallway egress. See also the specific architectural requirements of program of requirements for more specifications.

Emergency Power System

The electrical engineer preparing the standard requirements document should provide a statement similar to the following:

Certain pieces of equipment used in the laboratory rooms need to be connected to the emergency generator or to the uninterrupted power supply (UPS) or to both. These pieces of equipment are identified in the "List of Special Equipment Used in This Room" of the RDS. See also the general electrical requirements of the program of requirements for more specifications.

Outlets

The following are examples of how to specify telephone, LAN, and LIMS outlets.

Telephone Outlets

One telephone outlet is required per single module space. The exact location for each outlet shall be determined by the user representative at an early design stage.

LAN (Local Area Network) Computer Outlets

One LAN computer outlet is required per single module space. The exact location for each outlet shall be determined by the user representative at an early design stage.

LIMS (Local Information Management System) Computer Outlets

One LIMS computer outlet is required per single module space. The exact location for each outlet shall be determined by the user representative at an early design stage.

Fixtures and Outlet Cover Plates

All telephone, computer, and electrical outlets shall have PVC or equivalent corrosion-resistant cover/face plates; metal cover shall not be used.

Chapter 5

Designing the Laboratory Wings

THE design of the laboratory wing is affected by the requirements stated in the RDS and by many other complex and sometimes conflicting requirements, including those related to operation, health, and safety, as well as to architecture, civil engineering, mechanics, electrification, and sustainable design. The combined and successful response to all these requirements should result in an efficient, safe laboratory.

The designer's first step in designing the laboratory wings should be to confirm his or her understanding of the project's requirements as well as those of the individual rooms that make up the laboratory wings. Once the designer is familiar with all the requirements, it is often helpful to draw each room to scale on a thin piece of cardboard and then cut out the individual rooms. The designer can color code the rooms according to function and to the entity's division, branch, or team with which the rooms correspond, as well as specify the required adjacencies or separations of each room. Naturally, the same result can be obtained by computer modeling. The designer then groups the rooms according to function and organization as required by the program of requirements. Each group becomes a "block": a group of rooms with the same or very similar functions, architecture, and mechanical and electrical requirements. Rooms in a block are also subjected to the same or very similar health and safety requirements. This type of grouping allows for efficient, effective servicing of the rooms, as well as the provision of health and safety protection. These blocks can be shifted around schematically until the designer finds an arrangement that responds to all stated requirements. Next, the designer regroups the blocks into design elements and laboratory wings.

Questions to be Answered Before Rooms are Grouped

Before the designer groups the rooms into wings, several important questions must be answered. While many of these questions are considered elsewhere in the design process, their resolution is necessary to the design of the wings. These questions are listed below. A more detailed description of the first five questions follows this abbreviated list. The last question is addressed in chapter 6.

- How many floors are allowed or required for the laboratory wing(s)?
- What are the required adjacencies and separations?
- What degree of flexibility and expansibility is required?
- Could or should the laboratory rooms have windows or other means of natural lighting?
- How will the laboratory rooms be provided with the required services and utilities?
- What is the location and orientation of the wing(s) in relation to other wings, rooms, and functions? What are the locations of fume hood exhaust stacks in relation to the building's air intakes and to the predominant wind direction(s)?

1. *How many floors are allowed or required for the laboratory wing(s)?* The number of floors allowed or required for laboratory wings is determined by local and national codes and standards or by the program of requirements of the entity for which the laboratory is designed. The operations and their risk and hazard levels, the potency and quantity of the chemicals used, the construction type and materials, and the fire separation and the subdivision of the spaces are among the factors that must be considered. The decision regarding the number of floors will affect both the way the laboratory rooms are serviced and how the laboratory building is designed.

2. *What are the required adjacencies and separations?* Adjacencies and separations may be explicitly required in the individual RDS or may result from functional or health and safety requirements or from practical considerations. For instance, a certain number of rooms may be grouped together simply because they perform the same kind of functions or belong to the same division or branch of the entity. Some rooms may be adjacent, perhaps with connecting doors, in order to promote operational efficiency or to reduce the hazard of transporting potent chemicals between rooms. Separation of rooms may be explicitly required for operational, programmatic, or legal reasons. Such separation may also be necessary to avoid contamination of samples or to prevent the operations conducted in one laboratory from affecting those conducted in another. In general, when dealing with low-level samples in one room, one should avoid having high-level samples in an adjacent or nearby room. Organic and inorganic laboratory rooms should be grouped in separate areas and their fumes must be segregated. Generally, the laboratory operators and scientists can guide the designer because they usually are aware of the adjacencies and separations their operations require.

3. *What degree of flexibility and expansibility is required?* Flexibility and expansibility are two of the major determinants in the design of the laboratory wing. They are also important factors in sustainable design and are major components of a green laboratory building.

Once the number of floors is decided, the designer should determine how to respond to the flexibility and expansibility required by the project. Responding to flexibility and expansibility needs is difficult and may be costly and wasteful if the user's present and future space modification and expansion needs are not fully understood.

Flexibility for a laboratory designer means convertibility, versatility, and interchangeability. *Convertibility* is the ability of a building to be reconfigured with minimum disruption. It is generally obtained by modulation. Structurally, modulation is accomplished through the placement of the structural elements at regular intervals; architecturally, modulation is achieved by repetition of areas or modules. In the design of the mechanical and electrical systems, modulation is obtained by having cut-off and connections outlets provided to each architectural module. Another factor that affects convertibility is the way services and utilities are provided. There are several ways to service laboratory rooms—among them are interstitial floors and utility or service corridors. These and other ways of servicing the laboratory wing will be further discussed later in this chapter. *Versatility* is the ability of the laboratory room's interior to be reconfigured with minimum disruption.Versatility is accomplished by modulating the service and utility lines in accordance with the space modulation. This means that every module must have access to each service or utility outlet or cut-off, regardless of whether the outlet or cut-off is actually needed at the present time. The location of services and utilities outlets and cut-offs should be consistent among modules.This facilitates the process of interchangeability. *Interchangeability* is obtained by using casework that can be disassembled and reassembled in different ways from one laboratory room to another. Several types of system casework or flexible laboratory cabinetry are on the market. These systems offer components that can be attached to and later separated from benchtops and service strips and reconfigured. These systems are discussed in chapter 6. The required flexibility of the laboratory wing and of other components of the laboratory building and how this flexibility is achieved have a major impact on the building design.

Expansibility of a building, a wing, or a system is its ability to grow or expand at a minimum cost and with minimum disruption to the surrounding activities. The manager should determine the laboratory's need for expansibility and define how and what components should have this capability. Expansibility must be incorporated in the design and construction drawings, and adjustments must be made to the building during initial construction. Therefore, when expansion becomes necessary it can take place smoothly, at a minimum cost, and with minimum disruption to surrounding activities.

4. *Could or should the laboratory have windows or other means of natural lighting?* Before design takes place, a determination should be made on the use of windows, skylights, clerestories, or other means of natural lighting for the laboratory rooms. There are advantages and disadvantages to all these lighting options. Most scientists like to have windows in their laboratory rooms as well as in their offices. However, if they must choose between having windows in their laboratory rooms or in their offices, most prefer having windows in their laboratory rooms. Many scientists feel that a room without a window is a bad place to do science. A window gives a view of the outside, relieving scientists' sense of isolation and entrapment. It allows the laboratory operator to enjoy the sky and sun and to feel the changes in the day. It provides a soothing feeling, which is conducive to thinking and creativity.

Laboratory rooms with specific airflow and temperature requirements should have fixed pane windows. When the use of windows in laboratory rooms is not feasible, skylights and clerestories should be seriously considered. They should be designed to allow maximum lighting without glare where lighting is needed, while preventing direct sun from entering the room. They also must not interfere with service and utility lines.

Mechanical engineers generally prefer laboratory rooms without windows or other types of openings. They feel that these types of openings expose the room to substantially different day and night temperatures. To maintain consistent temperature in rooms with windows, skylights, or clerestories, the mechanical engineer and the architect must increase the insulation of the openings and use or design special engineering and architectural systems to shade the sun. The construction of these systems inevitably increases the initial and operation costs of the building. Thus, the laboratory owner must compare the value of his scientists' comfort and productivity with the slight increase in initial and operation costs. This assessment and decision should take place during the predesign phases.

Whether laboratory rooms are required to have windows and/or other types of openings will have a major impact on how the building is designed, how the laboratory rooms are serviced, and how the building components relate to one another. Windows always should be located on external walls, and laboratories with windows must be supplied with services and utilities differently than laboratories without windows. If the laboratory building is limited to one floor, the laboratory rooms can be located at the center of the building around a service corridor, and natural lighting can be provided by skylights and clerestories. This situation, as noted before, requires special design accommodation for the provision of services and utilities. This will become obvious in the discussion of the wing design later in this chapter.

It should be noted that the use of fiber optics as a means of introducing natural lighting to interior spaces is being researched in several countries. When this technology becomes economically viable, it can be used for office or laboratory rooms.

Answering these first four questions will help direct the provision of services and utilities to laboratory rooms and wings. It will also help determine where laboratory wings should be located in relation to other wings with other functions and in relation to the direction of predominant winds (discussed in chapter 6). Additionally, it will help establish how wings should be oriented to maximize efficiency and reduce health and safety risks.

5. *How will the laboratory rooms be provided with the required services and utilities?* There are several ways to provide a laboratory room with services and utilities. These options are presented and explained in the context of the historical

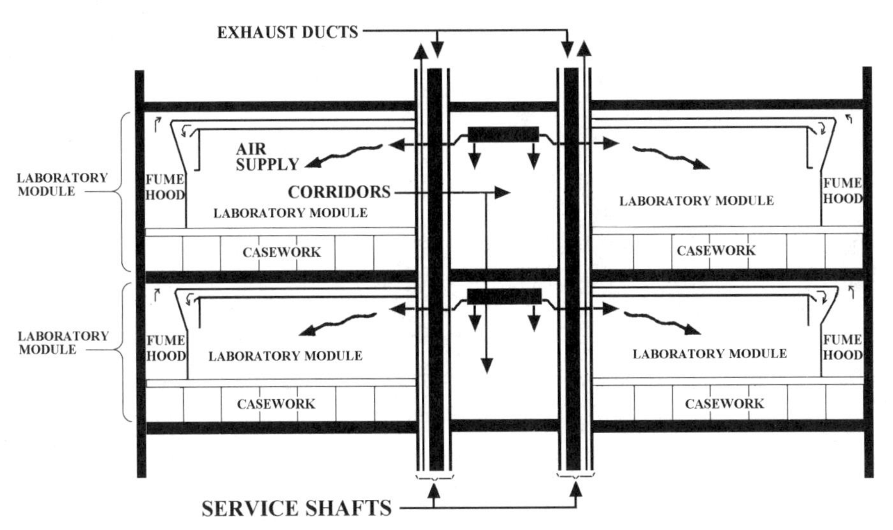

FIGURE 5-1
Section through double-loaded corridor laboratory

evolution of laboratory wing design and will be discussed in the context of the laboratory wing design.

Evolution of Laboratory Wing Design and the Provision of Services and Utilities

After World War II laboratory design patterns focusing on services and utilities started to seriously emerge. The first attempt was the introduction of the module. Then, a module was simply a self-contained, fully equipped cell. The modules were lined up on each side of a corridor. Vertical service shafts were located either on each side of the corridor or on the outside walls of the building. Figures 5-1 and 5-2 illustrate the design solution with service shafts placed on each side of the corridor. The service shafts were used to deliver the services and utilities to adjacent modules. In figure 5-1, with the vertical service shafts on each side of the corridor, each floor had access panels.

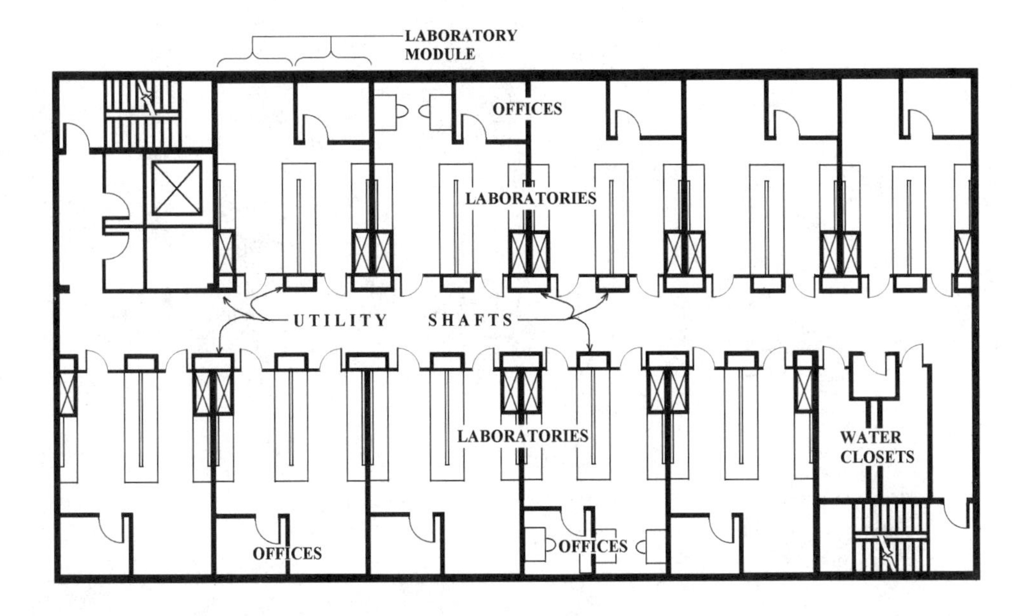

FIGURE 5-2
Floor plan of a laboratory wing with double-loaded corridor

There are several types of laboratory shafts and corridors: people corridors, supply and equipment corridors, and utility and service corridors and shafts. The people corridor is used by the laboratory operators: scientists, chemists, and people who work mainly in the laboratory rooms. Often, supplies are transported in a people corridor. The supply and equipment corridor is mainly (if not always) used to move heavy supplies and equipment from the dock to the hazardous materials storage building, and from storage rooms to the utility or service corridors or shafts and to laboratory rooms.

The utility and service corridors and shafts directly service the laboratory rooms. The corridors are generally located on the back of the laboratory rooms and service two rows of laboratories, one on each side of the corridor. The shafts are located so as to service two or more laboratory rooms. The utility corridor is generally narrower than the service corridor and essentially lodges the utility and service lines feeding and servicing the laboratory rooms. All cut-offs to the lines are also located in this corridor. These corridors are generally from 6′-0″ to 9′-0″ (1.8 to 2.7 m) wide and are accessed only at their two ends. Their accesses are restricted to service personnel only. The service corridor has the same function as the utility corridor but is significantly wider. The width is from 12′-0″ to 20′-0″(3.66 to 6.1 m). These corridors can be accessed at either end (by service personnel), as well as from the laboratory rooms on either side (by laboratory operators). In addition to lodging the utility and service lines, service corridors can be used to house large equipment, such as incubators or refrigerators.

Figure 5-3 illustrates a design solution in which service shafts are placed on the outside walls of the laboratory wing. This solution (also shown in figures 5-1 and 5-2) offers the advantage of providing laboratory rooms with direct utility and service lines with cut-offs and connections at each access panel. The utility and service lines are generated in either the basement or the attic of the building. This solution offers ease of servicing at the two ends and of connection at each of the laboratory rooms. It also allows the building to have several floors and most, if not all, laboratories to have windows.

The disadvantage of this solution is that if work must be done on the shaft, the access panel must be removed where the work is performed. This means that the service personnel may clutter the corridor with equipment, piping, or ducts,

FIGURE 5-3
Section through laboratory wing with utility shafts outside the external walls

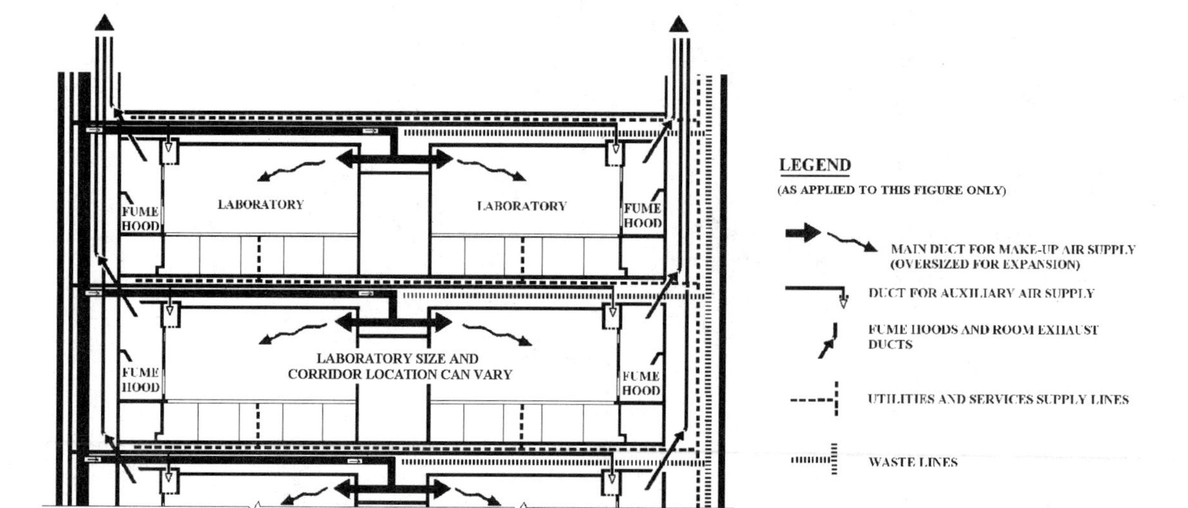

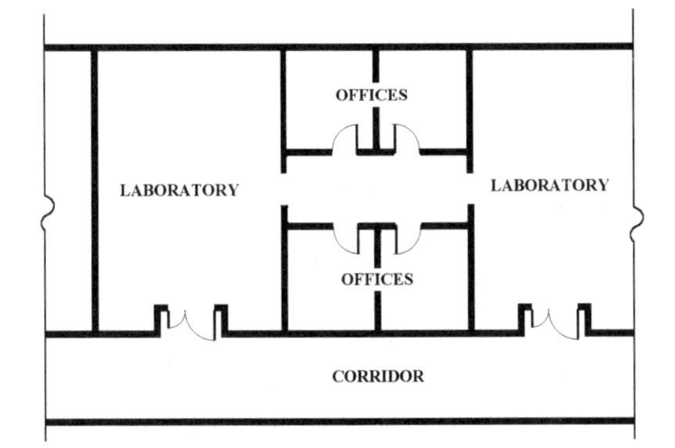

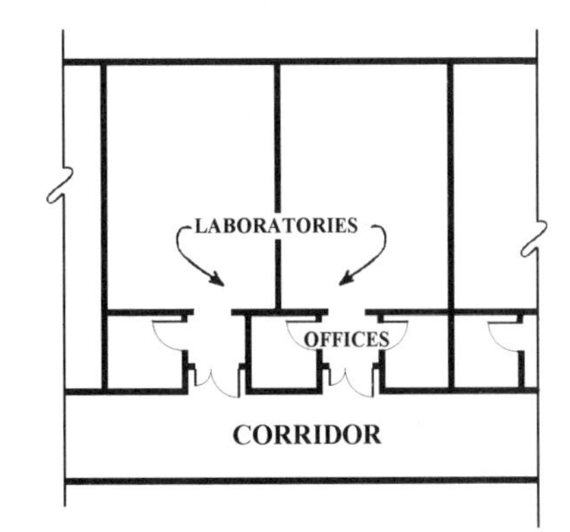

FIGURE 5-4 (left) Laboratory-related offices using modules between laboratory rooms (present USA codes do not allow this arrangement)

FIGURE 5-5 (right) Laboratory-related offices located in strips between corridors and laboratory rooms

which interferes with traffic and the transportation of chemicals. Another disadvantage of this solution is that the laboratory-related offices (offices for scientists directly involved in laboratory operations) would have to be located either in other parts of the building, far away from their assigned laboratory rooms, or occupy expensive laboratory space. If the offices are located in the laboratory wing, the scientists' choice is to create a small office space in their laboratory (figure 5-2) or to place it between laboratory rooms (figure 5-4).

Offices placed between laboratory rooms offer the advantage of being near the laboratory where the scientists work, but their use of expensive laboratory space is not economical. Additionally, the lack of independent access to these offices means that future alterations will be costly if the office space must be converted to laboratory space. Although part of the evolution of laboratory wing design, this option may be unavailable for future designs, as today's codes do not permit exiting from offices to laboratory rooms.

The solution shown in figure 5-3, with the vertical utility shafts outside the external walls, offers the same advantages to the laboratory rooms as the solution shown in figures 5-1 and 5-2. It also allows a more flexible relationship between the laboratory rooms and the corridors. The disadvantage of this solution is that performing maintenance or changing the service corridor is problematic. If the panels are located outside the external walls, service personnel must work from an access ladder or from a small temporary platform. This makes using material and equipment difficult. When panels are located inside the external walls, they are often in a laboratory room, which means maintenance work or changes interfere with operations and may affect the quality of these operations. Such work also inconveniences the laboratory scientists.

In spite of these problems, however, locating service shafts outside exterior walls can be advantageous because it provides a strip of space between the corridor and the laboratory rooms where offices can be located. This arrangement, shown in figure 5-5, permits direct access from the laboratories to the offices and minimally affects wing expansion or reconfiguration. The offices' access to natural light is through partitions with limited glazed areas, particularly if the fire code does not allow the corridor and laboratory walls to have large areas of glazing.

The solution illustrated in figure 5-3 can be further developed by consolidating the individual service shafts into large common shafts. These large common shafts are located either outside the laboratory wing or within the wing in areas that do not affect room modulation (figures 5-6 and 5-7). This solution reduces the number of access panels needed at each floor and offers the advantage of providing the laboratory rooms with direct utility and service lines with cut-offs and connections at the access panel locations. It also allows more flexibility in expanding or decreasing the size of the laboratory rooms.

In this solution, the utility and service lines are generated either in the basement or in the attic of the building, which offers ease of servicing at the two ends and of connections at each of the laboratory rooms. It allows the building to have several floors and all laboratory spaces to be shaped as needed and to have windows.

In order to perform changes or maintenance work in a laboratory wing using this solution, the service personnel have to reach the access panels of the service shaft either from the inside or from the outside of the building. If the access panels are located inside the building shell, they are often in a laboratory room or corridor, which means that maintenance work or changes interfere with operations and may affect the quality of these operations. Such work also inconveniences the scientists. If the access panels are placed outside the exterior walls, built-in access ladders and/or mini-platforms must be provided, neither of which, unfortunately, greatly improves working conditions for the service personnel.

Figure 5-6 is an imaginative solution that illustrates how the consolidation of utility or service shafts allows quite a bit of flexibility in the use of the wing floor space. The entire floor can be used for one large laboratory room or can be subdivided for multiple laboratories and laboratory-related offices. Another advantage of this solution is that it allows the laboratories to have natural lighting through windows.

Another configuration of this floor plan, with the utility shafts placed as shown in figure 5-6, could locate some or all the offices alongside the external

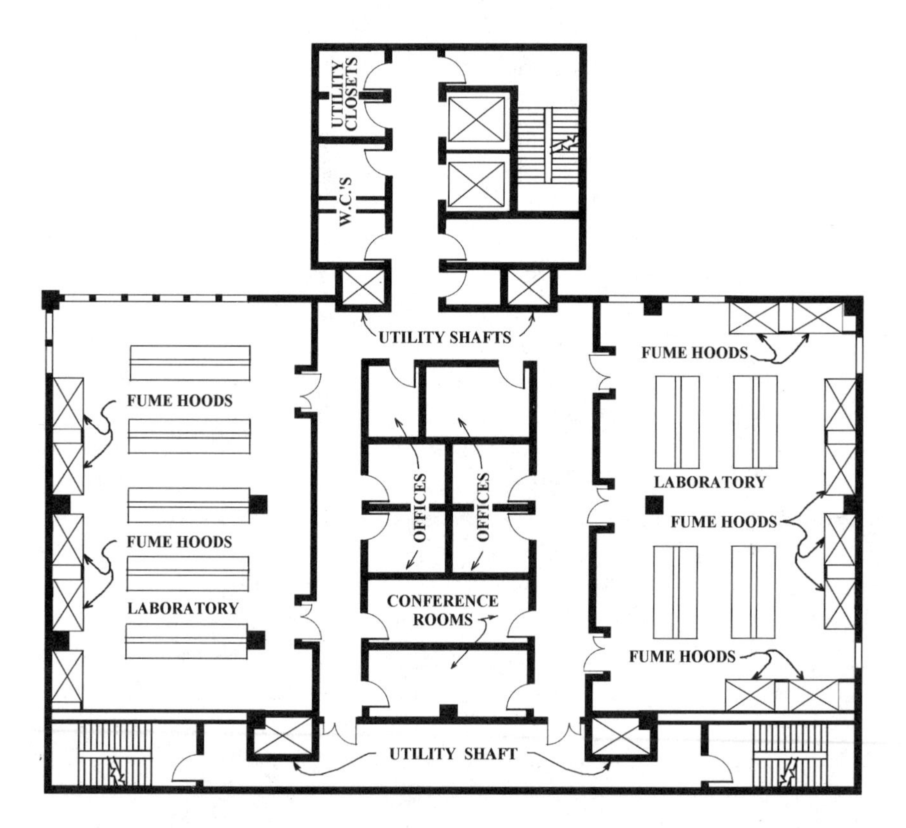

FIGURE 5-6
Floor plan of a laboratory wing with consolidated utility shafts

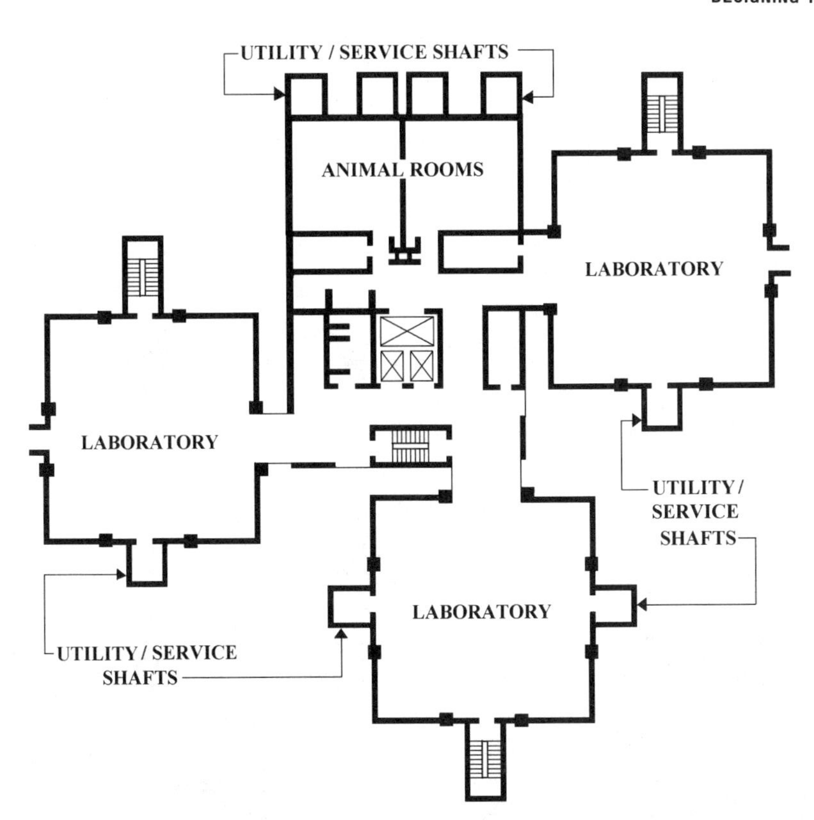

FIGURE 5-7
Floor plan for the Richard medical research laboratory

walls. This makes it possible for the laboratory-related offices and the laboratory rooms to have windows.

In this scheme, the utility or service shafts are accessed through the people corridor (instead of through a utility or service corridor) or through a laboratory room, depending on how the space is configured. This has the same disadvantage of disrupting traffic and operations discussed earlier.

Figure 5-7 shows a multifloor laboratory designed by Louis I. Kahn. In this design, the architect used three distinct laboratory wings and placed the consolidated utility and service shafts on the perimeter of each wing. This imaginative solution allowed quite a bit of flexibility for internal changes in space use. Each of the wings could be one large laboratory room or could be subdivided into laboratories and laboratory-related offices. This solution also allows maximum natural lighting through windows, either for the laboratories or for the offices along the external walls. The access to the utility and service shafts is the same as described for the solution shown in figure 5-6 and would have similar advantages and disadvantages.

Another way to provide flexibility in laboratory servicing is to locate the laboratory rooms along a central utility corridor. This approach is illustrated in figure 5-8. This and other similar approaches are useful when the flexibility required is limited to the extension or reduction of the number of modules that a laboratory room has at any time during the life span of the building. The utility corridor lodges all air ducts, supply and exhausts, all water supply and waste lines, all piped gases, and all electrical and telecom lines. These lines can be generated in the basement, in the attic, or at one or both ends of the corridor in the same floor, and connections and cut-offs are provided at each module. The central utility corridor access is restricted to the service personnel and maintenance

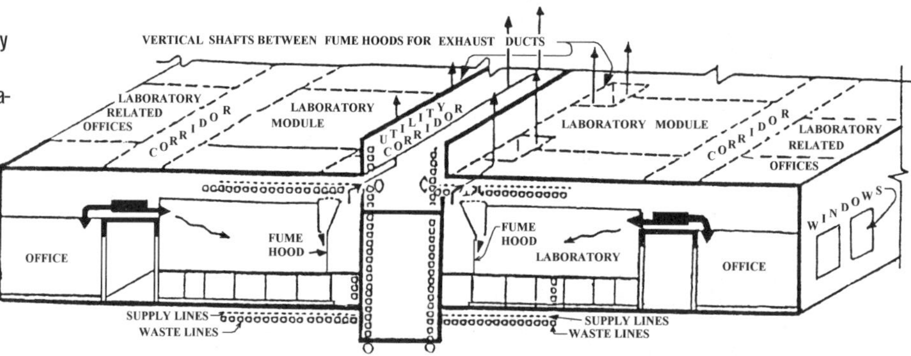

FIGURE 5-8
Section through laboratory wing with utility corridor servicing one set of laboratories on each side and with laboratory related offices on the external wall.

crews—no laboratory scientist or operator should be allowed to perform any function in the utility corridor under normal conditions. This grouping of all the service and utility lines in one accessible space allows for repairs, servicing, and changes of the laboratory rooms along the utility corridor with minimal disruption to operations.

Figures 5-9 and 5-10 are pictures taken in the utility corridor of the USEPA regional laboratory in Athens, Georgia. These pictures show how useful the corridors are for service and maintenance as well as how easily changes can be made without undue restrictions or disturbances of activities in the laboratory rooms not affected by the work.

When laboratory rooms are placed on each side of a people corridor (figure 5-11), extension lines must be used to provide the laboratory row on the far side of the corridor with services and utilities. These extension lines run horizontally above ceilings or under floors. The connections and cut-offs remain in the utility corridor, but any repairs or servicing performed between the utility corridor and the point of use cause disruption of the laboratory operations and clutter the corridor used by the scientists and through which chemicals are transported.

When a utility corridor is used, the laboratory wing may have several floors. The laboratory rooms alongside the utility corridor cannot have windows, but, as shown in figure 5-11, those on the other side of the people corridor can have them.

All air supplied to the laboratory wing is one-pass air and the laboratory rooms' air pressure is negative to that of the people corridor, as well as to that of the supply and equipment corridor. This means that when a laboratory room

FIGURE 5-9 (left)
FIGURE 5-10 (right)

door opens, the air pressure in the corridor forces air into the laboratory room, thereby preventing the contaminated air in the laboratory room from flowing to other parts of the building.

Obviously, the disadvantage of the solution shown in figure 5-11 is that the laboratory-related offices are not adjacent to the laboratory rooms used by the offices' occupants. These offices would be located in other parts of the building, generally at the end of the laboratory wing. As previously mentioned, this arrangement forces scientists to either do a lot of walking back and forth or to create a small office space in their laboratories. The latter makes inefficient use of expensive laboratory space and is unhealthy for the scientists.

One variation of this approach is to group the laboratory rooms alongside the utility corridor, thereby providing them with all the aforementioned advantages, and to place laboratory-related offices on the other side of the people corridor. This solution, illustrated in figures 5-8 and 5-12, allows the building to have several floors. Naturally, the plenum above the office space is much deeper than needed.

This solution prevents the laboratory rooms from having windows, but the offices on the far side of the people corridor can have them. These windows, however, must be fixed pane windows, because open windows would disrupt the wing's air balance as well as the air pressure, temperature, and humidity of the laboratory rooms. (It should be noted that if the approaches illustrated in figures 5-8 and 5-11 through 5-13 are used for one-story laboratory wings, it is possible to introduce daylight into the laboratory rooms along the utility and service corridors by means of skylights and clerestories. However, the designer must be careful to shade the skylight or orient it so that no direct sunlight enters the room.)

The solution shown in figures 5-8 and 5-12 can use recirculated air to supply the offices, but the offices' air pressure must be higher than that in the people corridor. Additionally, the corridor must have one-pass air but could use part or all of the offices' recirculated air. Its pressure should be higher than the air pressure in the laboratory rooms, which also should have one-pass air. This means that when an office door opens, the air pressure in the office forces air into the corridor, and when a laboratory room door opens, the air pressure in the corridor forces air into the laboratory room. This prevents contaminated air in laboratory rooms from flowing into the corridor and from there into other parts of the building.

Another variation of the approach shown in figure 5-8 keeps the configuration of laboratories and laboratory-related offices shown in the figure but substantially enlarges the utility corridor and expands its functions. This is illustrated

FIGURE 5-11
Section through laboratory wing with service corridor servicing one set of laboratories on each side and with laboratory related offices on the external wall

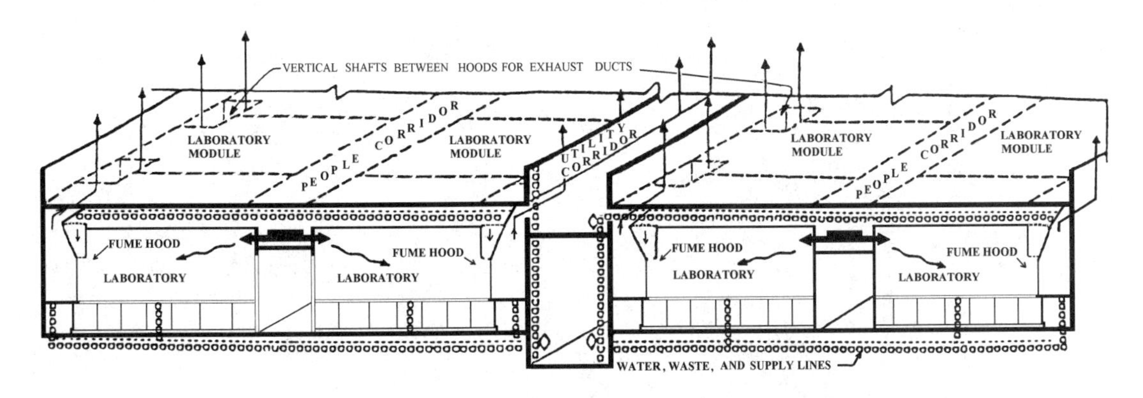

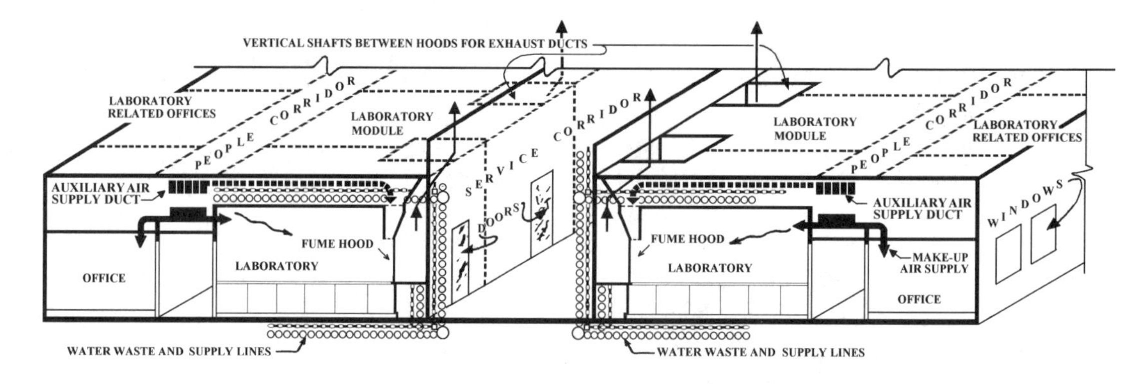

FIGURE 5-12
Section through laboratory wing with utility corridor servicing two sets of laboratories on each side of people corridor (combination of vertical and horizontal services)

in figures 5-12 and 5-13. The utility corridor becomes the service corridor, with doors providing access from laboratories on each side.

This corridor still lodges all air ducts, supply and exhausts, all water supply and waste lines, and all piped gases, as well as all electrical and telecom lines. This corridor can also be used by the laboratory operators to store instruments and equipment (e.g., small compressors, sample storage refrigerators, and incubators) that would be out of place or too cumbersome in the laboratory rooms, as long as such storage is not hazardous. This option has the same disadvantages but more advantages than the option shown in figure 5-8.

The solutions shown in figures 5-8 and 5-12 allow flexibility and expansibility of the laboratory rooms located next to the service shafts or along the utility, service, or people corridors. A laboratory room may have as many modules as needed as long as they can expand along the service or utility corridor. The dimension of the room between the corridor servicing it and the people corridor or between the people corridor and the external wall always remains the

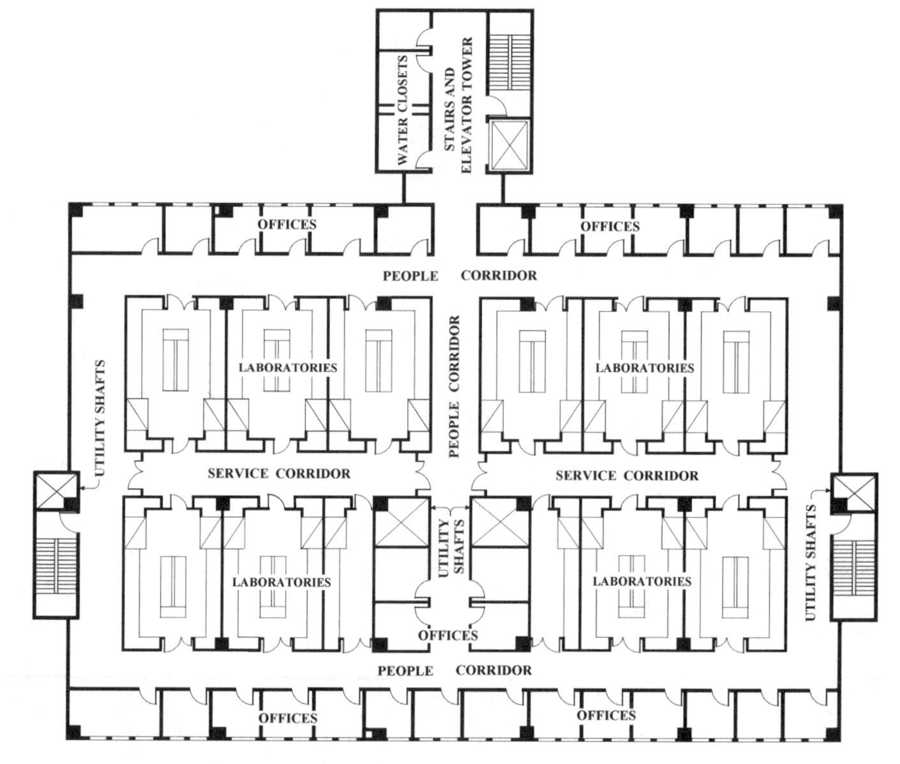

FIGURE 5-13
Floor plan of a laboratory with service corridor servicing one set of laboratory rooms on each side with the laboratory related offices on the external walls

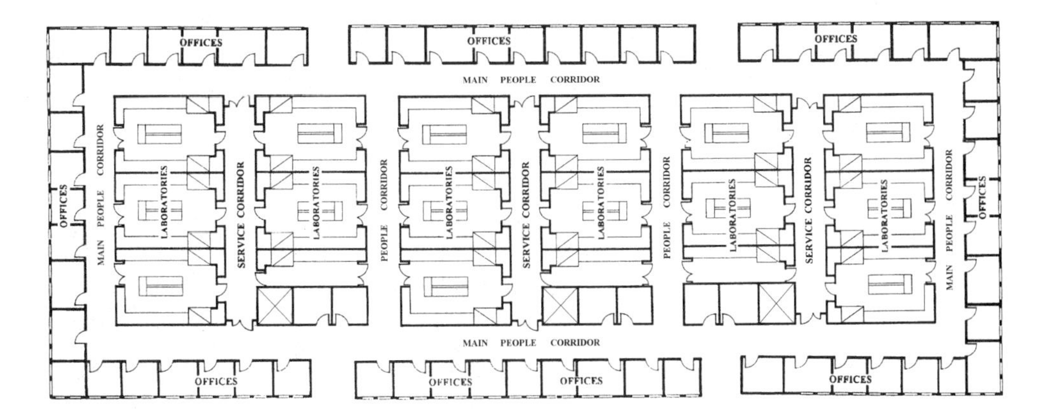

FIGURE 5-14
Floor plan of a laboratory where the people corridors leading to the laboratory rooms and the service corridors are at right angle with the two main corridor

same. Despite the disadvantage of expanding in only one direction, this type of flexibility and expansibility along the service or utility corridor may be sufficient in many cases. Expansion can be accomplished by the addition of laboratory rooms at the end of the laboratory block.

When program requires numerous rooms, the solutions shown in figures 5-8, 5-12, and 5-13 produce very long, wagon train–like, corridors. Such solutions become unpractical, not only because the distance between the building's extremities is too great, but also because they require an elongated site that is difficult to orient. The solution shown in figure 5-14 solves this problem. While derived from the solutions shown in figures 5-12 and 5-13, this arrangement places the laboratory rooms, the service corridors, and the people corridors directly leading to the laboratory rooms at a right angle to the two main corridors. The laboratory-related offices are located alongside the two main corridors. Obviously, this arrangement allows more design flexibility. Laboratories that perform functions that require them to be close to each other can be grouped in blocks. The laboratory-related offices of a given block are closer to the laboratories of that block but scientists from different blocks are not isolated from each other.

Further, to a certain extent this solution allows for the lateral expansion of the laboratory rooms located alongside the utility, service, or people corridors. A laboratory room may have as many modules as required as long as they can expand laterally. The dimension of the room between the corridor servicing it and the people corridor always remains the same. However, this lateral expansion is limited because the blocks are framed by the main corridors. This laboratory is normally expanded by adding blocks of laboratories at the end of the wing. This type of flexibility and expansibility, despite its disadvantages, may be sufficient and attractive in many cases.

In some cases, laboratories may change the scope of their work and require frequent and complete reconfiguration of the space. In such cases it may be more cost-effective in the long run to keep the floor on which the laboratory rooms are located free from service or utility corridors and service shafts. All services and utilities are provided through a slab separating the laboratory room's floor from either the floor above or the floor below the room.

The floor from which the services and utilities are provided is often referred to as an interstitial or service floor. In buildings with laboratory rooms on several floors, the interstitial or service floors can serve both the floor above and the

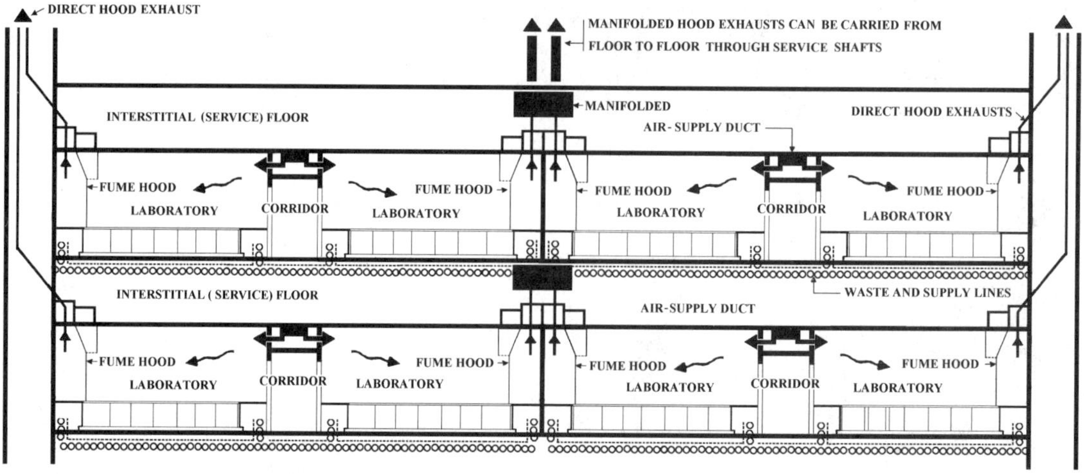

FIGURE 5-15
Section through laboratory wing showing horizontal distribution of services by use of interstitial space

floor below it. Naturally, vertical circulation of people as well as equipment and chemicals must be provided for. One or more vertical utility or service shafts, strategically located, may also be needed for the vertical distribution of services and utilities to each interstitial space or service floor. The hoods' exhaust ducts—manifolded or individual exhaust ducts—can also be located in these or other vertical utility or service shafts. This arrangement is illustrated schematically in figure 5-15 and was used by Louis Kahn in the design for the Salk Institute in La Jolla, California. Croquis of this laboratory's plans and sections are illustrated in figures 5-16 through 5-18. These figures respectively show a laboratory/office floor, a mechanical equipment floor (referred to as interstitial space or floor), and a section through the building showing the relationship of floors.

Access to the services and utilities follows a modular design that may resemble a checkerboard, as there is not necessarily a determined location for corridors in the laboratory floors. Obviously, this arrangement offers extensive

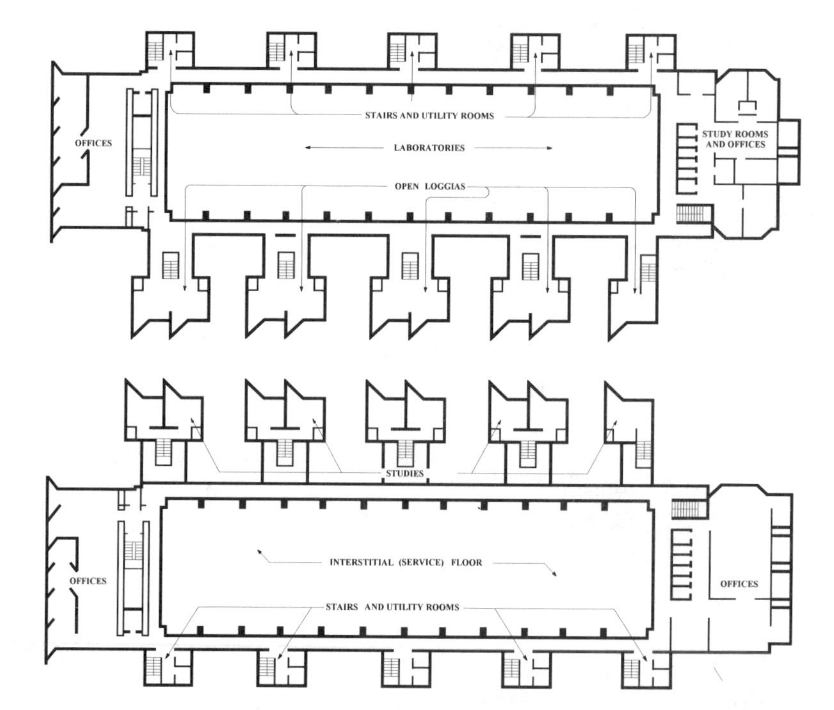

FIGURE 5-16
Interstitial (service) floor plan–Salk Institute building

FIGURE 5-17
Laboratory and offices floor plan–Salk Institute building

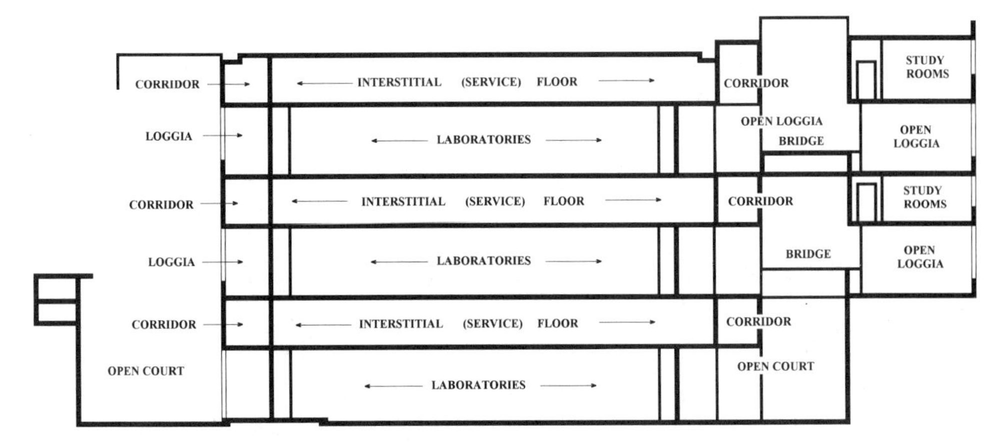

FIGURE 5-18
Section through the Salk Institute building

flexibility in all senses of the word. It also mandates a high initial cost since it requires the addition of an entire floor to service the laboratory floor or floors. In the long run, however, the flexibility that this solution offers may be economically desirable. When changes are needed, as often happens in research laboratories, construction work is limited to two levels: the laboratory floor (or portion of the floor) that needs to be changed and the interstitial floor above or below it. This minimizes displacement of researchers during the construction phase, which consequently saves time and money.

Figure 5-19 pictures an interstitial space of a laboratory located in the southeastern U.S. This space lodges most of the equipment required by this laboratory to function properly.

FIGURE 5-19

Since its inception, the use of the interstitial or service floor design concept has created intense interest among laboratory users and designers. Both groups tried to find ways to achieve the flexibility offered by this design concept, but at a construction cost closer to that of less expensive solutions. As a result, several new design concepts emerged.

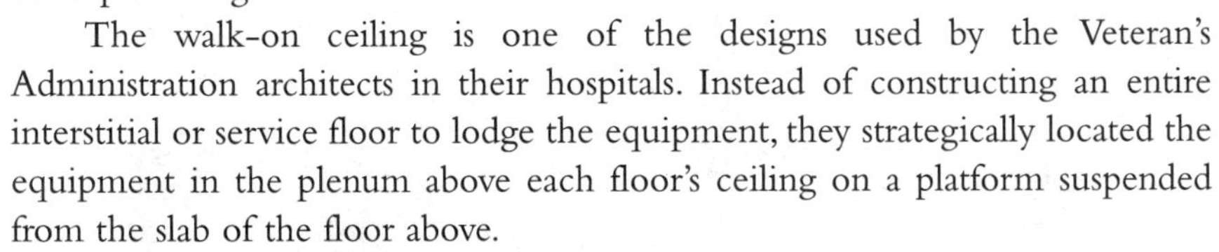

The walk-on ceiling is one of the designs used by the Veteran's Administration architects in their hospitals. Instead of constructing an entire interstitial or service floor to lodge the equipment, they strategically located the equipment in the plenum above each floor's ceiling on a platform suspended from the slab of the floor above.

The ceiling structure was reinforced and the ceiling panels were designed to carry the live load of the maintenance people. The cost of the walk-on ceiling is about 15–20 percent higher than the cost of a regular ceiling. This cost is much lower than the cost of an entire interstitial or service floor. However, there are certain disadvantages to this system. The suspended equipment must be relatively light and must not require frequent servicing. Walking over the suspended ceiling may be noisy and disturbing to the users of the floor below. It is inconvenient and complicates the tasks of the maintenance people.

The catwalk is also often used to circumvent the construction of an entire interstitial or service floor. In this design, the architect strategically houses the equipment in the plenum above each floor ceiling on a platform suspended from

FIGURE 5-20
Section through a laboratory wing with back-to-back laboratory rooms serviced by a service penthouse with laboratory-related offices across the corridor

the slab of the floor above. They connect this platform with a catwalk to a point of entry that is generally located in a corridor or in a storage room, not too far from where the equipment is suspended. Access to the point of entry is generally provided by a built-in ladder.

Another alternative is shown in figure 5-20. This figure shows a section through a one-floor laboratory wing. This design was used by the firm of Bullock, Tice and Associates for a laboratory built in Montgomery, Alabama, for the U.S. Environmental Protection Agency. In this scheme, the laboratory rooms are placed back-to-back and are serviced from a service penthouse. This penthouse is placed above the central part of the wing and extends in length over the laboratory areas that are likely to require changes. The penthouse's width is sized to envelop the mechanical, plumbing, and other equipment required to service the laboratory rooms below. It is a metal shell, fully insulated and lightly heated during the cold season, to protect the equipment and the maintenance crew from the elements.

Figure 5-21 is a bird's eye view showing the location of the penthouse over the laboratory areas. Figures 5-22 and 5-23 are external views in which the penthouse location is clearly apparent. Figure 5-24 shows the interior of the penthouse.

FIGURE 5-21

FIGURE 5-22 (left)
FIGURE 5-23 (right)

The reduced size of the penthouse, as well as the materials used to build it, made the construction cost of this penthouse much lower than that of a full interstitial floor. However, this solution obviously makes the future addition of one or more laboratory floors more difficult and lacks the maximum flexibility a full interstitial floor provides. For the laboratory pictured in figures 5-21 through 5-24, expansion and flexibility were not required or expected within the foreseeable future. However, if the laboratory did need to expand—for instance, if the laboratory-related offices across the corridor were converted to laboratory rooms—it would be cost-effective to expand the service penthouse, as the space is available and the expansion construction costs would be low because of the materials used.

FIGURE 5-24

Figure 5-25 shows a section through a one-floor laboratory wing that is a variation of the previous concept. The cross section shows four laboratories on the lower floor, two back-to-back, and two across the corridors from them. These four laboratories are serviced by a service penthouse similar to the penthouse of the previous scheme. This space was placed above the central part of the wing and extended over its entire length. Again, the width was sized to envelop the mechanical, plumbing, and other equipment required to service the laboratory rooms below. The hood exhaust stacks of the two back-to-back laboratory rooms are manifolded together. The hood exhaust stacks of the laboratories located on each of the external walls of the wing are also manifolded together. The exhaust duct is placed low in the space above the laboratory rooms and is located next to the edge of the external wall. These ducts are screened with a parapet. Air is supplied to all the laboratories by ducts placed in

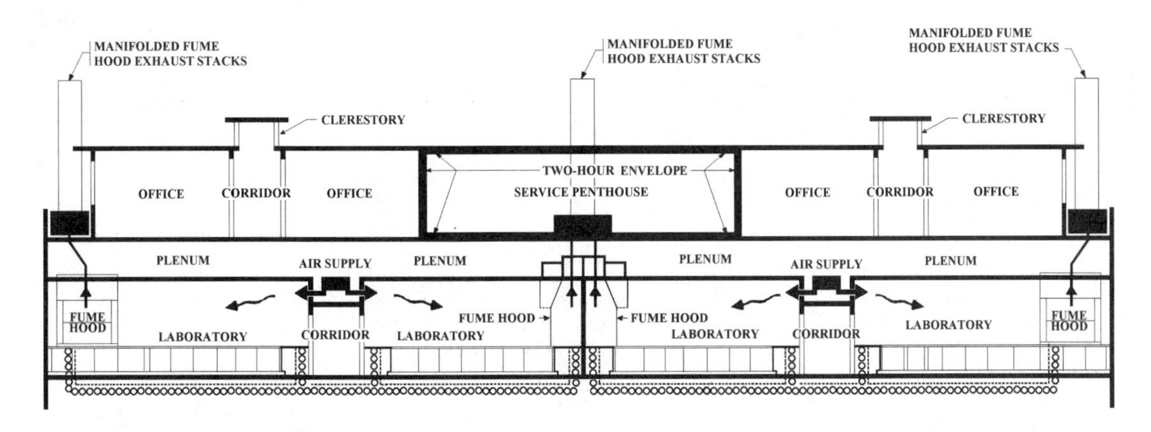

FIGURE 5-25
Section through laboratory wing with four laboratory rooms on the lower floor and with laboratory-related offices located on each side of the service penthouse

the plenum below the second-floor slab. The air handlers generating this air can be placed in the service penthouse or in another mechanical room in a lower floor.

A row of laboratory-related offices is inserted on each side of the service penthouse, between it and the exhaust ducts located on the external walls. These offices are placed on each side of a central corridor that connects them, by mean of stairs, to the laboratories (at the lower level). The offices on the outside walls have windows. The offices next to the service penthouse receive daylight through skylights. The use of such a scheme requires rigorous separation between spaces for fire and ventilation protection.

This solution, although very compact and economical in terms of space, is even more difficult to expand than the previous one. It also does not provide the flexibility that a full interstitial floor allows. The cost to add additional floors and the extensive disruption which will result makes it an unrealistic option. Reconfiguration of the laboratory floor, however, is possible and cost-effective, but it may also seriously disrupt the laboratory's operations.

Figure 5-26 shows a section through a one-floor laboratory wing that uses the interstitial or service floor design concept in a new and imaginative way. In this design the laboratory rooms are placed alongside a people corridor and are serviced by a service corridor placed immediately above and extending over the people corridor. This service corridor, referred to in the section as service/mechanical corridor, takes advantage of the difference in ceiling heights between the laboratory rooms and the people corridor. It is like a penthouse, and it receives natural light through clerestories along its length. The width of the service/mechanical corridor encompasses the width of the people corridor and intrudes into the laboratory rooms over the door recesses. All supply ducts for make-up air and all piping servicing the laboratories are placed along this service/mechanical corridor.

The main air supply ducts are placed directly above the door recesses. This allows the make-up air to be provided to all laboratory rooms from the area above the room's door. As the air is introduced into the room, it has a broom-

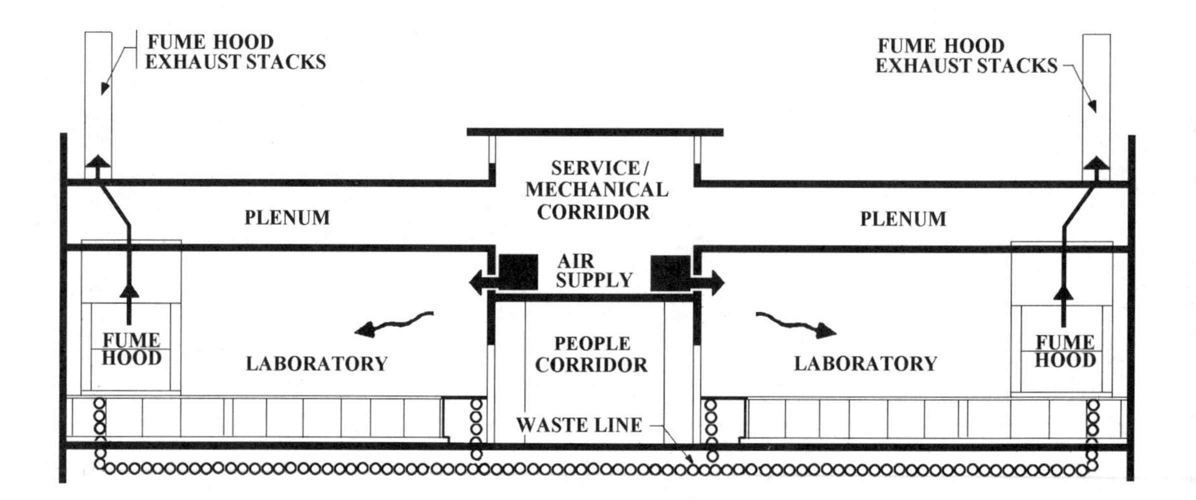

FIGURE 5-26
Section through laboratory wing laboratory rooms on each side of the people corridor and with the service/mechanical corridor above the people corridor

FIGURE 5-27

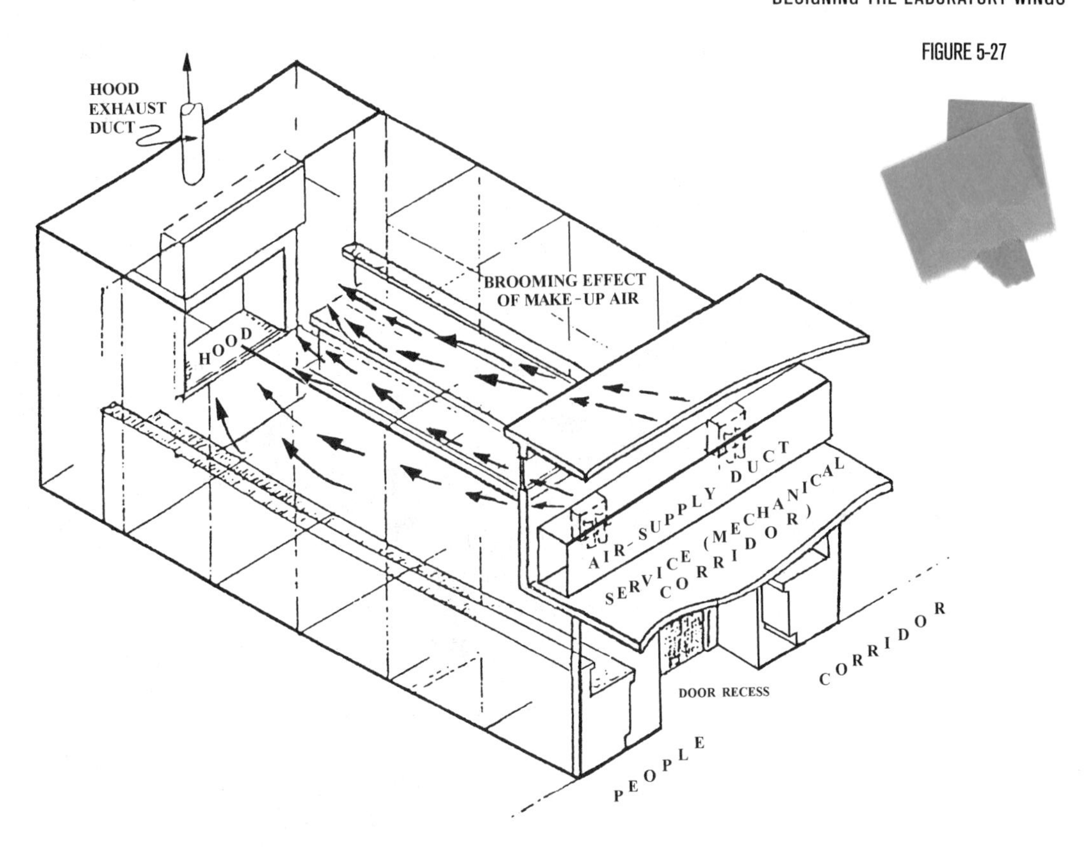

ing effect, which directs all fumes and particles floating in the room toward the hood, which absorbs them. This is illustrated in figure 5-27.

All gases, liquid, and vacuum piping are also located either along the walls of the service/mechanical corridor or on its ceiling. From there they are either dropped in a small utility shaft placed on each side of the laboratory entrance door and from there directed behind the casework to the place of use, or directed through the plenum above the laboratory room and directly dropped at the place of use.

All air supply ducts and all gas, liquid, and vacuum piping have a cut-off for each module located in the service/mechanical corridor. This allows servicing, alterations, and changes to take place without disturbing operations in rooms not affected by the work performed. The design of the service/mechanical corridor can allow it to house all or most of the mechanical equipment needed to service the laboratory wing. A permanent hoist can be installed in a small mechanical room located at one end of the laboratory wing. This room should be accessible to large service vehicles.

This arrangement, pictured in figure 5-28, was used in the USEPA laboratory in Gulf Breeze, Florida. The upper-floor service corridor can be reached by stairs and ladder from the inside of the building. Heavy equipment can be lifted to the service/mechanical corridor by a hoist in the secondary entrance area shown in this picture.

This type of service/mechanical corridor offers a great number of possibilities for one-floor laboratories. Some of these layout possibilities are illustrated

FIGURE 5-28

schematically in the two-laboratory block designs shown respectively in figures 5-29 and 5-30 and figures 5-31 and 5-32. The laboratory block design shown in figures 5-29 and 5-30 has three laboratory wings and a long row of laboratory-related offices. This laboratory block could be connected, with a proper fire separation, to the administrative block of this laboratory. This block may be located at one end of the main people corridor. It could also be located parallel to the laboratory block and connected to it by way of a corridor, using the open space shown between the laboratory-related offices and extending from it. The ground floor of this design (figure 5-29) shows that almost all the laboratories and all the laboratory-related offices have windows. The laboratory wings can expand at their free end, assuming that no structure or other impediment is placed there. The laboratory-related offices are close to the laboratory rooms. This office arrangement does not lend itself to expansibility, except if part of the upper floor is used, as is discussed later. This arrangement can be used when there is no immediate need for additional laboratory-related offices. Stairs at the two ends of the main people corridor lead to the interstitial (service) floor above.

The upper-floor plan of this laboratory block design is shown in figure 5-30. This floor is an interstitial (service) floor with open plenums over the laboratory rooms. Access to this floor is through stairs at the two ends of the interstitial (service) floor and through a platform providing access to a service space and a service road at the ground floor. It is through this platform that heavy equipment can be lifted to the upper floor.

The floor area available at this level is much too large for a service floor. Therefore, the space located just above the ground-floor laboratory-related offices can be left open for future use as laboratory-related offices. These offices would be placed over the ground-floor offices. A corridor dedicated to the upper-floor offices would lead to two or more sets of stairs independent from the one servicing the service space. The stairs should be located strategically so as to reduce distances between these offices and their occupants' assigned laboratories.

Another variation of this design, with more room for expansion, is shown in figures 5-31 and 5-32. The laboratory block design shown in these figures has three laboratory wings, and three two-floor wings for laboratory-related offices, instead of the long row of laboratory-related offices offered by the previous design. This design's entire laboratory block, as in the previous solution, can be connected, with a proper fire separation, to the administrative block of this laboratory. The ground floor of the design shown in figure 5-31 shows that almost all the laboratories and all the laboratory-related offices have windows. The laboratory wings as well as the wings of laboratory-related offices can expand at their free end, assuming no structure or other impediment is placed there. The

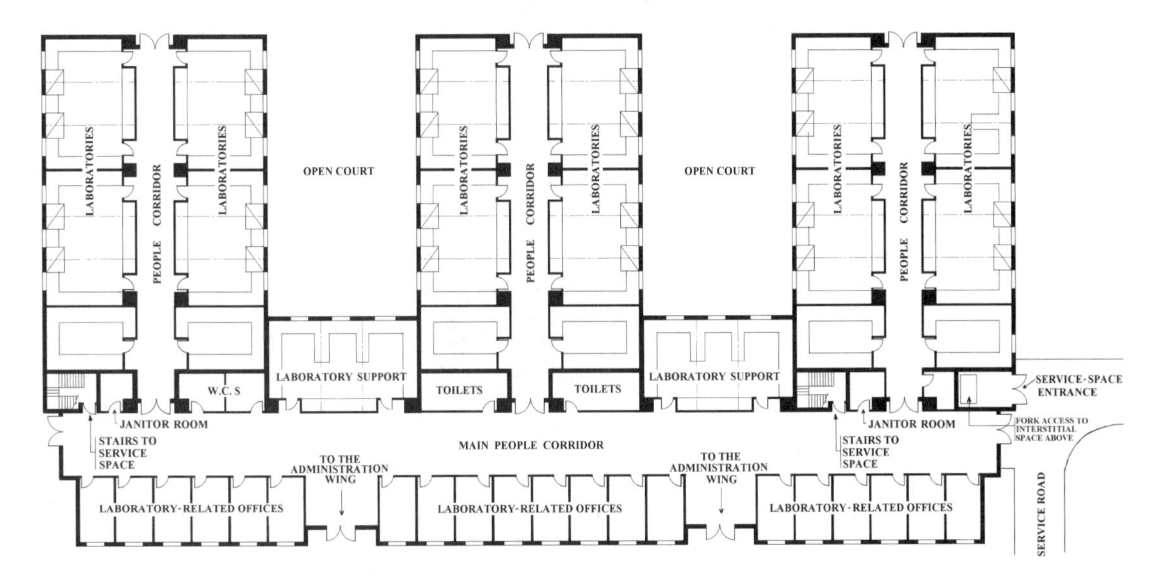

FIGURE 5-29
Ground-floor plan showing laboratories and laboratory-related offices

laboratory-related offices still remain relatively close to the laboratory rooms. This office arrangement offers an easy way to expand. This arrangement can be used when there is an initial need for a large number of laboratory-related offices, especially if laboratory operations may later require an even larger number of such offices. Stairs at one end of the wings of laboratory-related offices lead to the main people corridor. An emergency fire escape staircase is located at the other end of each wing.

The upper-floor plan of this laboratory block design is shown in Figure 5-32. This floor is an interstitial (service) floor with open plenums over the laboratory rooms. The service area is reached through stairs at the two ends of the interstitial (service) area of this floor and through a platform providing access to a service space and a service road at the ground floor. As in the previous solution, it is through this platform that heavy equipment can be lifted to the upper

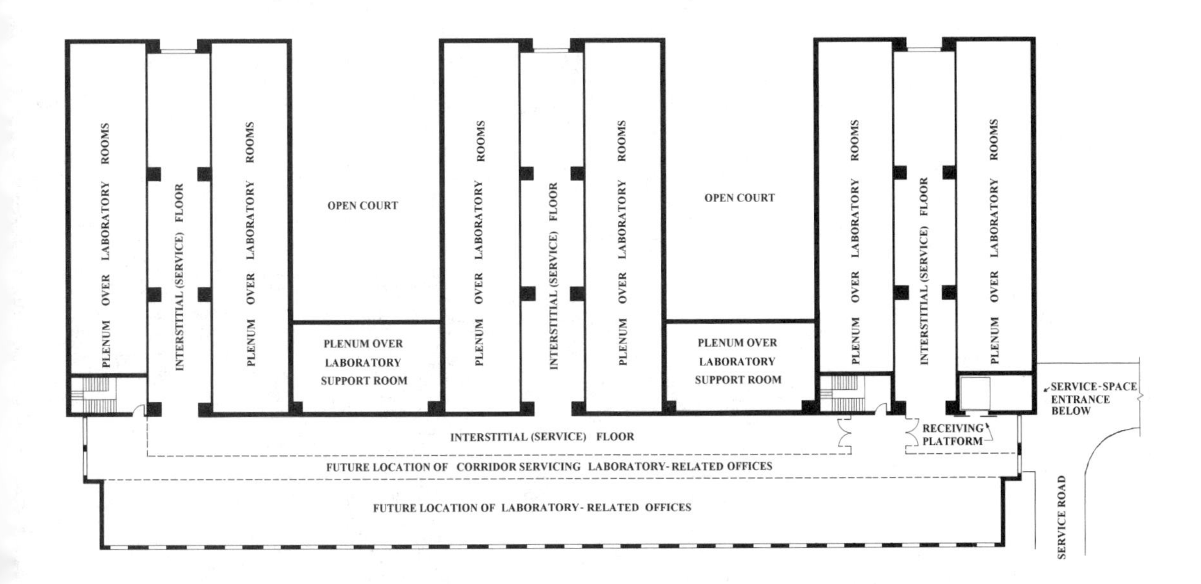

FIGURE 5-30
Ground-floor plan showing laboratories and laboratory-related offices

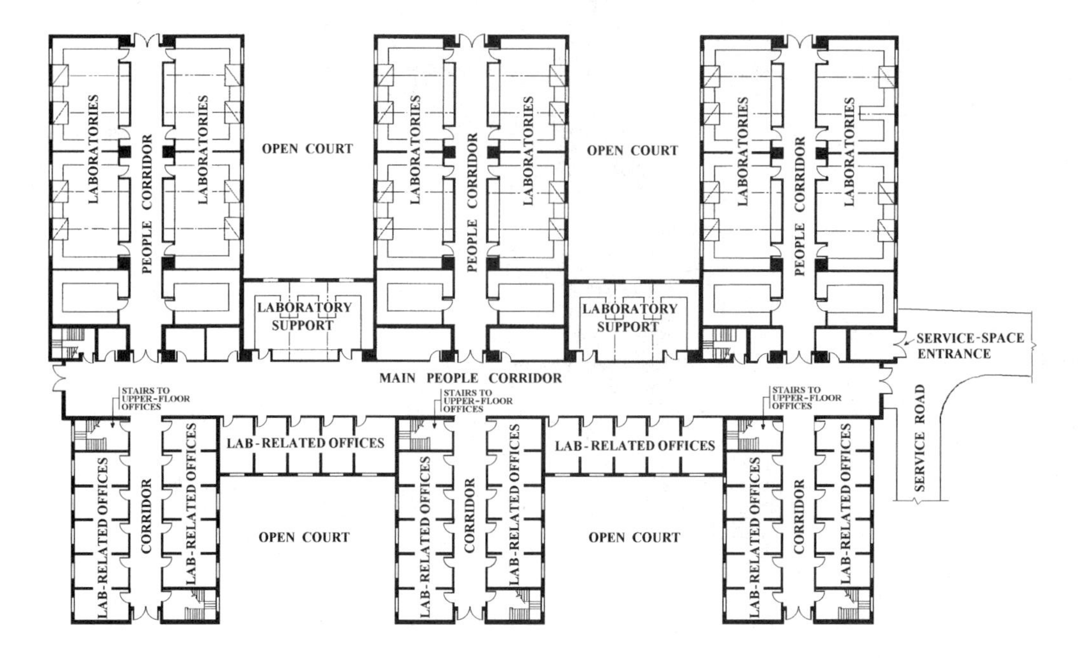

FIGURE 5-31
Ground-floor plan showing laboratories and laboratory-related offices

FIGURE 5-32
Upper-floor plan showing interstitial (service) space and laboratory-related offices

floor. The service area available at this level and the location of this area is very efficient for the provision of services and utilities to the laboratory rooms below.

One of the great advantages of the last two solutions is that the laboratory rooms as well as the laboratory-related offices have windows. Because of the isolation of the laboratory-related offices, if the right air pressures are used, the windows of the laboratory-related offices may be operable. The two sets of wings—the laboratory wings and the wings of the laboratory-related offices—have windows looking out onto open courts. If properly landscaped, these courts provide a very pleasant environment to the rooms' users.

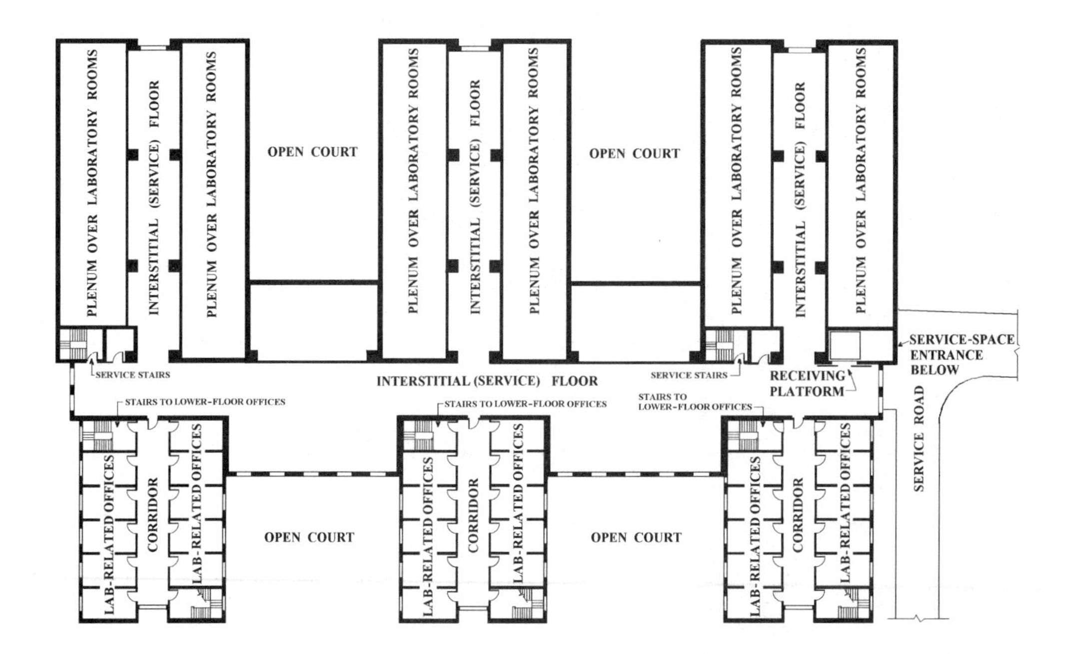

Chapter 6

Designing the Laboratory Facility

THE previous chapters addressed laboratory room and wing design. This chapter deals with the guidelines that should govern the design of a laboratory facility.

GUIDELINES FOR DESIGN

A laboratory facility generally encompasses several functions, each requiring different types of space. Therefore, the facility space should be subdivided, and the different activities performed in the facility should be grouped according to function and to their hazard and risk level. Separate space or blocks should be provided for administrative personnel and for the administrative supporting functions. The block that incorporates the laboratory and the laboratory-related offices (which are adjacent to or near the laboratories) should occupy a space separated from the space allocated to the other facility functions. Separate space should also be allocated for the mechanical equipment and for servicing laboratory wings and rooms (e.g., mechanical rooms, service and utility corridors, interstitial floors, catwalks, and walk-on ceilings).

A popular grouping of laboratory components is shown in figure 6-1. This grouping can be achieved through use of separate structures or of a single, subdivided structure with fire walls and other adequate precautions. The building or buildings composing the laboratory facility and lodging its different functions should be built according to the applicable codes and regulatory standards. These buildings should be flexible in their interior arrangement and lend themselves to expansion with minimum disruption to activities.

FIGURE 6-1

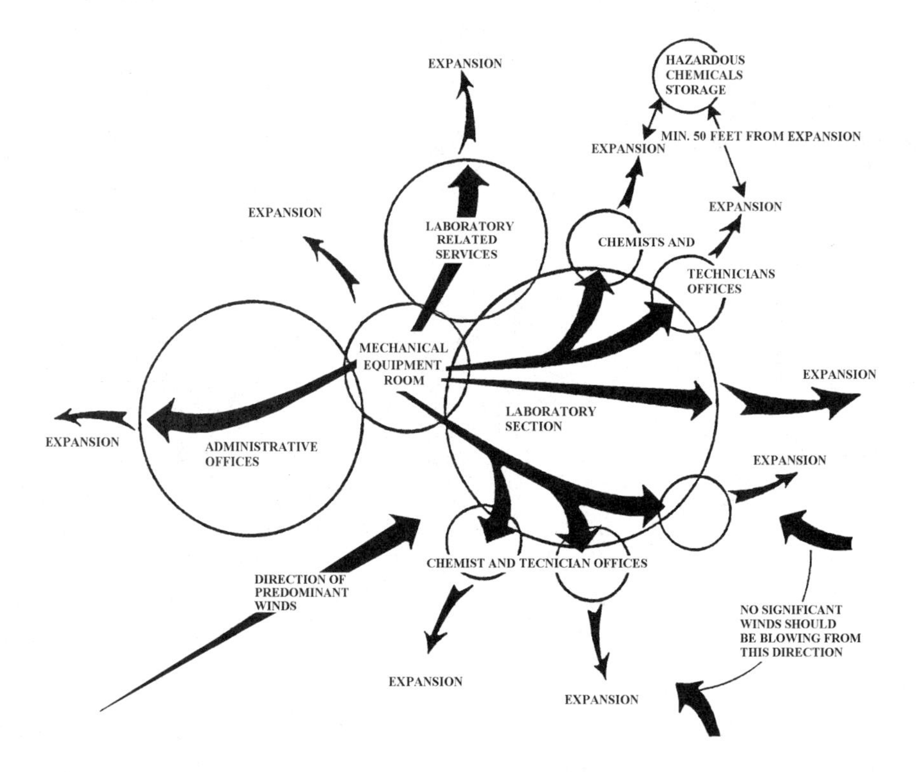

The laboratory wings, which include the chemical and biological laboratories and their related spaces, form the laboratory block. The laboratory block should have its space—the arrangement of its wings and rooms, and the relations of its wings and sections to one another—designed in such a way as to preclude the transmission of chemical fumes, vapors, and other contaminants from any one room, wing, or area to other rooms, wings, or areas, either from the inside or by cross-contamination between exhaust stack and intake ducts.

The laboratory block should be downwind from the other blocks of the facility. This allows the chemical fumes, vapors, and other contaminants exhausted by fume hood stacks to dissipate away from the other blocks or buildings of the laboratory facility. The prevailing winds should reach the air intake ducts free of any fumes or contaminants. Thus, exhaust stacks and air intake ducts should be segregated from each other and strategically located so that the exhaust fumes are not sucked into the intake ducts. Figure 6-2 shows airflow patterns when predominant winds hit the laboratory wing. The air intake ducts should be placed upwind and away from parking lots, loading docks, and other sources of fumes and contaminants so that prevailing winds reach the air intake without entraining motor vehicle and other types of fumes.

Figure 6-3 schematically illustrates the location of laboratory components in relation to each other and to the other functional blocks of the laboratory building. This figure also shows the location of the main air intakes, the loading dock and parking lots, and the direction of the prevailing winds. To be fully efficient, the arrangement shown in the figure should be supplemented by other mechanical considerations, such as the balancing of airflow and of pressures in and out of the building.

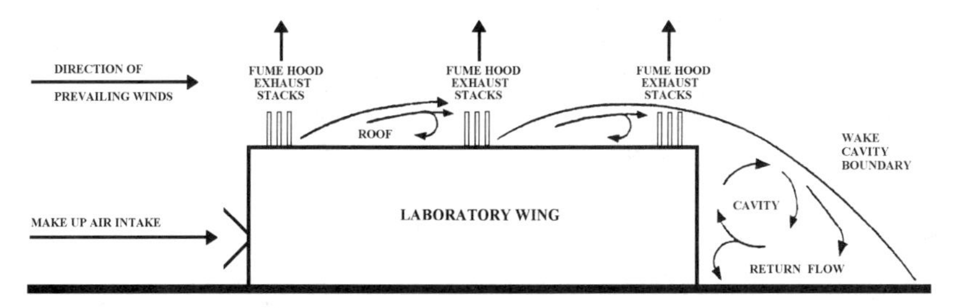

FIGURE 6-2
Air intake, fume hoods' exhaust and air flow patterns

SELECTION OF LABORATORY MODULE

In chapter 4, the interior design of laboratory rooms and the principles governing such designs were discussed. A laboratory room was said to be composed of one or more modules. In chapter 5 space, mechanical, and electrical modularity was defined. This chapter discusses how to determine the size of a module. Mechanical requirements are controlling factors in the selection of the module. Because the mechanical factors control the architectural design of the laboratory, they must be considered at the conceptual stages of the project.

As discussed earlier, the adoption of a modular plan is a must for increased efficiency and flexibility. It certainly decreases the cost of future changes. The selection of a module size tailored to the laboratory needs is one of the first tasks that the designers—the architect and the mechanical engineer of the design team—should address. A room 11 to 12 feet by 22 to 33 feet (3.35 to 3.66 m by 6.71 to 10.06 m) with a height of 9 1/2 to 10 feet (2.90 to 3.05 m) would be an acceptable module for a laboratory with one 6-foot-wide (1.83 m) or one 8-foot-wide (2.44 m) fume hood in it. In a module with these dimensions, the vol-

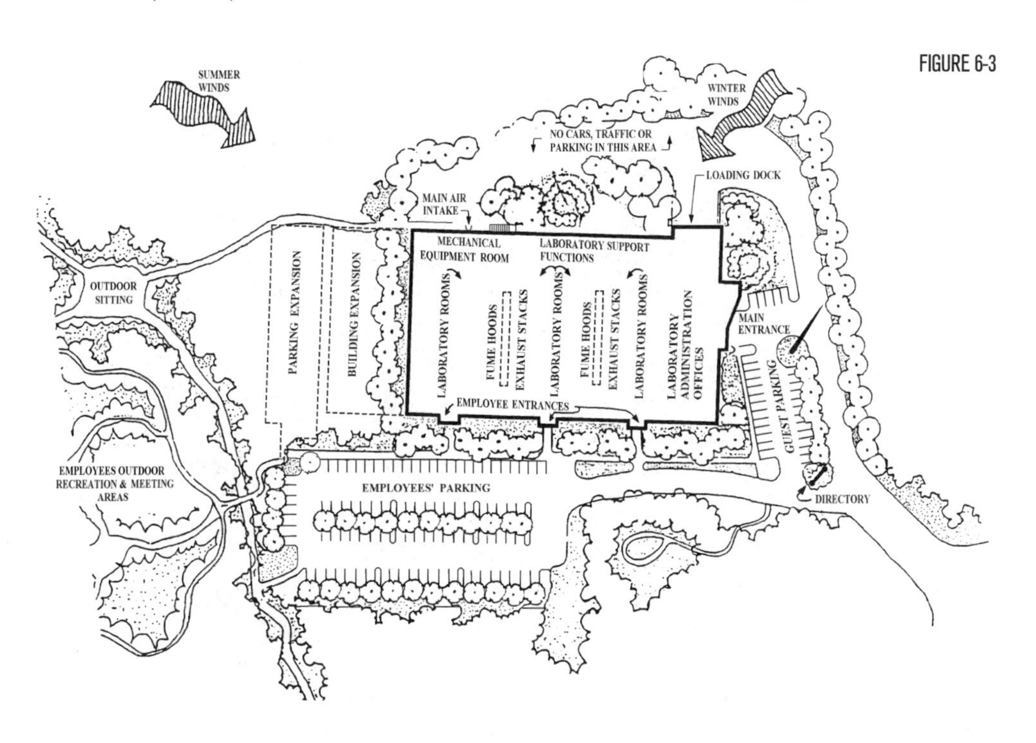

FIGURE 6-3

ume of air required for a fume hood's safe operation would flow safely without causing undue turbulence. As was said earlier, rooms or spaces can be composed of one or multiple modules. Each module can locate the fume hood so that rooms can keep the capability for partitioning along modular lines with minimum alteration to existing conditions. The design team should, therefore, be conscious of the current and future desired hood density per given area and of ways to satisfy their ventilation needs.

DETERMINATION OF MEANS TO PROVIDE SERVICES AND UTILITIES TO THE LABORATORY ROOMS

One of the first issues that the architect and the mechanical engineer should address at the conception phase is how to service the laboratory rooms in the context of the laboratory wings and how to provide these rooms with the needed utilities and services. The utility and service lines are the "life system" of the laboratory. They should be designed and located so as to always efficiently and safely perform functions and to be easily maintained. These lines will always need to be handled, either for repair or servicing, or to execute changes in the laboratory arrangements. Easy access to them with minimum disruption to the normal laboratory operations is thus of utmost importance.

In chapter 5, six factors were cited as affecting the arrangement of rooms into wings. These factors were:

1. The number of floors allowed or required for the laboratory wing(s)
2. The required adjacencies and separations
3. The required flexibility and expansibility
4. Whether the laboratory rooms could or should have windows or other means of natural lighting
5. How laboratory rooms should be provided with services and utilities
6. The wing(s) location and orientation in relation to other wings, rooms, and functions, as well as the location of fume hood exhaust stacks in relation to the air intakes and to the predominant winds direction(s)

The first five factors were discussed abstractly in chapter 5. When a laboratory building or facility is being conceived, the project's specific factors should be among those analyzed and used to determine how the laboratory wings will be serviced. The above-mentioned sixth factor requirements are addressed in this chapter and must be considered, in addition to other factors such as the site characteristics, the impact of and on neighboring uses, existing land uses and zoning, direction of predominant winds and other climatic considerations, the economics of the project and management inclination, and possibly other relevant factors.

CIRCULATION IN THE LABORATORY SECTION

Circulation in the laboratory section should be restricted to the people who are working there. The arrangement of laboratory and service rooms should provide paths allowing for tasks to be performed efficiently as well as for the circulation of chemicals, gases, and people. In principle, the circulation of chemicals and gases should be segregated from the circulation of the people working in the laboratory section. This is particularly necessary when the operations conducted in the laboratory require the frequent use and transportation of large quantities of potent chemicals, gases, and samples to and between the laboratory rooms. If, on the other hand, the operations conducted in the laboratory require only small quantities of chemicals and samples to be handled and transported between the laboratory rooms, an enlarged people corridor could handle the two types of circulation. The type and pattern of the corridors selected should be such that people working in laboratory rooms are not disturbed by the traffic. Corridors should be simple and direct and should be as short as possible when connecting points between which there is frequent communication.

Three types of corridors are normally used in the laboratory section: people corridors, utility corridors, and service corridors. People corridors connect the laboratory rooms to each other and to other sections of the building. They are generally only used by the scientists, the laboratory technicians, and other selected personnel.

Utility corridors are placed in back of the laboratory rooms and contain all utility and service lines servicing these rooms. They should be accessible only to maintenance and service personnel.

Utility corridors can be widened to provide additional services and utilities to the laboratory rooms. They then become service corridors. These corridors can then be used for the transportation of chemicals, gases, and samples to and between the laboratory rooms. This separates the service traffic from the people traffic. They also can be used to house certain types of equipment that are too bulky to be kept in the laboratory rooms that use them. Scientists, laboratory technicians, and other selected personnel should have access to the service corridors.

Vertical Circulation

Laboratory buildings with multiple floors should have at least two sets of stairs for the laboratory operators. In principle, chemicals, gases, and samples should not be transported via the stairs used by the laboratory operators. If, however, the chemicals and samples that need to be transported from floor to floor are not potent, are in small quantities, and are in containers small enough to be carried by hand, they can be transported via the stairs used by the laboratory operators.

If the laboratory building is conceived to have only one elevator, as required by the American Disability Act, and the chemicals and samples needing to be transported are potent, in large quantities, and in containers too large to be carried in the elevator, one or several dumbwaiters should be incorporated in the design. These dumbwaiters should be located strategically to allow the transportation of chemicals in a manner that does not interfere with the traffic of the people corridor.

If the laboratory building has two or more elevators, one of these elevators can be used for passengers as well as for transporting supplies and other freight. If the materials' volume, weight, and frequency present a hazard that prohibits the use of that elevator, a separate freight elevator should be provided.

When passenger elevators are used for transporting chemical supplies, samples, gases, and other freight, no passenger (other than those involved in transporting these items) should be allowed to use the elevator during the transportation. If service corridors are not used, a service path should be selected and clearly defined. This service path must be from the point of entry of the building, or from the location where these items are stored, to the location of the elevator and to the destination point. This path should be selected as the one with minimum traffic and minimum safety risks. When a separate freight elevator is provided and service corridors are not used, a service path should be selected and have the same characteristics as just described.

CORRIDORS IN THE LABORATORY SECTION

Walls and floors of people and service corridors in the laboratory section are subjected to more abuse than the other parts of the building. Obviously this is not the case for the utility corridors. Frequent use of hand trucks, the presence of emergency showers in corridors, and the possibility of chemical spills require that the wall surfaces and floor coverings of people and service corridors be adequate. Materials should be selected according to how efficiently they respond to these risks.

Figure 6-4 shows the utility corridor of the Emory Woodruff Memorial Research Building. Notice that this corridor's floor, walls, and ceiling are unfinished.

The corridors' wall finishing should be selected according to the functions performed and the chemicals used in the laboratories they service. Walls can be unfinished CMU or drywall. Wall surfaces are generally painted with a gloss or semigloss latex paint, when special treatment is not required. Certain low-level analysis and experimentation or similar types of operations can be affected by the fumes emanating from the corridor painting. When this is the case, the type of painting used should not emanate fumes. Guard rails or wall guards should be located on both sides of every corridor where carts and hand trucks may be used. Corner guards should also be used.

FIGURE 6-4

FIGURE 6-5

Floor coverings must adequately reduce the risk associated with the use of the space. Flooring may have to be seamless and resistant to abrasion; it always should be of sanitary and nonslip materials when resistance to chemicals is not required. In most instances vinyl floors will do. Use of 12-by-12-by-3/16-inch thick (305-by-305-by-5 mm) vinyl squares is a good choice, as they are easily and cheaply replaced should they become damaged. When chemical-resistant floors are required, it may be necessary to use epoxy finished floors or quarry or ceramic tile floors with or without sealed joints. Floors at corridors and between laboratories and rooms related to the laboratories and corridors must be smoothly sloped when their level changes and must be without impediments such as thresholds at doors of bumpy expansion joints.

Figure 6-5 shows the service corridor to the Georgia Public Health Laboratory in Atlanta, Georgia. Notice that this corridor's floor and walls are finished and that wall guards are located on each side of the corridor.

BUILDING FRAMING FOR THE LABORATORY SECTION

Framing systems in the laboratory section should be selected based on functional requirements and expected flexibility and expansibility of the laboratory, as well as on cost. Steel framing offers the advantage of smaller vertical dimensions for equal strength and allows layouts to be more easily modified to accommodate clearance problems for piping and ductwork. A concrete design offers the advantage of speed in construction. A construction of steel and reinforced concrete can be used in a multistoried structure—the lower floor has a reinforced concrete frame and the upper levels have steel framing.

EXTERIOR AND INTERIOR WALLS OF THE LABORATORY SECTION

Exterior and interior walls and partitions in the laboratory section can be of a variety of materials, the selection of which is well beyond the scope of this book. Their common denominator, however, is that their construction and finishing need to make them resistant to fire and smoke. Further, the chemical composition of the materials used for walls, ceilings, floors, and wall coverings, should be reviewed, as it may affect laboratory experimentation. All wall coverings in laboratories should be stain- and mildew-resistant.

DOORS IN THE LABORATORY SECTION

The number of doors in a laboratory room should depend on the size, number of occupants, and the distance to exits. Fire codes and other code requirements, as well as the classification of activities undertaken in the room, determine the labeling of the doors, the direction in which they open, and their hardware (panic and otherwise).

Regardless of code requirements, every laboratory room should have one door opening of at least 4 feet. This allows heavy equipment and large instruments to be moved in and out of the room. As stated in chapter 4, the door for this opening is often composed of two leafs—a 1-foot-wide (0.30 m) fixed leaf and a 3-foot-wide (0.91 m) active leaf that is used for entering and exiting the room. The 1-foot-wide (0.30 m) leaf is kept closed with a latch when there is no need to move equipment or large instruments in or out of the room. Although few codes mandate recessed doors, it is highly recommended that all doors in the laboratory section's corridors be recessed in an alcove so that they do not protrude into the corridor more than 6 inches (0.15 m). This is to avoid a collision between an opening door and a cart or person carrying chemicals.

Labeled doors are generally either C or B doors. C doors can resist fire for twenty minutes and B doors can resist fire for 1 1/2 hours. C doors are allowed to have a viewing window (called a "lite") of 1296 square inches (0.84 m^2). B labeled doors are allowed to have a viewing window of 100 square inches (0.06 m^2). The glass used is tempered glass for C doors and wired glass for B doors. Door lites in C doors are large enough for anyone to easily see through. The door lites in B doors should be vertically elongated and placed on the doorknob side, with the lower part of the lite about 3 1/2 feet (1.07 m) from the floor and the upper part of the lite about 5 1/2 feet (1.65 m) from the floor. This is to prevent someone from opening the door into someone else on the other side.

For laboratory rooms with more than one door, it is good to keep the location of the door consistent—for instance, the wider door should be on the right (or left) side of every laboratory room. This means that the emergency shower (if it is outside the room) is on one side of the door alcove and the electrical panel box servicing the room is on the other side of the alcove. (The exact location of emergency showers is discussed in chapters 4 and 9).

The material selected for external doors and door frames is also important. There is a large selection of these doors on the market, and most of them are acceptable if the selection is made according to their use, location, and security and aesthetic requirements. It should be noted, however, that steel doors and door frames quickly corrode when exposed to salty, humid, marine climates.

WINDOWS IN THE LABORATORY SECTION

Laboratory room windows have long been a controversial subject for several reasons. Everyone agrees that in most laboratories windows are not acceptable as a means of ventilation because the air pressure inside the building must be precisely maintained. From there, opinions differ on the use of fixed pane windows.

As discussed in chapter 5, most mechanical engineers prefer windowless laboratory rooms. This is because windows have a high thermal conductivity factor (the "u" factor), even if double glazed glass is used. This means that changes in outside temperatures would affect the laboratory room's temperature and humidity. Depending on the seasons, the geographic location, and the orientation of the building, such temperatures differences could be substantial and would affect the HVAC system.

Some architects and laboratory owners found it convenient to adhere to these views. A laboratory building with windowless rooms is more compact and therefore costs less to design and build. A shoebox design is easier to create, and it generally pleases the cost-conscious laboratory owner, as the utility costs are slightly lower. However, most laboratory users, being technicians or scientists, prefer to work in laboratory rooms with windows. For them, windows are a source of natural light as well as a means of communicating with the outside world. Windows in laboratories diminish the isolation felt by the laboratory users and substantially increase their efficiency. The recognition of this fact prompted laboratory owners and management who considered the productivity of their staff more important than money saved in building construction and utility costs to ask designers to conceive laboratories with fixed pane windows.

Windows in laboratory rooms should be sized and located in each module so that they do not affect the typical arrangement of chemical fume hoods and casework. These arrangements are discussed in chapter 4. The window width should be such as to allow room for of a chemical fume hood on each side of the window within each module. The window height should be such as to allow the maximum natural lighting to penetrate the room. The window sill should be about 4 feet above the floor for safety reasons and to allow room for casework against the window wall.

Double or triple glazing can be used to reduce the thermal conductivity factor. If the windows face south, built-in horizontal blinds should be used to screen the sunlight. If the windows face east or west, built-in vertical blinds should be used.

The material selected for window frames is important. There is a large selection of wood, aluminum, and steel windows and window frames on the market. Many of these windows can be used if the selection is made according to the specifications determined by the rooms' requirements. It should be noted, however, that steel window frames quickly corrode when exposed to salty, humid, marine climates.

LABORATORY FURNITURE AND CASEWORK

The choice and arrangement of laboratory cabinetry and countertops is governed by the operations and tasks expected to be performed in each laboratory and by the funds set aside for this purpose. The nature of the operations or tasks (e.g., research, development, quality control) determines the type of materials being investigated (odorous, flammable, unstable, etc.), the chemicals and gases used, and the space needed for storage and for each scientist, each student, and each task. The task's nature also determines the degree of maintenance for and permanence of the furniture and casework.

When choosing the casework for a laboratory room, one should consider the effect the casework might have on the chemicals handled as well as the effect the chemicals would have on the casework. Metal casework (and doors and windows) should not be used in laboratories dealing with trace metal analysis, as the metal of the casework could affect the analysis. Similarly, metal casework (and fume hoods) should not be used in laboratories where large amounts of condensed acids and oxidants are used because the chemicals would corrode the metal casework (or fume hood). The designer should have a list of the chemicals that will be used in each room. This list should indicate the level of dilution and potency of each chemical and the quantity used per unit of time (per day, week, or month). The designer should then provide safety cabinets in sufficient number and size—and enough space for them—to safely store all chemicals requiring storage for the needed amount of time. A description of safety cabinets is provided later in this chapter.

Usually, the architect gives too little attention to the design of the casework. Often the decision is left to the user or to the cabinet supplier, both of whom try to fit in as many cabinets as possible without seriously considering functional and equipment needs. The result is often chaotic, as figure 6-6 depicts. Poorly designed laboratory rooms may not look too different from this picture.

Good casework design requires patient collaboration between the architect, the facility manager, and the room user or users. It takes into account the work culture of the organization as well as the habits of the scientists and technicians who will be using each room. The operations that will take place in the room, as well as the space needed by equipment and its services and utilities, must also

FIGURE 6-6

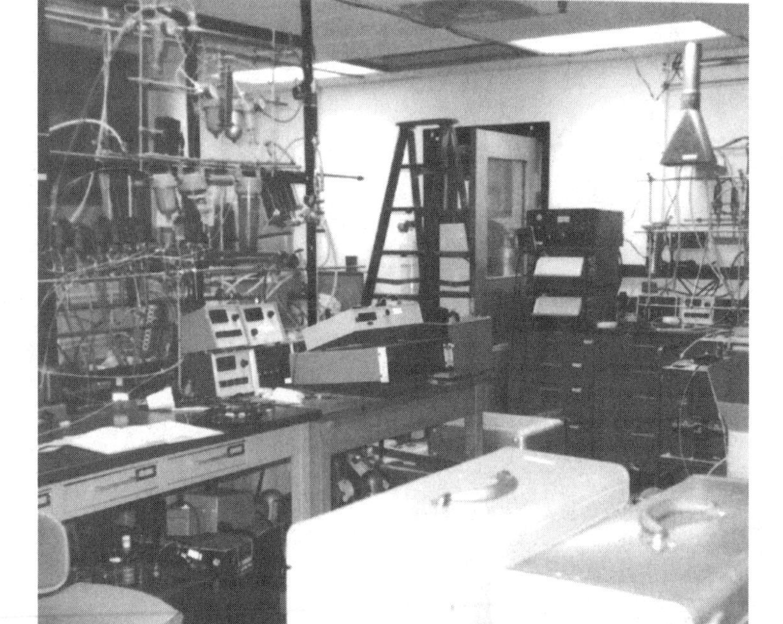

FIGURE 6-7

be considered. When all these factors are taken into account, the environment is not only functional but also pleasant, as shown in figure 6-7.

The available choices for cabinetry are steel, wood, plastic laminate, and polypropylene. Each has advantages and disadvantages and each may be fixed or movable.

Steel Cabinetry

Steel cabinetry is most often used in general chemistry laboratories, except when special types of operations dictate otherwise. Steel cabinetry has great fire resistance, requires less maintenance, has great flexibility by offering interchangeable parts and components, and, when baked with epoxy-type finishes, becomes mostly impervious to many acids, bases, and solvents. This type of cabinetry is not well suited for marine climates or when exposed to a corrosive environment, man-made or natural. Steel cabinetry, as well as any untreated metal equipment, including doors and windows, quickly corrodes when exposed to salty, humid, marine climates.

Wood Cabinetry

Wood cabinetry creates a lower noise level than steel cabinetry. It is easier to damage but also easier to repair and to refinish. Wood is a warmer material, both in feel and appearance, and makes the environment more aesthetically pleasing. Wood cabinetry is appropriate for marine climates and for laboratories used for operations that may be affected by the presence of metal in the room or operations that intensively use chemicals that corrode metals.

Plastic Laminate Cabinetry

Plastic laminate cabinetry also creates a lower noise level than steel cabinetry. It is easy to damage and difficult to repair. Some types have a warm feeling and attractive appearance. Plastic laminate cabinetry should not be used in laboratories where chemicals are used frequently. Also, when exposed to fire, plastic laminate cabinetry may cause a toxic atmosphere.

Polypropylene Cabinetry

Polypropylene cabinets are very expensive. They should be used for laboratories in which operations require a high level of cleanliness or when large quantities of corrosive chemicals are used. For instance, polypropylene cabinets are highly recommended for clean rooms and laboratory rooms where low-level or trace analyses are performed (such as a trace metal analysis laboratory) because such rooms require a high level of cleanliness. Polypropylene cabinetry is also ideal for rooms where perchloric acid and oxidizers are used because the cabinets' surface is not affected by strong acids. Another benefit of polypropylene is that it requires less maintenance. It is a rigid material and its smooth, nonporous surface does not collect dust or residues. Because it does not corrode or oxidize, this type of cabinetry is well suited for marine climates and all types of corrosive environments, man-made or natural. A disadvantage is that polypropylene hardware is weaker than metal hardware. To compensate for this weakness, polypropylene hardware is larger, thicker, and bulkier than metal hardware. It is also much more expensive than metal hardware.

Fixed Versus Flexible Cabinetry

Certain laboratories are expected to change tasks and operations often. These changes may require a reconfiguration of the cabinetry layout. When this situation occurs, flexible cabinetry or a combination of fixed and flexible cabinetry is a better choice than the conventional fixed cabinetry. The components of flexible cabinetry can be separated from the benchtops and the service strip and then reconfigured as needed. The cabinets are hooked from the back and can be moved from one location to another in the same room or in any other room. This relocation does not require too much expertise and can be performed by the laboratory maintenance staff, without the involvement of formal cabinet installers. This allows the cabinets to be rearranged quickly and with minimum disruption.

There are some disadvantages to flexible cabinetry. It does not have the stability or durability that fixed cabinets have, and, because each component has its own, separate benchtop, the countertop surface is seldom perfectly level. Further, the countertop has as many joints as it has components. These joints must be sealed, and if they are not perfectly level (because of minor defects in the floor finishing or misadjustment by the cabinet installer), the countertop will be wavering and unsafe.

A combination of fixed and flexible cabinet systems is possible and in some situations preferable. It might consist of a fixed service strip with modulated cabinets and tables locked to each other and to the service strip. The cabinets, although fixed in this situation, could be easily moved and their components (doors, shelves, and drawers) interchanged. The cabinetry described thus far would be used to store laboratory instruments and equipment, tubing gloves, and other laboratory paraphernalia. It could also store small supplies of chemicals used in the laboratory room as long as these chemicals are not reactive, flammable, combustible, or toxic and the storage of such chemicals adheres to codes and regulations.

Countertops

The available choices for countertops are natural stone, modified epoxy resin, wood, high pressure laminate, and stainless steel.

NATURAL STONE COUNTERTOPS

The most common natural stones used for countertops used to be alberene and sandstone. Alberene is a silicate with or without talc. It is structurally strong, has a low coefficient of absorption, and is resistant to most reagents. Alberene countertops have a long life expectancy but are expensive. They are, however, rarely used in today's laboratories because of their cost the difficulty of shipping and installing them. The so-called natural impregnated stones are impregnated hard sandstones. Countertops made of these materials are highly resistant to chemical attacks but should not be subjected to high, continuous localized heating. If these countertops are impregnated with organic materials, they are not suitable for perchloric acid use. Impregnated stone countertops exposed to violent oxidizers should be impregnated and sealed with nonorganic materials.

MODIFIED EPOXY RESIN COUNTERTOPS

The most common artificial stones used for countertops are made of modified epoxy resin. Counters made of this material have an extremely hard surface. They offer excellent resistance to a wide range of acids, solvents, and alkalies. They have good resistance to heat, abrasion, stain, impact, moisture, and bacteria. They have a low shine finish and are easy to clean. They should be used in areas requiring chemical and heat resistance.

WOOD COUNTERTOPS

Wood, natural and plastified, makes an excellent countertop for physical test tables and in electronic laboratories. Carbonized wood finishes and wood treated with acid- and alkali-resistant finishes also provide good and relatively inexpensive countertops.

HIGH PRESSURE LAMINATE COUNTERTOPS

These countertops have a wood particles core and a plastic laminate surface. They have a smooth finish and are manufactured in a variety of colors. Their edges, when banded, are easy to damage and difficult to repair. They offer poor chemical resistance but good stain, bacteria, and moisture resistance. They should be used in general purpose areas requiring limited heat resistance and no chemical resistance.

STAINLESS STEEL COUNTERTOPS

Stainless steel surfaces are excellent for radioactive work, bacteriological work, and for work where organic solvents are used. It can be used with certain acids such as perchloric and nitric acids but should not be used with hydrofluoric, hydrochloric, sulfuric, or hydrobromic acids.

LABORATORY STORAGE

Sufficient and segregated space should be provided for the storage of items necessary for the conduct of laboratory activities, as well as for chemicals, reagents, samples, and laboratory wastes. Storage space for such items should be considered hazardous material storage space and must respond to specific health and safety requirements.

Laboratory Support Space

Laboratory support space should consist of several rooms of different sizes evenly distributed throughout laboratory areas and centrally located to the laboratory rooms they service. Naturally, these rooms cannot be used for storage of chemicals, reagents, or laboratory wastes. The size of each room should be adequate for the activities for which the space is designed and for housing what should be stored there. As a rule of thumb, approximately 5 square feet (0.46 m^2) of laboratory storage should be made available for every 100 square feet (9.29 m^2) of laboratory space. Laboratory support space generally includes glassware washing and storage rooms; sample receiving, log-in, and storage rooms; sample preparation rooms; and field and other types of laboratory-related equipment storage rooms.

Storage rooms distributed throughout laboratory areas efficiently provide storage for items not conducive to within-laboratory or remote-warehouse storage. They free up within-laboratory storage areas for items most frequently used, and provide lower cost storage for items previously occupying expensive laboratory space. Another major advantage of such rooms is that they prevent inappropriate use of laboratory floors, benches, carts, and cabinet tops as storage spaces. Obviously, the use of storage rooms reduces safety hazards.

When these rooms are located centrally to the laboratory rooms they service, they reduce time-consuming trips to remote warehouse. By keeping aisles and floors clear of obstructions, they also promote safe and efficient movement of personnel within each laboratory. The relationship between laboratory rooms and these storage rooms should be somewhat similar to the relationship between a kitchen and a walk-in pantry, where traffic between the two areas may not be heavy but where quick access may be required several times a day.

These rooms store many useful items necessary for laboratory operations. Among these items are: large pieces of equipment and instrumentation that are used occasionally or are being stored temporarily before being repaired or discarded; new instruments and hardware covers and casings normally removed after the instruments and hardware are installed; customized shipping containers used to return equipment; instrumentation and accessories for repair or replacement; bulky consumables too large to fit in the laboratory's under-counter storage, such as large rolls of bench protection mats, poly bags of disposable and plastic ware as well as large, multiunit boxes temporarily stored before breakdown into smaller units for in-laboratory storage; carts and movable tables temporarily out of service; new instruments and equipment awaiting vendor inspection and installation; demonstrator instruments and equipment awaiting evaluation;

oversized apparatus such as aluminum support rods, glass rods, (empty) reagent carboys, tubs and new waste cans and liners; hand trucks; step stools; ring stands; glassware; and chairs, movable tables, and other similar types of equipment.

Traditionally, bulk storage areas for laboratory use were located in basements or in fairly remote warehouses with little air quality control and where delicate instruments were affected by the air quality. Access to these types of storage areas is often under the control of facility personnel and does not correlate well with laboratory work hours, including overtime and weekend work. Further, the utilization of these types of bulk storage areas was observed to be less frequent because of the time and distance it took to get from the laboratory wing to the storage areas.

Easy and proximate access to the storage rooms minimizes the hassle for laboratory operators and scientists by circumventing nonlaboratory assistance or authorization and restrictions on access times. When storage rooms are well integrated into the laboratory environment, laboratory operators and scientists are encouraged to take control of the rooms' organization, content, and management. The storage rooms should be designed to respond to specific needs, if such needs are known during the design phase. If they are not known during the design phase, storage space should be designed to respond to general storage needs. This means they should:

- Have large door openings with either a 4′-0″ (1.22 m) single door or a 1′-0″ (0.30 m) fixed panel and a 3′-0″ (0.91 m) operable door panel with a vision panel and a doorstop to keep door open when large items are being moved.
- Have either card readers or locks to limit the access to people authorized to enter.
- Be equipped with heavy-duty adjustable and removable shelving, composed of laboratory grade laminate (if a nonmetallic environment is required) or of epoxy coated metal (if a regular, nonrestrictive environment is acceptable). Also, optional "space saver" vertical carousels can be utilized in about one third of the available wall space. This provides flexibility and increases the storage capacity.
- Be provided with adequate lighting to illuminate all shelves. Room lighting should be activated by motion sensors. If rooms are provided with light switches, the switches should be located outside the room, close to the door.
- Have a ventilation system designed to provide protection from moisture, heat, cold, and gross particulates.
- Have sufficient space to allow one person to easily move large tables through the entrance and within the room.
- Be provided with outlets for phone, intercom, central LIMS/LAN networks, and be connected to the paging system.

Storage of Chemicals in the Laboratory Building

Several types of segregated spaces may be needed for the storage of chemicals, samples, and laboratory wastes. These spaces, depending on what they are, are located either in the laboratory rooms or in or nearby the laboratory wing(s). Space for the storage hazardous chemicals must respond to specific health and safety requirements. Chemicals, samples, and laboratory wastes should be stored in the space provided in the laboratory wing only for short periods of time.

There are three types of spaces generally used for storage of chemicals, samples, and chemical wastes (a more comprehensive description follows this list:

1. *Chemical storage cabinets, refrigerators, and freezers.* These are specially designed to store different kinds of chemicals and are generally located in the laboratory rooms. They are used to store the small quantities of chemicals needed for daily consumption in the laboratory room.
2. *Stockrooms.* This is a special room used to store a certain quantity of the chemicals needed for the operations taking place in the laboratory rooms. These chemicals should be stored in approved, small containers. Larger quantities of chemicals should be stored in a remote building known as the hazardous materials storage building. Stockrooms, when used, are generally located in the laboratory wings. Depending on the size and number of floors of the wing, and on the quantities of chemicals needed, a stockroom may service an entire wing, an entire floor of the wing, or a set of laboratory rooms within a wing or a floor.
3. *Hazardous materials storage building.* When large quantities of chemicals are used in a laboratory facility or when large quantities of laboratory wastes are generated, these chemicals and wastes are generally stored in a remote building called the hazardous materials storage building. This building must respond to specific health and safety requirements, in addition to the requirements dictated by the users' needs. The quantities of chemicals stored in the hazardous materials storage building may be sufficient to provide the laboratory operations for a month or more. These chemicals should be stored in approved containers of any size up to 50-gallon (189.25 l) drums. The requirements for a hazardous materials storage building are stated in chapter 7.

1. CHEMICAL STORAGE CABINETS, REFRIGERATORS, AND FREEZERS KEPT IN LABORATORY ROOMS

Scientists and laboratory operators like to keep a certain amount of chemicals in their laboratories. These chemicals must be stored in sealed containers and in a prescribed place. Except in very small quantities, these chemicals must not be placed on benchtops or open shelves where they would be unprotected from fire or mishandling. Nor should they be placed near fume hoods, where they could interfere with airflow. If the chemicals need to be refrigerated or frozen, they should be stored in explosion-proof refrigerators or freezers. If they do not, they can be stored in cabinets. These chemical storage cabinets must respond to specific health and safety requirements, in addition to the requirements dictated by the users' needs.

Tables 2-2(a) and 2-2(b) in the NFPA-45 set the maximum quantities of flammable and combustible liquids and gases that can be stored in sprinklered and nonsprinklered laboratory units. The tables also specify the laboratory classifications resulting from the quantities of flammable and combustible liquids or gases stored in each laboratory. A class A laboratory has a high hazard level, a class B laboratory has an intermediate hazard level, and a class C laboratory has a low hazard level. Quantities of chemicals or chemical wastes larger than what is permitted by these two tables must be stored in either a stockroom or a hazardous materials storage building.

Special cabinets, refrigerators, and freezers should be used for storing reactive, flammable, and toxic chemicals or chemicals having any combination of these hazards. These special cabinets, refrigerators, and freezers are named according to their content: hazardous chemicals storage cabinets, refrigerators, and freezers; flammable materials storage cabinets, refrigerators, and freezers; and safety cabinets, refrigerators, and freezers. They must comply with specified requirements and codes. Quite a variety is available on the market.

These cabinets, refrigerators, and freezers must:

- Be built in compliance with OSHA and NFPA-30 requirements
- Be labeled on the outside door to indicate the nature of the chemical stored and any related warnings (e.g., "Flammable—Keep Fire Away" or "Caution—Toxic Substance")
- Have raised sills or spill trays covering all shelves in their entirety
- Be equipped with locks
- Be spaced 2 feet (0.61 m) apart from each other

The cabinetry must:

- Be mechanically vented with 5–20 cfm (0.00236–0.01 m^3/s) of air supplied at the top of the cabinet and exhausted from the bottom so that it sweeps across all the shelves, and
- Have its ventilating inlet duct fitted with a flame arrester and the exhaust duct connected to a roof-mounted spark-proof exhaust fan.

Figure 6-8 pictures an under-counter flammable materials storage cabinet and figure 6-9 pictures an under-counter acid storage cabinet. Each of these cabinets was manufactured to respond to all the applicable safety requirements.

When stored in a laboratory room, mutually reactive chemicals, such as strong oxidizers, promoters, curing agents, and activators, should be stored separately in special cabinets, refrigerators, or freezers. When a chemical belongs to more than one of these groups, it should be stored in compliance with all the requirements of the groups. Detailed requirements for storage of chemicals is provided in NFPA-30 and NFPA-45.

FIGURE 6-8 (left)

FIGURE 6-9 (right)

STORAGE OF TOXIC GASES WITHIN A LABORATORY BUILDING

Toxic gases should be stored in gas cabinets or exhausted enclosures that have positive exhaust ventilation. Such gas cabinets or exhausted enclosures must be connected to treatment systems. A few exceptions to the storage requirements may be permitted by the authority having jurisdiction. Such exceptions, however, must provide an equivalent level of safety.

2. STOCKROOMS

As stated in the previous paragraphs, the quantities of hazardous chemicals stored in a laboratory room's work area must be kept to a minimum. Separate storage rooms located in or next to the laboratory wing should be used when the quantities of hazardous chemicals are larger than the maximum quantities allowed to be stored in the individual laboratory rooms. These storage rooms are called stockrooms.

Stockrooms are located within or next to the laboratory wing requiring certain quantities of chemicals for operations. Only one stockroom may be needed for the entire building if the operations do not require very large quantities of chemicals. There may be one stockroom per floor if the needed quantities of chemicals are large enough or if the chemicals cannot be safely stored in individual laboratory rooms. The use of stockrooms reduces the laboratory operators' trips and access to the remote hazardous materials storage building. The stockrooms must respond to specific health and safety requirements, in addition to the requirements dictated by the users' needs.

The hazardous chemicals inventories stored in a stockroom should be within the prescribed capacity of the room. Incompatible chemicals and mutually reactive chemicals should be segregated—either by safety cabinets within the stockroom or by being placed in different stockrooms. Sometimes it may be necessary to have solvent stockrooms as well as acids and bases stockrooms, such as when the chemicals are flammable, combustible, or explosive and are in quantities that make it unsafe to store them in the same room.

Flammable liquids stored in stock rooms must be in containers made of glass, metal, and/or approved plastics. The size and type of the containers depends upon the flush point of these liquids. There are three classes of flammable liquids: 1A, 1B, and 1C. The container type and the maximum amount of liquids that can be stored in a given type of container is specified in paragraph 7-2.3.2 and table 7-2 of NFPA-45.

STOCKROOM'S ENVELOPE AND MATERIALS OF CONSTRUCTION

Stockrooms where hazardous materials are stored should be separated from adjacent spaces by at least a 2-hour resistive envelope. The materials of construction used in the floors, walls, and ceilings, as well as those used for ducts, piping, and vessels should be compatible with the chemicals stored, transferred, or handled in the room.

The floor of a stockroom should be either depressed or segregated from the other spaces and rooms by use of a curb. Such floor should be sealed and impervious. The depression depth or the height of the curb should be such as to contain a spill of the same volume as the largest container stored in the room. The

FIGURE 6-10

changes in floor level must be ramped, so that a cart carrying chemical containers can go in and out of the stock room without being bumped.

Shelving in a stockroom should be built from materials not affected by the chemicals stored in the room. Each stockroom should have at least one set of double doors with panic hardware. The recommended minimum door opening for a stockroom is 4′-0″ (1.20 m).

Figure 6-10 shows a stockroom with a grid over a depressed floor. The grid is at the same level as the corridor floor. Shelving is built from materials not affected by the chemicals stored in the room, and doors are much wider than 4′-0″ (1.20 m). Notice also that the room air is exhausted at the floor level as well as at the ceiling level.

When flammable chemicals are stored in a stockroom in certain quantities identified by code, the room should be equipped with blast (or blow) panels to conform with the requirements of NFPA-30, section 4-4.2.5. These panels should be located on an outside wall so that if an explosion occurs, its thrust will propel the panels within the laboratory grounds and within an open area. This area should be designed as buffer (e.g., a wooded area) and should be a place where people are unlikely to be.

If the stockroom is on the upper floor of the building, the blast panels can be located in the roof of the room. In any case, the location of the blast panels should abide by the same requirements described above.

Blast panels are not required in stockrooms where class 1A liquids are stored in containers of 1 gallon (3.79 l) or less. Panels are also not required where dispensing of class 1A or 1B liquids (or any unstable liquids) in the stockroom is prohibited.

STOCKROOM'S VENTILATION

A stockroom containing hazardous (such as flammable or toxic) chemicals must be vented to the outside atmosphere by means of a mechanical exhaust system that provides a ventilation rate of 1 cfm per square foot of floor area with a minimum of 150 cfm (0.07 m^3/s) per room. The room should have supply and exhaust registers placed in the upper part of the room and lower part of the room, on opposite walls. The air supplied to the room must enter the room within 12 inches (0.30 m) of the floor and within 12 inches (0.30 m) or less of the ceiling on one side of the room. The air exhausted must be exhausted from a location also within 12 inches (0.30 m) of the floor and 12 inches (0.30 m) or less of the ceiling, but on the wall opposite from where the air is supplied. Ducts in stockrooms and the entire ventilation system supplying air to the room must comply with NFPA-91. The air exhausted must not be vented into a fume hood or into a fume hood exhaust system.

LIGHTING, MECHANICAL, AND ELECTRICAL EQUIPMENT USED IN STOCKROOMS

Explosion-proof light fixtures, switches, and outlets are required in stockrooms housing flammable chemicals, if the quantities and condition warrant it. If heating or cooling equipment is used in the room, all equipment—wiring systems, duct systems, switches, etc.—must be explosion-proof when inside the room or directly in contact with any gases or fumes that may emanate from the chemicals in the room.

Chapter 7

STORAGE OF CHEMICALS AND CHEMICAL WASTES IN A LABORATORY FACILITY

SEVERAL types of segregated space should be provided in a laboratory facility for the storage of chemicals, samples, and laboratory wastes. This space is considered hazardous material storage space and must respond to specific health and safety requirements.

Large quantities of chemicals and laboratory wastes are generally stored in a remote building called the hazardous materials storage building (HMSB). It, too, must respond to specific health and safety requirements, as well as to the requirements dictated by the users. When the laboratory wing or wings are large and multistoried, each floor may have a chemical stockroom that stores an amount of chemicals sufficient for the needs of the floor for a certain amount of time. This reduces trips to the remotely located HMSB. The chemical stockroom must also respond to specific health and safety requirements as well as to the requirements dictated by the users.

Often, scientists and laboratory operators like to keep a certain amount of the most frequently used chemicals with them in their laboratories. These chemicals must be stored in segregated chemical storage cabinets. The chemical storage cabinets must also respond to specific health and safety requirements and to the requirements dictated by the users.

Table 2-2(a) and Table 2-2(b) in NFPA-45 specify the maximum quantities of flammable and combustible liquids and gases that can be stored in class A, B, C, and D laboratory units with sprinkler systems and without. It should be noted again that a class A laboratory has a high hazard level, a class B laboratory has an intermediate hazard level, and a class C laboratory has a low hazard level. Classification is based on the quantity of flammable and combustible liquids or gases stored in the laboratory. Quantities of chemicals or chemical wastes larger than those permitted by the two tables must be stored in either a HMSB or in a chemical stockroom.

THE HAZARDOUS MATERIALS STORAGE BUILDING

The first task to be undertaken before designing the HMSB is to identify the chemicals and wastes that will require storage in the different chambers of the building. The quantities and nature of these chemicals must also be identified.

Chemicals are classified as class I, II, or III. Class I encompasses all flammable gases, vapors, and liquids; class II consists of combustible dusts; and class III comprises ignitable fibers or flyings. As a general rule, any room, space, or area where any of these chemicals is stored, handled, or transferred must meet the requirements of the latest edition of NFPA-70, Article 500.

Of the above-named chemical classifications, class I chemicals are the most frequently used in general chemistry laboratories. Therefore, we will focus on the storage of class I chemicals. Class I storage locations are subdivided into division 1 and division 2. In class I division 1 storage locations, ignitable concentrations of gases or vapors exist under normal conditions. This includes areas where volatile liquids are dispensed or transferred from one container to another. In class I division 2 storage locations, ignitable concentrations of gases or vapors do not exist under normal conditions. This may be because the chemicals are stored in sealed containers or because of the provision of mechanical ventilation. It is possible that a class I division 2 storage location may experience an ignitable concentration of gases under accidental or unusual conditions, such as the accidental rupture of a storage container or a failure of the mechanical ventilation system. Such an extraordinary condition should not be the basis for reclassification of the class I division 2 storage locations as a class I division 1.

As stated earlier, relatively large quantities of class I chemicals must be stored in the HMSB. Class I chemicals will therefore be emphasized in the design criteria for the HMSB.

Chemicals should be separated for storage in accordance with NFPA-30 and local codes. Further, these chemicals should be segregated in accordance with NFPA-80A. Incompatible chemicals must be stored separately to prevent their intermingling. As a general rule, solvents and other flammable chemicals should be stored in one room, acids and basis in another room, toxic chemicals in a third room, and so on. When a large number of drums of a given type of chemical is needed for the work conducted in a laboratory facility, a special, separate room can be used for their storage. The drums should be segregated according to the chemicals stored in them, and in accordance with the requirements of NFPA-80A. Gas cylinders can be stored either in a room or in an open-air restricted area. This room or area can be adjacent to the other storage rooms. Each of these rooms must meet specific codes and health and safety requirements, most of which are described below.

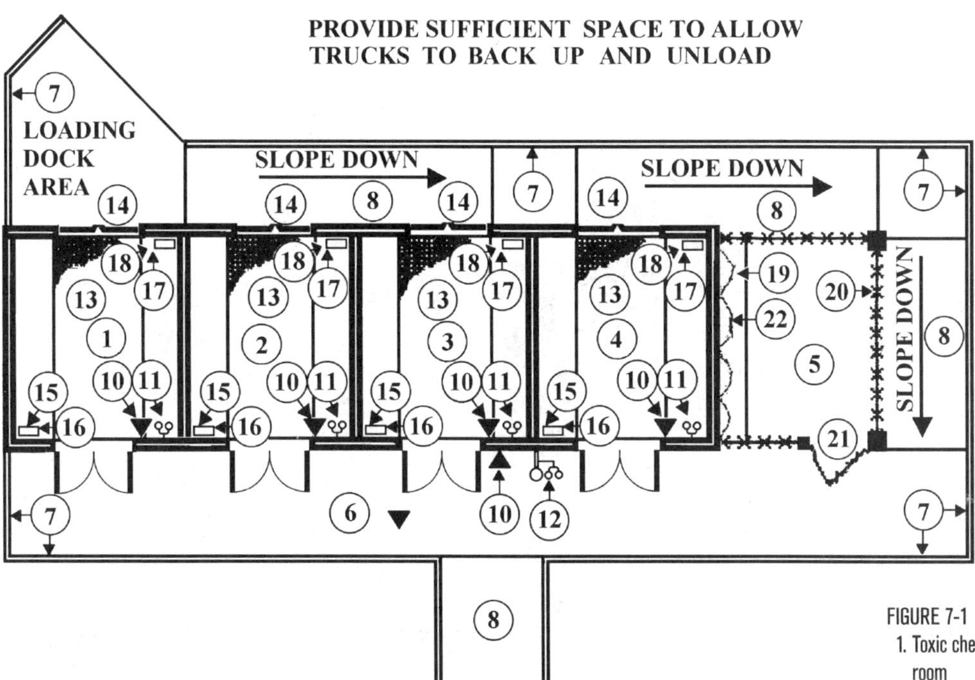

FIGURE 7-1

1. Toxic chemicals storage room
2. Acid storage room
3. Flammable chemicals storage room
4. Drums storage room
5. Gas cylinders storage area
6. Curbed area 8″ (00 cm) above road level
7. Metal railing
8. Smooth ramp
9. Market paved path to main laboratory
10. Telephone and panic button
11. Eye wash
12. Emergency shower & eye wash
13. Depressed floor area with chemical resistant grill
14. Blow panel
15. Upper supply register
16. Lower supply register
17. Lower exhaust register
18. Upper exhaust register
19. 8″ curb
20. Chain link fence
21. Chain link gate
22. Gas bottle restraining device

LOCATION AND ACCESSIBILITY OF THE HAZARDOUS MATERIALS STORAGE BUILDING AND RELATED REQUIREMENTS

The HMSB, a schematic plan of which is shown in figure 7-1, should be located at least 50 feet away from the facility buildings where people work and from the facility's property line. Structures such as water tanks and emergency generator sheds may be located less than 50 feet from the HMSB, as long as the blast panels of the HMSB are not facing or directed toward these structures.

The chemical delivery trucks should have easy, unobstructed access to the HMSB. There should be sufficient paved space at the loading area of the building to allow the truck to easily turn around and unload. A protected area should surround the HMSB. For instance, this area could be a wide sidewalk 8 inches above the road level with metal railing or posts that would prevent trucks from getting too close or accidentally hitting the HMSB.

A smooth ramp should lead from the unloading spot at the sidewalk to the floor level of the HMSB. Figure 7-2 illustrates this spot as well as the railing just described. All floors from the unloading spot to the final location of the chemicals in the storage rooms must be unobstructed and without bumps.

A paved path must be provided from the HMSB to the main laboratory building(s). This path must be clearly marked and, if in the way of vehicular traffic, must have priority over the vehicular traffic by use of stop signs. In geo-

FIGURE 7-2

graphic areas of extreme weather (cold, rainy, or hot) this path must be protected from the elements. All floors of the paved path from the HMSB, to the point of entry of the laboratory building(s), where these chemicals will be used, to their final location in the laboratory rooms of this building, must be unobstructed and without any bump. When going from one level to another smooth ramps must be used.

General Safety Requirements

HANDLING AND STORAGE REQUIREMENTS

The hazardous chemical inventories stored in any storage room must be within the prescribed capacity of the room. Incompatible chemicals and mutually reactive chemicals should be segregated to prevent accidental contact. Flammable liquids can be stored in containers made of glass, metal, approved plastics, or metal drums. The size and type of containers used for flammable liquids depend upon the flush point of the liquids. There are three classes of flammable liquids: 1A, 1B, and 1C. The container type and the maximum amount of liquids that can be stored in a given type of container are specified in paragraph 7-2.3.2 and table 7-2 of NFPA-45.

The safe handling of hazardous chemicals is described by several NFPA codes. The procedures described by these codes should be followed to avoid serious accidents. How the hazardous chemicals are transferred from one vessel to another and the size of the vessels used affect the design of the storage room where the chemicals are kept or used. For this reason it should be noted that:

- Class 1 liquids should not be transferred between metal containers without the containers being electrically interconnected by direct bonding or by indirect bonding through a common grounding system in the room where the transfers are made. The maximum impedance of the bonding must not exceed 6 ohms.

- Transfer of class 1 liquids from containers of 5 gallons (18.9 l) or larger must be carried out in a separate area outside the laboratory building. If this amount of liquid has to be transferred inside the laboratory building, it should be done in a separate area that meets the requirements of the Flammable and Combustible Liquids Code for Inside Areas in NFPA-30.
- Transfer of class 1 liquids from containers smaller than 5 gallons (18.9 l) can be carried out inside a laboratory building or in a laboratory room or work area as long as the transfer is effectuated in a laboratory hood. The transfer can also be performed in an area provided with enough ventilation to prevent the accumulation of flammable vapors and of air mixtures that may exceed 25 percent of the lower flammable limit. Transfer of class 1 liquids from containers smaller than 5 gallons (18.9 l) can also be carried out inside a storage area or building as described in the Flammable and Combustible Liquids Code for Inside Areas in NFPA-30.

GENERAL BUILDING PROTECTION REQUIREMENTS

The HMSB should be fully protected with a sprinkler system and a full fire detection and alarm system. The design must incorporate a telephone in each room and another outside each room. These telephones allow communication between the person at the HMSB and any laboratory within the facility. All telephones in the HMSB must be equipped with a panic button that activates an alarm at the HMSB and the control panel and that puts the caller in direct communication with the control panel operator.

The HMSB should have an eye wash device near the entrance of and inside each room. An emergency shower/eye wash must be located near the entrance of and outside each room. In figure 7-2 the emergency shower/eye wash is located on the wall at the center of the elevation, between the two rooms it services. See chapters 4 and 9 for a detailed description of emergency shower/eye washes.

STORAGE ROOMS FOR FLAMMABLE, COMBUSTIBLE, AND EXPLOSIVE CHEMICALS AND THEIR WASTE

The quantities of hazardous chemicals stored in the open in a laboratory room's work area must be kept to a minimum. Separate storage rooms are required when the quantity of hazardous chemicals needed is larger than that allowed to be stored in the individual laboratory rooms. The maximum quantities allowed to be stored in an individual laboratory room are specified in table 2-2 of NFPA-45. The rooms covered by this table are storage rooms and stockrooms for solvents, acids, and bases that are flammable, combustible, or explosive.

The Envelope and Construction Materials

Each of the rooms storing hazardous materials must be separated from adjacent spaces by at least a 2-hour resistive envelope. There are several ways to construct a 2-hour resistive envelope for these rooms; the project architect should be familiar with them. The construction materials used in the floors, walls, and ceilings, as well as those used for ducts, piping, and vessels, must be compatible with the materials to be stored, transferred, or handled.

Floors

The floor of a hazardous materials storage room should be depressed, sealed, and impervious. The depression depth should be such that it can contain a spill of the same volume as the largest container or drum in the room. Generally, 4 to 6 inches (0.10 to 0.15 m) is adequate. The floor of the room should not have a drain. The room floor should permit loaded carts to be carried into the room without being bumped. Therefore, the transition between different floor levels should always be smooth.

One way to accomplish this is to have the changes in the floor levels ramped. One ramp should lead from the road level to the sidewalk level. Another ramp should lead from the sidewalk level to the depressed level of the storage room. The depressed area of the storage room can also be covered with a removable grille. The grille must be constructed from a material that will not be affected by the type of chemicals stored there. If the grille's top is not level with the adjacent space, a ramp will be necessary. Figure 7-3 shows the floor of a storage room covered by a removable grille.

Storage Arrangement

Drums containing up to 50 gallons (189.25 l), and sometimes up to 60 gallons (227.10 l), can be placed directly on the floor, against the wall. A storage room where such drums are stored must have a grounded bonding system. All drums placed there must be electrically interconnected by direct or indirect bonding to the system.

Shelving in the storage room must be built from materials not affected by the chemicals stored in the room.

FIGURE 7-3

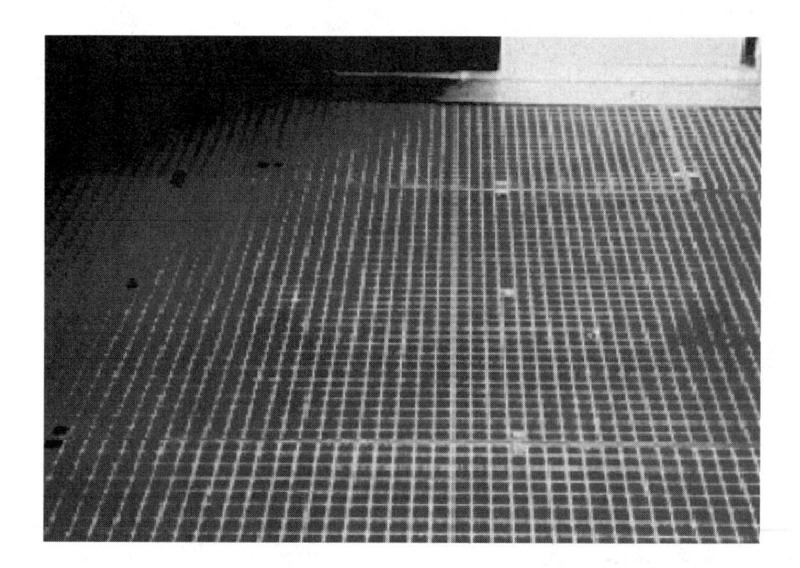

Regardless of whether they are located in a remote building or in the main laboratory building, storage rooms for flammable chemicals and for flammable wastes must be equipped with blast (or blow) panels to conform with the requirements of NFPA-30, section 4-4.2.5. These panels should be located so that if an explosion occurs, its thrust will propel the panels into an open area within the laboratory grounds. This area should be designed as a buffer (it may be a wooded area) and should be such as to discourage people from being there. The blast panels can be located on one of the walls or in the roof. If the storage room is a stockroom, the blast panel should be placed on an outside wall. If the stockroom is located on the top floor, the blast panels can be located in the roof. Figure 6-10 shows a stockroom with a blast panel in the back wall. This figure also shows three rows of shelving built with a solvent-resistant material.

Blast panels are not required in storage rooms where class 1A liquids are stored in containers of one gallon or less. Nor are panels required in storage rooms where dispensing class 1A or 1B liquids or any unstable liquid is prohibited.

Doors

Each room must have at least a set of double doors with panic hardware. The doors' rating must conform to the 2-hour envelope. The doors must open toward the outside of the room. Each door should have a small view window at eye level.

Ventilation

Each storage room should be vented to the outside atmosphere by means of a mechanical exhaust system that provides a ventilation rate of 1 cfm per square foot of floor area, with a minimum of 150 cfm (0.07 m^3/s) per room. Each room should have supply and exhaust registers placed in the upper part of the room and lower part of the room, on opposite walls. The air supplied should enter the room within 12 inches (0.30 m) of the floor and 12 inches (0.30 m) or less of the ceiling on one side of the room. The air exhausted should leave the room also within 12 inches (0.30 m) of the floor and 12 inches (0.30 m) or less of the ceiling, but on the wall opposite from where the air is supplied. The locations of several supply and exhaust ducts are apparent in figure 6-10. Ducts—and the entire ventilation system supplying air to these rooms—must comply with NFPA-91. The air exhausted must not be vented into a fume hood or into a fume hood exhaust system.

Lighting and Mechanical and Electrical Equipment

Explosion-proof light fixtures, switches, and outlets are required in rooms housing flammable, hazardous chemicals in the quantities and condition described earlier. If heating or cooling equipment is required, all the equipment used (wiring systems, ducting systems, switches, etc.) must be explosion-proof when in direct contact with the gases or fumes that may emanate from the chemicals stored in the room.

GAS CYLINDER STORAGE ROOM OR SPACE

Codes and Standards Governing the Storage of Compressed and Liquefied Gases

Methods of storage of compressed and liquefied gases are generally stipulated by several NFPA standards as well as by the American National Standards Institute (ANSI) standards. NFPA-50 standards govern the handling of bulk oxygen systems at the consumer site. NFPA-50 A standards govern the handling of gaseous hydrogen systems at the consumer site. NFPA-50 B standards govern the handling of liquefied oxygen systems at the consumer site. NFPA-51 standards govern the handling of k oxygen-fuel gas systems for cutting and welding. NFPA-54 gives the National Fuel Gas Code, NFPA-55 standards govern the storage, use, and handling of compressed and liquefied gases in portable cylinders, and NFPA-58 standards govern the storage and handling of liquefied petroleum and gases. The ANSI standards (ANSI B 31.1.1.0, ANSI B 31.2, and ANSI B 31.3) govern power piping and the piping of fuel gas and petroleum refinery.

The following are guidelines for the storage of portable cylinders containing compressed or liquefied gases. These recommendations do not cover every requirement for every situation, nor do they apply to every compressed or liquefied gas. Therefore, the previously mentioned standards should be carefully studied.

Storage Room/Space Requirements

The location and size of the gas cylinder storage room or space are governed by the type of gases contained by the cylinders as well as by related code requirements. Gases are categorized according to how toxic, pyrophoric, flammable, and oxidizing they are. Storage rooms or spaces containing hazardous gases must always be secured against unauthorized entry.

When the gases are flammable, their storage location in relation to adjacent combustible materials is rigorously regulated by the previously stated codes. Important relevant standards are identified in NFPA-50, chapter 2, paragraph 2-2.1.6 through paragraph 2-2.1.14. When flammable gas cylinders are stored within a protective structure rated as a 2-hour fire-resistant envelope, or separated from any combustible materials by a 2-hour fire-resistant envelope, most types of flammable gas cylinders can be stored as close as 1 foot from the fire-resistant envelope.

Different flammable gases can be stored together in the same room or space. However, incompatible gases should not be stored in the same room or space without adequate separation and protection. When flammable gas cylinders are stored together, they must be separated in accordance with table 2-1.5 of NFPA-55 standards. When a gas is categorized in more than one category, all compatibilities must be checked and the most stringent separation must be used.

The storage location of nonliquefied flammable gas cylinders is indicated in table 2-2.1 of NFPA-55.

When more than a few liquid hydrogen cylinders need to be stored, prudent practice and common sense dictates that they be stored alone in a separate room or space or with cylinders of nonflammable and nontoxic or hazardous gases. Storage locations containing liquid hydrogen cylinders should have a sign reading: "HYDROGEN—FLAMMABLE GAS—NO SMOKING—NO OPEN FLAMES" or an equivalent warning. The location of a storage room for liquid hydrogen cylinders and the maximum allowable quantity of liquefied hydrogen are indicated in NFPA-50B, chapter 5, tables 1 and 2.

For calcium carbide, the storage space must be dry, waterproof, and well ventilated. Calcium carbide in quantities less than 600 pounds and in sealed packages can be stored in a room inside a building or in the same room with fuel gas cylinders, as long as the storage is in accordance with the previously stated requirements.

The protective structure for the storage of flammable gases will consist of the floors, walls, and roof of a storage room or of the floors, walls, and roof separating the storage space of the flammable gases from other spaces. The main factors determining the number of gas cylinders that can be stored in the structure are: the square footage, configuration, and dimensions of the structure; the separation required between different gas cylinders (distance and type of separation for each flammable gas); and the way the cylinders must be stored.

When a separate building, or a special room within a building, is used to store flammable gases, its floor, walls, and ceiling must be constructed of noncombustible or limited combustible materials. If the volume of gas contained in the cylinders is between 2501 scf and 5000 scf (a scf is 1 cubic foot of gas at 70°F [21.1°C] and at 14.1 psi [0.97 bar]), the entire interior envelope—floor, walls, and ceiling—must have a fire resistance rating of at least one hour. If the volume of gas contained in the cylinders is greater than 5000 SCF, the entire interior envelope—floor, walls, and ceiling—must have a fire resistance rating of at least two hours. The walls should extend from the floor to the ceiling or bottom part of the roof and should be securely anchored vertically and horizontally. When a special room within a building is used, one of its walls must be an exterior wall. It also must not have an opening to another part of the building, and its window(s) and door(s) should be placed in the exterior walls and must be accessible in case of emergency.

The building or room where gas cylinders are stored must be equipped with a sprinkler system designed in accordance with NFPA-13 standards. Shelves used for the storage of cylinders must be of noncombustible construction and designed to bear the weight of the cylinders. Liquefied flammable gas cylinders must be stored in the upright position, so the storage space must be designed accordingly. Spill control, drainage, and secondary containment is not required for the storage of compressed gas cylinders.

The building or room must be provided with natural or mechanical ventilation vented to the outside. The ventilation should provide a ventilation rate of 1 cfm (0.000472 m^3/s) per square foot of floor area. Inlet/outlet openings should be located within 12 inches of the floor and within 12 inches (0.30 m) of the roof or ceiling, on exterior walls only. Inlet and outlet openings should each have a minimum total area of 1 square foot (0.0929 m^2) for 1000 cubic feet (28.32 m^3) of room volume, if natural ventilation is provided to the space. If mechanical ventilation is used, the room or building should have supply and exhaust registers placed in the upper part of the room and lower part of the room, on opposite walls. The air supplied should enter the room within 12 inches (0.30 m) of the floor and within 12 inches (0.30 m) or less of the ceiling on one side of the room. The air exhausted should leave the room also within 12 inches (0.30 m) of the floor and 12 inches (0.30 m) or less of the ceiling, but on the wall opposite from where the air is supplied. When toxic gases are stored, mechanical ventilation is preferred, and it must operate continuously and not recirculate exhausted air. Further, a gas detection system, with alarm, must be installed to warn of the presence of toxic gases.

Discharge from outlet openings must be directed to a safe location where the air and gases would not affect people should a leak occur. The discharge location must be at least 50 feet (15.24 m) from air handling systems, air-conditioning equipment, and air compressors. Also, the entire exhaust system used in the room and building housing flammable or toxic gases must comply with NFPA-91.

The storage room or space for gas cylinders should be designed to store the maximum number of cylinders requiring storage. Codes regulate the storage of different types of gas cylinders and recommend prudent practices. The room or space should provide the appropriate type and number of gas cylinder stands and have a fixed and approved mechanism to prevent the gas cylinders from falling.

The design must provide adequate fire separation between incompatible gases. Further, flammable and combustible gases or liquids must be separated from oxidizing gases. A log of the gases stored in the gas storage room or space should be regularly updated. Gases corrosive to cylinders or cylinder valves or that may become unstable while stored in the cylinder should have a maximum retention period of 6 months.

Cylinders of gases with a health hazard rating of 3 or 4—and cylinders of gases with a health hazard rating of 2, with no physiological warning properties if stored in a room or other confined space—should be kept in a hood or other enclosure that is continuously mechanically ventilated. No more than three cylinders of gases with a health hazard rating of 3 or 4 should be stored in a hood or other enclosure.

Explosion panels should be installed in the room or building housing flammable gases. These panels must be located in the exterior walls or roof. A panel or a combination of panels should be provided to relieve at a maximum internal pressure of 25 lbs (0.01 bar) per square foot.

Finally, all electrical equipment used in a room or building housing flammable gases must conform with the provisions of NFPA-70, National Electric Code, Article 501 for class I, division II locations.

Outdoor Storage Space for Gas Cylinders

The best storage location for a large number of gas cylinders is an open space adjacent to other chemical storage rooms and part of the hazardous chemicals

FIGURE 7-4

storage building. Figure 7-4 shows a gas storage location at the right end of a building. To qualify as an outdoor storage space, the area must have a minimum of 25 percent of its perimeter open to the atmosphere. Its floor and roof or ceiling should be noncombustible. The cylinders must rest on a curb or other floor surface where water cannot accumulate. Further, the storage space must be protected from vehicles by curbs and metal posts. The space can be enclosed by chain-link fence, open blocks, or other code-permitted, noncombustible types of fencing. The fencing can extend to the full height—a minimum of 8′-0″ (2.44 m)—and the entire width of the space. Figure 7-5 is a close-up of the outdoor storage space in figure 7-4. It shows with more detail how some of the requirements are met.

FIGURE 7-5

Naturally, any outdoor storage area must respond to the previously described code requirements and have adequate separation and protection. The storage area must be kept clear of dry vegetation and of combustible materials. Such vegetation and materials must not be closer than 15 feet (4.57 m) from the storage area's perimeter.

If toxic gases are being stored, the space must be located at least 75 feet (22.86 m) from the property line, from public ways and places of public assembly, and from buildings not associated with the gases being stored. Like any other gas storage space, outdoor space must be secured from unauthorized access.

If calcium carbide is being stored, it must be in unopened metal containers that are in good condition and are water- and airtight. The containers must be stored horizontally in single or double rows. The bottom row should be placed on wooden or woodlike planks to prevent the containers from coming into contact with the ground or ground water.

Chapter 8

RECOMMENDATIONS FOR THE RELATIONSHIP BETWEEN ARCHITECTS AND ENGINEERS DESIGNING LABORATORIES

LABORATORY design necessitates a relationship between architects and engineers that is much closer than that required by the design of less complex buildings. The need for such a relationship will become evident later in the chapter, but let it be said briefly that if architects and engineers do not communicate closely during the design of the laboratory, their work will be inefficient (if not totally faulty), which could ultimately be quite costly. This chapter is intended to help architectural and engineering firms understand how to correlate tasks and avoid the pitfalls inherent in the programming and design of laboratories. It will also allow laboratory owners to familiarize themselves with the process of programming and designing a laboratory. Tasks that need to be performed are restated, with an emphasis on how and by whom they should be completed. Only those activities that require coordination between the professions are described, as it is unnecessary to detail how and what an engineer or architect should do in the normal performance of his or her profession.

Architects and engineers have had to work together as a team for as long as the two careers have existed. Their relationship is like a marriage, in which both partners must work to make the marriage a success. Individuals from these two professions may be quite compatible socially but find that their perspectives and approaches are very different when they work together on a project. This may lead to a lack of cooperation and often negatively affects the quality of the design. Understanding the differences between architects' and engineers' viewpoints can ease tension and promote the cooperation needed for the successful design of a laboratory.

Architects are generalists, or "macro" designers. They like to first conceive an overall picture and then deal with its specific parts. They will conceive and produce a design either themselves or by delegating various parts of the design to other architects or to specialists, generally engineers. In this way, architects act as orchestrators and coordinators. In almost all cases, the architect focuses on the quality of the whole rather than the quality of the various parts.

Engineers, on the other hand, are specialists, or "micro" designers. When presented with a problem, they generally identify the parts of the problem that need to be solved and the different solutions to each of these parts. They then analyze and prioritize according to definite criteria and, finally, select the best, most efficient way to solve the problem. The engineers then do all the necessary calculations to complete their task. This description of the process is, of course, a generalization. The main point is that the engineer tends to approach a problem by looking at every detail and making sure it is taken care of. For the engineer, the quality of each part has to be assured.

If the design and programming of a laboratory is to be completed smoothly, architects and engineers must substantially alter the conventional way they coordinate and distribute tasks. The architect must still play a role similar to that of an orchestra conductor, telling each performer—the engineers and consultants—when and how to perform. However, this role of conductor is very different in that the architect must act as a second fiddle to most of the engineers during the conduct of their work. The real conceptual design must start with the mechanical engineer, not with the architect.

A look at how a conventional building project is developed will clarify the roles of the architect and engineer. In most instances, an architectural firm is hired to design a specific building. Once the architect is familiar with the building program, he or she immediately conceives a building, drafts sketches and preliminary drawings, and shows them to the client. This is done either to assure the commission or to reassure the client. Sometimes the architect will take liberties with the program and attempt to convince the client of the value of the proposed changes. Once a design concept is approved by the client, the architect secures the assistance of other professionals to help with the design of the building or buildings.

The architect asks the mechanical engineer to help select mechanical components for the building and to recommend adequate HVAC and other systems. The architect and the mechanical engineer evaluate the economics and efficiency of the proposed systems and eventually submit them to the client. Only after the architect secures the client's approval is the mechanical engineer allowed to begin designing and working on the systems and details.

The architect then asks the electrical and structural engineers and other professionals to contribute to the project as needed. From there on out, the architects, consultants, engineers, and others develop the project's construction drawings and specifications.

Following this commonly used procedure in the design of a laboratory or other advanced or high-technology building would result in major problems, if not an outright catastrophe because laboratories and other high-technology buildings are far more complex than conventional buildings and have many conflicting and complicated requirements. What exactly is an advanced or high-technology building? The ability to recognize such a building is imperative, especially because the drawings and specifications of the project may not distinguish it as such.

One of the most common definitions of an advanced or high-technology building is a building for which the program includes a combination of complex and conflicting architectural, mechanical, and electrical requirements. By this definition, hospitals and computer chip manufacturing facilities are examples of high-technology buildings. These buildings generally have complicated problems that are often without precedence. These problems may be created by the exigencies of the activities to take place in the building, by the use of new technologies with specific characteristics and requirements, or by both. Obviously, the laboratories discussed in this book are considered high-technology buildings.

The owner (or whoever represents the owner) generally serves as the project initiator. At the beginning of a laboratory project, the owner should simultaneously undertake these two additional tasks in addition to those identified in chapter 1:

1. Hire a specialized firm to start the National Environmental Policy Act (NEPA) environmental review process for the project. This process must comply with the specific requirements of the NEPA. The firm conducting the NEPA environmental review will rely heavily on engineers of several disciplines and will coordinate their work and findings.
2. Simultaneously, hire a firm experienced in programming the type of laboratory desired. This firm's engineers and architects will address and resolve the programmatic issues. This firm may or may not be the firm to design and prepare the construction drawings and specifications for the project. Generally, it is recommended that the same firm not be used for both programming and design. This is to prevent the architect(s) from being tempted to change certain aspects of the program to accommodate the design, which is not good for the project. The firm in charge of programming should monitor the design development if the entity for which the laboratory is designed does not include architects and engineers. This assures the design's conformance to the program of requirements.

The first task that must be addressed by architects and engineers is the identification of problems generated by the program of requirements. Conflicting program needs and requirements must be reconciled, and compliance with codes and standards must be insured. This may already seem obvious. Unfortunately, however, it is often accomplished by only very experienced engineers and architects working together in a coordinated manner.

Most, if not all, programmatic problems must be solved in the predesign phase. A few of these problems, when clearly defined, can be solved early in the design phase. The solution of these problems should be the result of a fully coordinated effort between the user (the client), the architect, the relevant engineers (with a major emphasis on the mechanical engineers), the industrial hygienists, and the interpreters of local and national codes. The outcome of this first effort is a program of requirements where most, if not all, the problems and conflicts are identified and solved. The final solutions may require a combination of compromises by the engineer, architect, and user. Often such solutions generate new and unique designs. The programmatic problems—architectural, mechanical, electrical, structural, and otherwise—should for the most part be solved when the design process starts. Once solved, these problems no longer need to be identified in the project program.

By this stage, the owner should have selected a firm to design the laboratory (see chapter 1 for a detailed description of this selection process). With the help of the firm and laboratory personnel who wrote the program of requirements, the firm to whom the contract is awarded should familiarize itself with the project requirements. It is only after this that it should start to conceptualize the project.

The conception of a laboratory can be described as the result of the combined efforts of the project architect and the mechanical engineer. The building's location on the site and its footprint, volumes, and exterior facades are only a few of the important characteristics that will derive from the cooperation between these two professionals.

As stated earlier, the real conceptual design must start with the mechanical engineer, not with the project architect. At the beginning of the predesign and conceptual design phases, the project mechanical engineer must assume the role of the project architect in response to several of the project requirements. The project mechanical engineer's first task is to look at the hoods requirements in the RDSs (see chapter 4 for a detailed description of the RDS) and to determine the needed volume of the modules. A module can be redefined from a mechanical standpoint as the floor space and surrounding volume that is necessary for the safe and efficient functioning of the fume hood. The size and type of the standard hoods will help determine a volume for the module that will satisfy the hood's appetite for air without creating an air tunnel. This information will naturally derive from the mechanical engineer's calculations. The module's ceiling height is generally set by the height of the fume hood—anywhere between 9′-6″ (2.90 m) and 10′-0″ (3.05 m). With this information—the module's volume and the room's height—the project architect can determine the width and length of the laboratory module.

The minimum width of a module with standard casework on the two long sides is 10′-8″ (3.25 m), if the total width of the casework does not exceed 5′-0″ (1.52 m). This will leave a space of 5′-8″(1.73 m) for circulation between the casework. This minimum width conforms to the Americans with Disabilities Act, which requires that a wheelchair be able to turn around in the space. Should the project architect choose a module width of 11′-0″ (3.35 m), the module's minimum length (where a 6′-0″ [1.83 m] regular fume hood with one-pass air and a face velocity of 100 cfm [0.64 m/s] are required) would be somewhere between 20′-0″ and 21′-0″ (6.10 and 6.40 m). This length would provide a volume sufficient to satisfy the hood's appetite for air without creating a wind tunnel. Naturally, the location of the hood and the air supply diffusers in the module, as well as the manner in which the required volume of air is supplied, will help determine the dimensioning upward of the module size, which will become the height of the room.

While the project mechanical engineer is the one to decide on the module's required volume (and therefore its dimensions), ultimately the project architect determines the final size of the module, as he or she considers all the factors dictated by the program of requirements and by staff needs. When deciding how many modules the room should have, the project architect must consider the operations to be undertaken in the room, as well as the equipment and instrumentation those operations require. Here again, the project mechanical and electrical engineers must assist the project architect in determining the room size, as many pieces of equipment may have special space requirements and may need to be placed in certain ways with certain connections and distances between them. The engineers will provide the architect with the space and location requirements for each instrument and piece of equipment. They will also indicate what utilities and services are needed, as well as where and how the instruments and equipment should be hooked up.

In the predesign phase, the project mechanical engineer specifies and/or interprets the needs of each instrument or piece of equipment. The engineer not only must decide where and how each instrument or piece of equipment should be located in the room, but also identify the instrument or piece of equipment's service and utility needs, as well as how these needs will be provided. Later, when the project architect completes the preliminary architectural design that includes locations for the mechanical and electrical equipment, the project mechanical engineer determines and indicates the exact location in the room where the outlets for these services and utilities should be placed.

Plumbing utilities, which include domestic hot and cold water, and laboratory gases, which include air, vacuum, natural gas, and other special gases, can also affect the arrangement of a laboratory module. Each of these utilities requires access to cut-off valves and to the piping itself to allow for future modifications. The design of the walls, ceilings, and casework will be affected by the layout of these piping systems. The project plumbing engineer should work with the project architect in the early stages of the design process to ensure that the laboratory layout supports access to these systems.

A laboratory generally requires a substantial amount of electrical power, without which it could not function. Thus, the project electrical engineer should determine—again, in the predesign phase—the approximate amount of electrical power that will be needed for the laboratory to properly function. The engineer should consult the power company to determine what amount can be made available and if existing powerlines can be used. If the required amount of power is available and existing lines can be used, the project electrical engineer must determine where connections to these lines can be made. These locations should be discussed with the project architect. The project electrical engineer should also determine the need for transformers, voltage regulators, and other devices needed to service the laboratory with the available power and should then identify the advantages and disadvantages of each location with the required equipment and estimate the cost of connection.

The power company may advise the electrical engineer that the required amount of power is not available through existing lines. If this is the case, the project electrical engineer should ask if the power company has the capability to generate the required amount of power. If it can, the engineer should request a cost estimate for extending the required amount of power to the site. The same analysis of advantages, disadvantages, and cost estimates should then be undertaken. Throughout this process, the project electrical engineer and the project architect share incoming information with each other and with the owner, since it may drastically affect finances and location.

In some cases, the electrical power company may not be capable of generating the required amount of power. The company may also, for whatever reason, not be willing to expand its capabilities to meet the laboratory's needs. In such a case, the project electrical engineer must look at other options, such as buying power elsewhere or generating power on site. The engineer must evaluate the legal and cost factors of such options and share this information with the project architect and the owner.

In the predesign phase, the project electrical engineer, project mechanical engineer, and owner should determine the electrical power needs of each instrument or piece of equipment. The owner will indicate how this power should be provided from the utility company to the building and from there to the instrument or piece of equipment. The owner, project mechanical engineer, and project electrical engineer, and possibly the project architect, should discuss the possibility of substituting certain instruments or equipment in order to reduce costs or increase efficiency. The project electrical engineer then determines the power required for each room's instruments and equipment and how this power will be supplied (again, from the utility company to the building and then to the instruments and equipment). Later, during completion of the preliminary architectural design, the project electrical engineer and the project architect will jointly determine the exact locations of each electrical outlet.

Once the mechanical and electrical engineers' predesign work is completed and coordinated, this work must in turn be coordinated with the architect's work. Once this is done, the project architect will again need to solicit the advice of engineers of different disciplines in order to start the design of the building and to locate it on the site.

In the early stages of the predesign phase, the project civil engineer should carefully study the site (or sites) being considered for the laboratory. The engineer should look at the site's topography and identify possible locations for the building(s). The project civil engineer should discuss these locations with the project architect and the mechanical engineer, and all three should work together to select locations that correspond to the direction of the predominant winds and that will not require too much adjustments in the grade. Then, the project civil engineer and the project architect should study the site's landscape, identifying the existing trees, their age, and condition. Together they should decide which trees to save and which can be easily replaced. The project civil engineer should also look at the location of the main utility and service lines and at the best and shortest ways for the project to connect to them. If there is more than one way to connect, the engineer should identify viable options, determine the costs, advantages, and disadvantages of the options, and provide the project architect with the findings and recommendations.

If there are no utility and service lines nearby, the project civil engineer should contact the companies that service the area and ask them if they plan to extend their lines to the location of the future laboratory. If extension is planned, the project civil engineer should ask if the extension could be expedited and, if there is a cost, what the cost of expedition would be. If the nearby utility and service lines are not sized adequately to service the new laboratory, the project civil engineer should contact the companies that service the area and ask them if they plan to increase the size of the lines servicing the future laboratory's site. If they do plan to increase the size of these lines in the future, the project civil engineer should ask if expansion could be expedited and, if there is a cost, what the cost of expedition would be. In some cases, the companies may not be willing to expand their capabilities to meet the laboratory's needs. If this is the case, the project civil engineer should look at other options, such as storing burning gas in large, on-site tanks, drilling on-site wells for water, and treating wastes on-site and disposing of them through percolation. Finally, the project civil engineer should evaluate the legal and financial factors of such options and advise the project architect and the owner of those findings. The ultimate decision may be to abandon the site.

The project mechanical engineer and the project architect must also join forces during the conceptual phase of the laboratory project. The participation of the project mechanical engineer during this phase is of paramount importance.

As the building is conceived, the combined design efforts of the architect, the project civil engineer, and the project mechanical engineer determine the footprint(s) and the exact building location of the building or buildings, as well as the arrangement of the laboratory rooms, the volumes and exterior facades of the building or buildings, and the location of the parking lots and roads servicing the facility. The project mechanical engineer should then study the site considered for the new laboratory and its surroundings (that is, if the site has not already been decided upon) and should orient the site and place on it the wind rose circle, which will help in the determination of the predominant wind directions during the different parts of the year.

Next, the project mechanical engineer and the project architect locate the off-site surrounding buildings, roads, and parking lots and identify the activities that take place in those buildings. If the activities generate fumes, the project mechanical engineer, assisted by the owner and by those familiar with the activities generating fumes, must estimate the volume and determine the nature of the fumes. The project mechanical engineer must also determine whether any activities taking place in surrounding buildings would be affected by the fumes generated by the future laboratory, and, if so, contact the person(s) responsible for the activities. Together they should work out an acceptable course of action, which should then be proposed to the project architect and the owner.

The project mechanical engineer must look at the roads and parking lots to calculate traffic patterns and density, as well as the volume and nature of the fumes generated by the traffic, and should then ascertain whether the predominant winds would direct these fumes toward the laboratory site. If they do, the project mechanical engineer must look at the kind and volume of the fumes and determine if these fumes would affect the kind of experimentation and operations that will take place in the projected laboratory. Once this determination is made, the engineer must propose an efficient corrective action or suggest that the site be abandoned.

If the proposed laboratory's operations do not conflict with surrounding activities, the project mechanical engineer can begin thinking about how to best use the site for the future laboratory building. The project mechanical engineer must ask the project architect to conceptually group the laboratory rooms in which fume hoods are needed. These rooms must be grouped according to function and required proximity. The project mechanical engineer and the project architect should determine the best way to service these rooms and to lodge the services, utilities, and stacks necessary for the laboratory to function properly. They may decide to group these rooms in one or several rows. These rooms may be back-to-back with a service corridor between rows, front-to-front with a service area in an interstitial floor, or in any other arrangement. The project architect, assisted by the project mechanical engineer, should locate the laboratory wing and other wings so that cross-contamination between wings is avoided. The project mechanical engineer then places the air intakes required to supply the building systems with fresh air. The location of these intakes should correspond with the most likely locations for the laboratory wings and the mechanical equipment rooms.

Along with the project architect, the project mechanical engineer must then look at the wind rose circle, previously located on the site plan, and at the directions of the predominant winds during the different parts of the year, and determine the locations and orientation (there may be several) that will best prevent cross-contamination between the laboratory wings and the air intakes. This information must then be confirmed by the engineer. In the same analysis, the project mechanical engineer should suggest the locations of the roads and parking lots that will be needed around the building's schematic footprint. These locations should not be between the predominant wind directions and the building's air intakes.

As the project architect develops the conceptual design, the project mechanical engineer needs to look at how the project architect is developing the buildings and their volumes—both in relation to the site and to off-site structures. He or she should reanalyze the location of the building inlets and outlets, as well as the hoods' discharge stacks and their assumed heights, to ensure that no exhausted fumes will be sucked back into the building through the intakes. At this stage it may be wise to ask the industrial hygienist to recommend an equation that will roughly identify the location of eddies that could provoke cross-contamination and then take the appropriate measure.

During the course of the conceptual design, the mechanical engineer should commission a computer air modeling study. This air modeling study must be conducted by a mechanical engineer specialized in such work. The study will show exactly where the potential for cross-contamination exists and will recommend corrective action. The corrective action may require some rearrangement of laboratory and other rooms or—especially if the project mechanical engineer is unfamiliar with laboratory design requirements—of the relationship of buildings within the site, which is a major design change. Once the cross-contamination corrections are completed, the predesign and conceptual phases are virtually finished. From there the design phase will continue in a conventional manner similar to that of any other complex building.

Chapter 9

GUIDELINES FOR DESIGN OF THE MECHANICAL SYSTEMS SERVICING THE LABORATORY WING

THE mechanical requirements in a laboratory are generally complex and of utmost importance if safe, efficient operations are to be conducted. These requirements are generally a direct response to both the present and anticipated needs of laboratory operations. They are also an essential means of implementing safety measures to protect the laboratory operators and occupants, on one hand, and the environment and population at large on the other.

Laboratory operations are usually expected to undergo significant change during the life span of the facility. Therefore, it is important to plan on how to effectuate future changes economically, rapidly, and with minimum disruption to the operations conducted in the different laboratory rooms. Architectural and structural modularity and flexibility (described in chapters 4, 6, and 8) provide for space rearrangement with minimum disruption.

The same principle can be applied to utility and services systems: modularity, adaptability, and flexibility allow for economic, rapid service to the rearranged spaces, with little disruption to operations. Modularity and flexibility are achieved by supplying all piped systems with convenient tee connections with isolation valves at each module, and the air systems with tapered branch connections at each module for supply, return, and exhaust. These arrangements also substantially improve serviceability and minimize disruption to operations. Further, the system must be designed for future expansion—in accordance with the management's projections for the next 10 to 15 years. It may be wise, however, for the mechanical designer to project an expansion of about 25 percent of the actual laboratory needs, as determined by the design loads.

Most of the mechanical requirements addressed in this book fall into the following three categories:

- *Plumbing.* This includes water supply and waste, all piping systems, emergency showers and eye washes, the sprinkling system, and steam service.
- *Laboratory services.* This includes the natural gas distribution system, the air and other gas distribution systems, the compressed air system, and the vacuum system.
- *HVAC.* This includes the supply and exhaust of air from the laboratory section, the treatment of the supply air (heating, cooling, and filtering) and all air-related protective devices (such as ventilated enclosures, hoods, and glove boxes). It also includes the treatment (filtering) of exhaust air before it is exhausted.

PLUMBING

Municipal Water

Special attention should be given to the municipal water that will be available for use in the laboratory experiments. This water should be analyzed and evaluated for its effect on laboratory experimentation. Corrections that may be needed to make the water suitable must be identified. The water may need to be deionized. Hard water may have to be softened; its level of purity may have to be increased; and its mineral, ferrous content and conductivity may have to be lowered or adjusted. Each of these corrections, if needed, will necessitate the use of specialized equipment and the installation of complex and delicate systems.

Backflow preventers of the reduced pressure zone type must be provided on all domestic waterlines. Further, water hammer suppressors and vacuum breakers should be provided at high points of supply lines or at the fixtures. Where pipe soldering is required, lead solder should not be used.

Laboratory Water Piping

All water piping systems that service the laboratory (hot and cold water, distilled water, deionized water, or any type of treated water) must be made of materials and assembled in a manner that will not affect the quality of water flow or be damaged by the system's contents. It is essential, therefore, that the specifications for equipment and systems (provision, installation, and preparation for use) be written by specialized professionals and that the actual provision, installation, maintenance, and operation of the equipment and systems be undertaken by reputable firms specialized in such work.

Facility Waste

Waste systems for laboratory facilities can be divided into two groups: laboratory waste systems, serving laboratories, and domestic waste systems, serving nonlaboratory areas such as toilets, locker rooms, and janitors' closets. Because laboratory waste requires special materials and treatment, the laboratory waste system and domestic waste system must be kept separate until after the laboratory waste has been adequately treated. The domestic soil, waste, and vent systems preferably should consist of cast-iron pipe, but PVC piping is also acceptable. These systems should be designed in accordance with plumbing design standards and applicable codes.

Laboratory Waste

Laboratory waste has to be disposed of separately to avoid needless inflation of the cost of the domestic waste system. The flow of laboratory waste lines should be designed so that the wastes will pass through holding tanks and treatment and neutralization chambers prior to entering the public sewer system.

The selection of laboratory drain lines is of crucial importance. The construction materials of these lines should stand up to attack from chemicals commonly used in the laboratory operations. Although chemicals are not disposed of in the laboratory drain lines, high-temperature acids and other chemicals often accidentally end up down the drain and in the lines. These chemicals remain lodged in the weak locations of the system until they are flushed away. This, with time, leads to leaks in the weak locations, from which toxic chemicals can spread and cause significant safety and liability problems.

Therefore, safety and reliability must be of utmost importance when waste line materials are being selected. The chemicals (and potential chemicals) that will be used should be identified and their impact on the waste line materials determined. In locations where many potent chemicals are used and the danger of high spills exists, glass should be extended to the neutralizer chamber, after which cast iron can be used.

Waste line pipes made of borosilicate glass are known to have the highest chemical resistance. These pipes are highly nonreactive, expand very little when exposed to high or low temperatures, and allow their contents to be visible because they are transparent. Waste line pipes made of high silicone cast iron, polypropylene, or PVC can also be used in waste lines, in combination with or even instead of borosilicate glass. How and when to use such pipes should, however, be the result of careful analysis of chemical uses and of designs that will protect the pipes from chemical attack.

Waste line pipes made of plastic materials tend to have more fire and smoke problems. They also have a high expansion coefficient when subjected to 140–180°F (60–82.2°C) liquids. They can crack when subjected to very low temperatures and sag and leak if improperly supported.

If initial costs are not a factor, it is best to use borosilicate glass pipes for all waste lines, particularly when extensive use of acids and other potent chemicals are used. Often, however, this is not possible. A good but relatively expensive design uses borosilicate glass pipes from the sink drain to an easily accessible, small (1 gallon) (3.785 l) holding tank. This holding tank acts as a trap for accidental spills and should be configured as shown in figure 9-1. The holding tank has a valve at its bottom and its connection to the drain line is at the midpoint of the tank depth. There is also an emergency cut-off valve at the beginning of the primary waste line that connects the holding tank to the main waste line. This holding tank should not be confused with the sampling ports and other types of holding tanks, such as those used for acid neutralization.

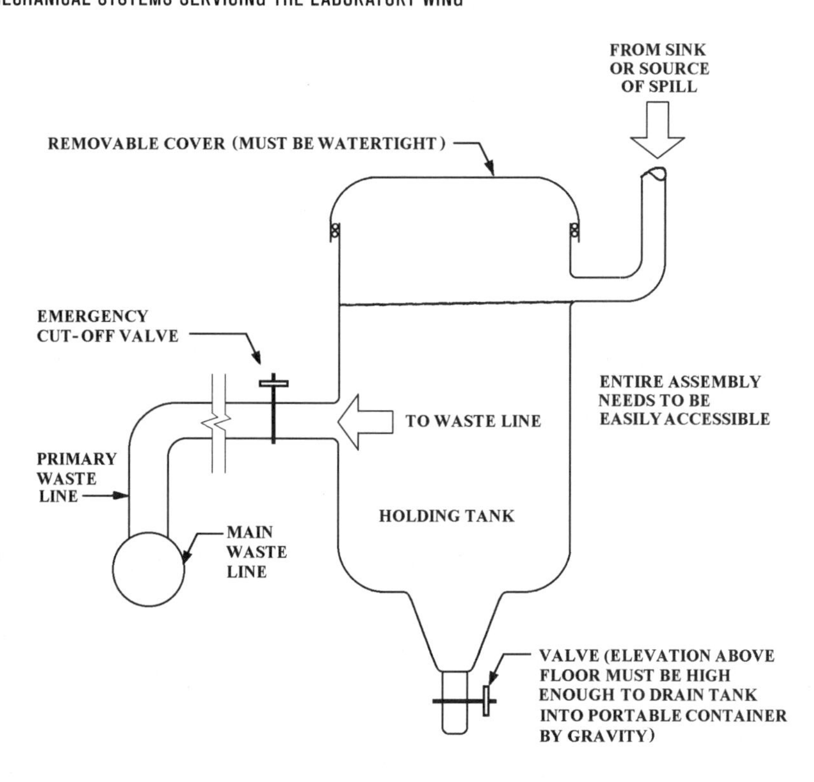

FIGURE 9-1
Schematic diagram for a holding tank

The primary and main waste line pipes can be made of less costly materials—that are still resistant to most of the chemicals used—as long as these lines are not buried underground. It is good practice to use heavy cast iron (schedule 80 or equivalent) or polypropylene for above-ground waste lines and very heavy cast-iron below ground. The below-ground waste lines should lead to an acid neutralization tank.

All soil, waste, and drain lines must be properly vented. The connecting segments of the vent pipes should be of the same materials as the lines they service. It is recommended that drains in the laboratory section be closed drains and that each be vented to the outside. Venting pipes should be made of the same material as the pipes of the line being vented. All floor drains and future capped drains must be equipped with traps or mechanisms to ensure a continuous seal.

Waste Sampling Access

Sampling ports must be installed in the laboratory's wastewater lines on both sides of the treatment equipment and in the combined (laboratory and domestic) wastewater lines prior to the connection to the municipal sewer system. These sampling ports allow monitoring of the facility's wastewater effluent. Each of these ports is a small holding tank 8 inches (0.20 m) in diameter, with removable closures and a trap access. The wastewater should enter the tank at least 12 inches (0.30 m) below the exit point of the exit line. Thus the tank will always have 12 inches (0.30 m) of wastewater stored in it.

Waste Treatment System

Laboratory waste should be required to pass through an automated acid neutralization system composed of two continuously controlled tanks prior to discharge into the local, publicly owned treatment system. The system should be readily accessible for sample collection and routine maintenance. Further, it is recommended that the system be designed to permit expansion for organic or inorganic treatment. The tanks in this system should be operated in parallel, one at a time and alternately. This will allow the cleaning and maintenance of one tank while the other is in operation. Each tank must be equipped with an audible and visual alarm system that is activated when the wastewater pH level reaches unacceptable limits. It is good practice to have all the alarms tied up into the building annunciator panel or under 24-hour watch through other means.

Emergency Showers and Eye Washes

Emergency showers and eye washes must be provided as needed in the rooms where potentially risky operations with hazardous chemicals will be conducted. Their location in relation to the hoods where the work will be performed and to the laboratory room's exit doors should be the result of a careful design agreed upon by the architect, the industrial hygienist, and the safety engineer. The water temperature of emergency showers and eye washes should be tempered, close to human body temperature, to avoid a second shock to the injured person using them.

EMERGENCY SHOWERS

Emergency showers must be installed just inside or outside the entrance of all laboratory rooms where hazardous work with chemicals may be performed. In some extreme situations, it may be necessary to have the emergency shower in the room, next to the location where the risky work is taking place. In any case, however, particular attention should be given not to have instrumentation or equipment using electricity without ground fault interruption provisions nearby or in the path between the shower and the location where the work is performed. In general, emergency showers should be installed either in the corridor outside the entrance door(s), if the laboratory is an instrument room, or in the room itself, next to the entrance door(s), if the laboratory room is a wet laboratory with no electrically powered instruments. In any case, its location should be consistent in all rooms.

If no floor drain is used under the emergency shower, the floor area under the emergency shower should be a different color from the surrounding floor and textured to allow a blinded individual to know that he or she has found the shower. It is good to use floor drains under the emergency showers in new construction work when possible. An area of about 3′-0″ by 3′-0″ (0.91 by 0.91 m) should be depressed under the shower head and should be pitched toward a floor drain located at the center of this area. The depressed area should be covered with a grid leveled with the surrounding floor. A shower rod for a decency curtain also should be installed.

The emergency shower must consist of a combination overhead shower and hand-held body wash/eye wash unit. The body wash/eye wash unit can be used for the eyes or any part of the body not effectively reached by an overhead shower. This unit should be held by a holder firmly installed in the wall underneath the showerhead. A rigid pull bar should be used to activate the shower. This pull bar should be installed with a wall-mounted guide and should hang 48

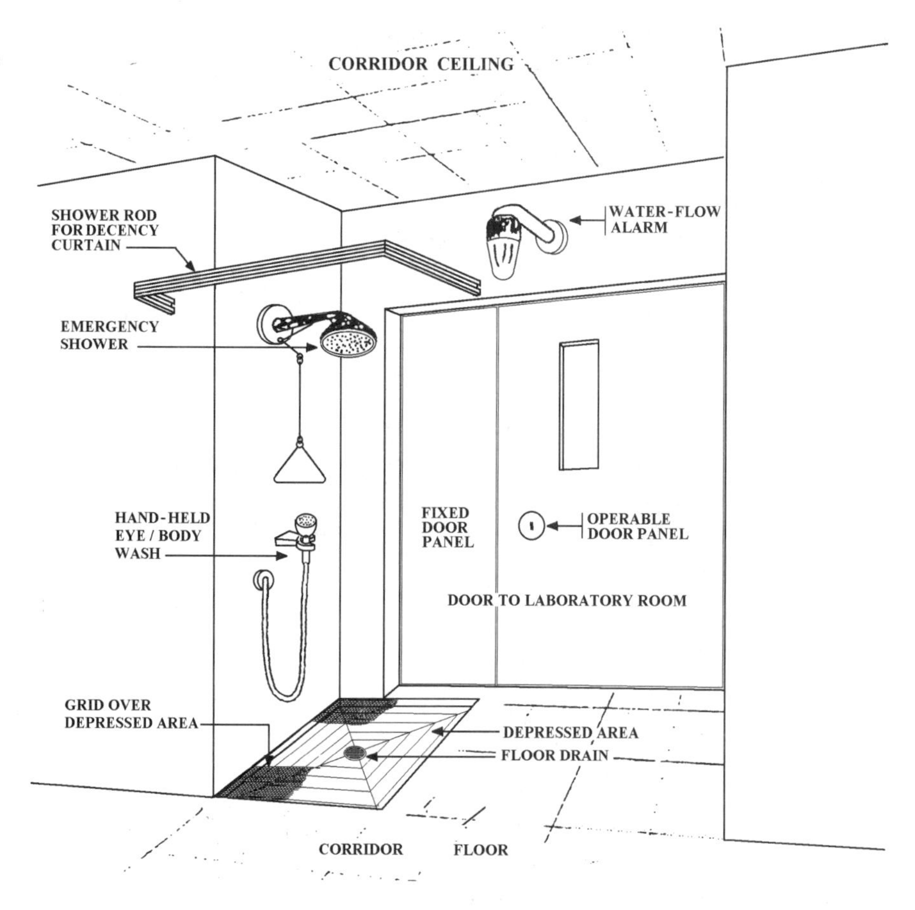

FIGURE 9-2
Emergency shower in corridor next to a laboratory room

inches (1.22 m) above the floor. The emergency shower should be connected to an audible and visual water flow alarm that activates when the shower is used. Figure 9-2 shows a good design for an emergency shower installed just outside the entrance door of a laboratory room.

EYE WASHES

Eye washes are generally installed within the room where risky work with chemicals is performed, near the location where the work takes place. Eye washes should be designed to flush both eyes at the same time. The unit should be equipped with a ball valve that stays open until manually closed and should have a foot treadle installed to supplement the hand lever. The model should provide protection of the nozzle area with pop-off covers to prevent contamination and should meet safety criteria established by American National Standard Institute (ANSI).

The location of the eye washes must be consistent in all laboratory rooms. There should be a minimum of one per module in all laboratory rooms where chemicals are used and where hazardous operations may be conducted. They should be placed near the hood they service, between it and the exit door. The location should be such that the eye wash is always accessible, that items cannot be stacked around it, and that it is as close as possible to the area of potential hazard.

Sprinkler Systems

Sprinkler system installations in laboratories are governed by codes and hence do not need to be discussed here. In the design phases, however, when decisions about fire protection are made, special consideration should be given to the instrumentation and equipment used—their cost and availability—and the effect of water and fire-suppressant chemicals on them. It may be more efficient and less costly, in its life cycle, to have the building built of noncombustible materials and compartmentalized for fire safety purposes, thus avoiding or limiting the need for sprinklering.

Steam

Steam is often needed for laboratory tasks. Depending on these tasks, the need for steam may be limited or substantial. If it is substantial and no source for steam is economically available, a central steam generator for the whole laboratory may be the best answer. If, on the other hand, the need for steam is limited, a small steam generator can be installed to service the room or rooms that need it.

LABORATORY SERVICES

Burning Gases

Natural gas is extensively used as a source of energy or heat in laboratory operations. When it is not available, bottled liquid petroleum gas can be substituted. Its use, however, is not always recommended. It should not be introduced or used in laboratory rooms where an open flame may be hazardous to the operations undertaken or to the operator. In such cases, electricity or steam, with required protection to avoid fire or explosions, should be used as a source of heat or energy.

Special Distribution Systems for Nonflammable Gas

A gas distribution system is often used to distribute nonflammable special gases (e.g., helium, argon, oxygen, nitrogen, carbon dioxide, methane, pure air, and other special gases) from portable cylinders to the different laboratory rooms that need them. This system may be central, servicing the entire laboratory; semicentral, servicing only a fraction of the rooms needing gases; or completely decentralized, with each room serviced by separate cylinders of gases. Each of these systems has merits and liabilities, depending on the tasks undertaken, the gases needed, and the level of risk.

When central or semicentral gas distribution systems are used, the gas cylinders for nonflammable gases (both in use and standby, full or empty) should be located either in a gas distribution room specifically dedicated to and designed for that use or in a utility or service corridor. The gas cylinders for nonflammable gases (both in use and standby, full or empty) can also be located in penthouses or interstitial floors, if access to these floors is safe and convenient. Regardless of what room or space is used for the gas distribution system, it must be designed in accordance with the applicable codes. The gas distribution room, when used, should be in a central location in the laboratory area served and should be accessible to the loading dock.

Regardless of the system adopted for gas distribution, piping for the system preferably should be of seamless copper with pressure-regulating valves and pressure gauges at both ends. In some cases, when extreme purity is required, a seamless Teflon- or Mylar-lined piping system may be used. When this is the case, all connection and valves should also be lined.

If the distance between the gas distribution room and the point of use is too great for seamless piping to be used, the connections between piping should be swage lock connections. The copper line pipe sizes should be coordinated to ensure proper and harmonious flow of gases from the cylinders to the point of use. The regulating valves should be selected for their capacity to reduce the delivery pressure from the bottled gas to the pressures required at the delivery ends. To be flexible and easily serviced, the system should have cut-offs for individual rooms. In some cases, when leakage would be harmful, it may be necessary to place the copper line inside a PVC piping system. The PVC system should be vented to the outside of the building. When liquid nitrogen or similar gases are piped from a gas distribution room, the delivery system should be designed and insulated to allow the gas to reach the point of use in liquid form.

The number, type, and location of gas outlets in each room must be determined during the programming phase and provided in the RDS. The exact location of each outlet should be finalized with the user during the design phase. Before installation, piping must be thoroughly cleaned in accordance with codes. Those used for oxygen should be specially prepared for by the manufacturer. Before accepting the distribution system, the user must have it pressure tested and purged by the provider. The testing and purging process will verify whether the required level of purity is being maintained at each point in the system.

Special Distribution Systems for Flammable Gas

Flammable gas cylinders (e.g. hydrogen, acetylene, etc.) may be located centrally or locally. Depending on the volume of gas used, cylinders should be kept outside the building or in fire-rated rooms. Piping materials for flammable gases require special attention. Flammable gas piping should not be routed in code-required exit corridors or in chases with environmental air-conditioning ducts.

Central Deionized Water (DI) System

When required, a central deionized water system should be used to provide the laboratory rooms with the required number of gallons (or liters) of high purity water per hour. The system's design may have to allow for an increase in the required number of gallons (or liters) per hour; e.g., the design may be for 50 gallons (189.25 l) per hour but expandable to 200 gallons (757 l) per hour. The user should help determine the storage tank capacity, which is calculated according to the required rate of use.

The water quality must always conform to the user-specified requirements for reagent-quality water and for water used in microbiological testing. Type I water is typically prepared by distilling feed water with a maximum conductivity of 20 umbos/cm (77°F) (25°C), followed by polishing with mix-bed deionizers and passage through a 0.01 inch (0.2 mm) membrane filter. Type I water cannot be stored without significant degradation; therefore, the designer should create a system that produces water for continuous use immediately after it is de-ionized and that includes a recirculating loop with a storage tank. Further, the designer must equip the system with cut-offs at individual rooms.

Pipes and fittings for deionized water systems traditionally were made of stainless steel; however, recently pipes and fittings in PVC schedule 80 have been proven quite effective and efficient. These systems are designed to perform a final polish of the water at the point of use. They also have a bypass capability in the event of a central system failure.

Central Vacuum Systems

Central vacuum systems generally evacuate air at a regulated suction—usually 25 to 27 inches (0.64 to 0.69 m) of mercury—to meet bench work requirements. Sometimes several vacuum pumps are needed to evacuate the required volume of air at the necessary suction level. It should be equipped with an exhaust silencer and have full automatic controls. Storage volume and number of pumps should be determined in the design stage and must meet the bench work requirements, as defined by the user.

Size, type, and specifications of equipment and piping should be determined by specialized professionals. It should respond to specific laboratory needs and be supplied and installed in accordance with codes and recommended practice.

Vacuum pumps must be secured over vibration pads and springs to substantially diminish the vibration and sound generated by the pumps. Concrete inertia bases may be needed to minimize vibration transfer to the structure. Vacuum pumps cannot be located in the roof or mezzanine, since the vibration will migrate to the rooms below regardless of padding or springs. They preferably should be located outside the building or in a service or utility corridor but, obviously, still within reach of the rooms they service. The location should be carefully selected to prevent transmission of vibrations or sounds to the building or rooms serviced.

Compressed Air System

Laboratory rooms often need a compressed air system to provide an adequate supply of compressed air at a regulated pressure. These systems generally have an oil-and-water trap and one or several air compressors providing compressed air at the required regulated pressure, generally about 100 psi (6.90 bar). This compressed air system, with all its piping, should be separate from other compressed air systems that may be needed for other purposes in other sections of the laboratory building. If pneumatic controls are used for the HVAC system, a separate and independent compressor and piping system must be furnished for it.

Like vacuum pumps, air compressors also must be secured over vibration pads and springs to substantially diminish the vibration and sound they generate. They also may need concrete inertia bases to minimize vibration transfer to the structure. Air compressors should not be located in the roof or mezzanine of the building, as the vibration will migrate to the rooms below regardless of padding or springs. They preferably should be located outside the building or in a service or utility corridor but, obviously, still within reach of the rooms they service. The location should be carefully selected to prevent transmission of vibrations or sounds to the building or rooms serviced.

HVAC

The HVAC system is to a laboratory what a lung is to a living being. It is, perhaps, the most complicated and most sophisticated ventilation system found in any type of building. Its performance and specifications are determined by the ventilation requirements of the chemical and biological tasks and operations undertaken in the building. These requirements have to do with:

- The needed level of purity of the air
- The needed temperature and humidity
- The likelihood of hazards altering the air quality

The type of laboratory affects the design of the HVAC systems. There are two major classes of laboratories:

- Those in which operations have general needs and can be performed safely in a general laboratory space (these operations can be conducted in parallel without affecting each other)
- Those in which operations have specific, unique needs and require special provisions or precautions

Each laboratory must be evaluated individually to determine the overall HVAC performance required for the building. This evaluation should be made by the architect, mechanical engineer, an industrial hygienist familiar with the future tasks and operations, and the scientist who will conduct the task. It needs to be thorough and factual, identifying and separating real needs and requirements from those arising from habits and whims.

Clean Room Air Purity

The level of purity of the air supplied to a laboratory space pertains to the amount of dust, bacteria, fumes, and other particles in the air. Normal filtering systems are sufficient for most laboratories. Some laboratories, however, need a reduction of the amount of particles or other matter contained in the air. Spaces that require this type of air are called clean rooms.

Clean room air purity is rated according to a class system. There are five classes, each referring to the number of particles of a certain size contained in a volume of air: 100.000, 10.000, 1.000, 100, and 10. The lower the number the cleaner the room. Purity is achieved by filtering the air. There are filters on the market that can achieve 99.999 percent efficiency in eliminating particles as small as 0.3 microns. Keeping a room clean may require 300 to 500 air changes per hour, depending on the class required. A class 100 clean room consumes 700,000 to one million Btu/sf/year.

Filtered air can be introduced in clean rooms in one of four ways: in vertical laminar flow, in horizontal laminar flow, in spot laminar flow, or in diagonal washing laminar flow. Air is introduced in laminar fashion to prevent turbulence, which could cause backflow of particles in the room. When air is introduced in vertical laminar flow, it enters the room from the ceiling and exits through low registers in the side walls or through perforations in a raised floor. Air introduced in horizontal laminar flow enters the room from registers in a wall and is exhausted through registers in the opposite wall. Air introduced in diagonal washing laminar flow is introduced through registers in the walls and the ceiling and exhausted on opposite walls and the floor. This promotes diagonal circulation and creates a brooming effect. Air introduced in spot laminar flow generally enters through a hood mounted over a work bench and creates localized air movement.

Access to clean rooms must be limited. Vestibules with interlocked doors should separate clean room space from normal space, and it is through these vestibules that people working in the clean room would have access to the space where they can change their clothes. Air pressure should be such as to prevent transmission of particles to the clean room. See chapter 11 for detailed clean room requirements.

Temperature and Humidity Control

The other major factors in designing the HVAC system are the temperature and humidity requirements. The most efficient and practical way to control the laboratory's temperature and humidity is to keep the laboratory airtight and control the temperature and humidity of the air supplied to it. This air is cooled in summer and heated in winter by a central HVAC system. Its humidity is adjusted through the cooling or heating process or, if tasks require specific or constant levels of humidity, is controlled centrally or at each room. Humidification requirements are large and variable due to the significant amount of outside air being conditioned. Duct-mounted heating coils can provide different temperatures in specific zones.

Some equipment and instrumentation, as well as certain tasks and experiments, produce noticeable amounts of heat and humidity. These loads must be estimated and considered during the design process.

Due to the large variance of heat loads from one laboratory to the next, each laboratory should have individual temperature control. This is normally accomplished by locating duct-mounted heating coils in the supply duct serving each laboratory. A temperature sensor in the laboratory modulates this heating coil to maintain room temperature. Humidity is normally controlled at the air-handling-unit zone level rather than at the individual room level.

It may be possible to group a certain number of nonlaboratory rooms into one or more common temperature-controlled zones. The temperature of each zone is managed by one strategically located temperature sensor. Rooms included in each zone must have similar heat load and temperature and humidity requirements. The final grouping of rooms in the temperature-controlled zones, and how their temperature and humidity will be controlled, will influence the selection of the HVAC system and the duct system used for air distribution and exhaust.

Level of Hazard

The levels of hazard of the chemical or biological operations (NFPA-45, class A, B, and C) are perhaps the most important factors affecting a laboratory HVAC system. These levels dictate the volume of air needed, how the air is induced in each space, and how it is exhausted. Each of the three classes has specific ventilation and other requirements, which are identified by code regulations and thus are not dealt with in detail here. They all aim, however, to protect laboratory users and the environment (in and out of the laboratory structure) from discharges, and the facility itself from the effect of the operation.

HVAC AIR SUPPLY AND EXHAUST SYSTEM

Air Circulation in Laboratory Facilities

In most laboratories facilities where chemicals are used, air should not be recirculated. Use of one-pass air, with 8 to 18 air changes per hour in each laboratory room and 6 to 8 air changes per hour in the chemical storage rooms, should be required where the possibility of moderate or high hazard exists. The U.S. government requires all its laboratories where chemicals are used to use one-pass air only. If contaminated by operations to a level above that permitted by code, one-pass air must be treated and rendered inoffensive prior to its exhaust.

Precautions Against Cross-Contamination

The main laboratory building, as well as all air intakes, should be located to provide the cleanest possible source of air for the building. The location of the air intakes in the building, the exhaust stacks, and the parking lots and other sources of contaminated air, should prevent the intake of car fumes and other contaminants, as well as of the laboratory's own exhaust fumes. Further, during the design phase precautions should be taken to arrange the supply and exhaust ducts to avoid the transmission of chemical vapors or air from room to room, from one section of the building to another, and from exhaust stacks to any part of the building or adjacent structures.

SELECTION OF A MECHANICAL SYSTEM

Selection of a mechanical system is one of the most important decisions the designer must make. There are two types of mechanical systems used for laboratories: a constant exhaust and supply airflow system, and a system that uses a variable air volume. The advantages and disadvantages of the two systems must be weighed according to how they affect the facility being designed in its given geographical location. Naturally, all the operational requirements and the characteristics of the geographical location must be identified and considered in the evaluation. The following is a brief description of the two systems.

Constant Exhaust and Supply Airflow

Laboratories with constant exhaust and supply airflow rates have hoods (when bypass hoods are used) that always exhaust a constant volume of air regardless of whether their sashes are open or closed. These are equipped with a bypass grille to allow the air to pass into the hood when the sash is closed, thus maintaining a constant exhaust rate. To make up for the air exhausted, an almost equal volume of treated outside air is supplied to the room where the hood is located. The treatment of this air (cooling, heating, filtrating, and moving) is energy-intensive and expensive.

Variable Air Volume System

The variable air volume system (VAV), as used in hoods, requires fewer energy and associated expenses than the constant exhaust and supply airflow system. The VAV system supplies and exhausts air at variable rates depending on the use of fume hoods. The VAV concept is based on the fact that fume hoods are used with full open sashes for only a few hours during the day, while the system (if it is a constant supply and exhaust airflow system) runs continuously, even when scientists are not using the hoods or when the hoods' sashes are closed. The major advantage of a VAV system is that it reduces the volume of air required for protection without reducing the level of protection. This substantially reduces the use of energy and associated expenses.

In a room where fume hoods are used, hazardous operations using hazardous chemicals must take place inside the fume hoods. When the hoods are closed there is really no need to exhaust any air from the laboratory room, except for the small amount that keeps the fumes from spreading into the laboratory and that pushes the fumes up the hoods' exhaust stacks. This volume of air can be as low as 25 to 30 percent of that required when the hood sash is open (assuming that this amount of air will maintain the required number of air changes in the room). This reduced volume of air should be introduced in the hood through a small (1- to 2-inch) (0.03 to 0.05 m) bypass at the top of the sash and another bypass one inch below the airfoil.

The reduction in the required air volume in turn reduces the energy required to cool, heat, or drive the air. Such air, even as reduced would decrease the danger of explosion and of interior hood corrosion. Further, these two bypasses at the top and bottom of the hood face ensure that some turbulence and interior mixing will occur within the hood at the lower exhaust condition and that the minimum stated volume will always be exhausted. This further decreases the danger of explosion and of interior hood corrosion.

The initial costs of the VAV system are significantly greater than that of the constant exhaust and supply airflow system. Sophisticated digital control systems are required to measure the airflow rates and to adjust the control dampers. Measuring and adjustment ensure that the laboratory's pressurization or depressurization is maintained at all sash positions and at all times. Further volume control devices must be added to all the supply fans to prevent overpressurization of ducts and to save energy. It is typically not cost-effective to add volume control devices for systems with individual fans for each fume hood.

The VAV system is not without its disadvantages. It is significantly more complicated than the constant exhaust and supply airflow system and requires more sophisticated, and constantly active, operating and maintenance personnel. Because of its complexity and its constant need for surveillance, this system is more likely to fail. Further, the peak energy usage of a VAV system is higher than that of the other system. This is due to the energy losses associated with the fan's static pressure control and to additional pressure losses of the measurement control devices. As long as the fume hoods have individual and independent exhaust stacks, the costs associated with VAV system controls will make the VAV systems economically unattractive.

The selection of laboratory mechanical system should be influenced first by safety factors (determined by codes, prudent practice, and the industrial hygienist and laboratory management) and secondly by four major economic factors:

- Frequency of hood use
- Heat loads (including the heat gain that is generated by equipment)
- Climate
- Cost of energy

These factors determine how much air should be moved through the building and the cost of conditioning that air. A life-cycle cost analysis may be the most rational and objective way to judge which system is more economical for the particular project. The factors considered in such an analysis can be identified and quantified or reasonably estimated. Climate for a given location can be accurately known, but the frequency of hood use must be estimated. As a rule, the important factors must be quantified with exactitude and realistic estimations should be made.

For a life-cycle cost analysis to be objective, all the operating costs (energy, maintenance, and incremental costs) pertaining to or impacting on the ventilation of the laboratory should be added and subdivided by the number of fume hoods. This conversion will provide an annual expense per fume hood. Then the factors should be grouped and converted so that the analysis can be conducted for a single fume hood. Disregard the factors that are common to both systems and that will not affect their ventilation. If utility costs are high and the climate is severe in terms of high heat gain, the life-cycle cost analysis will probably tell us that the correct choice is a VAV system. If the hood's diversity and the interest rates are high and the life expectancy of the laboratory is short, the best choice may be a constant airflow system.

Two-Level Constant Exhaust and Supply Airflow System

Some laboratories may have all the requirements to qualify them for a constant airflow system during operating hours, but not during off-hours. It may be cost-effective for these laboratories to set up a system for heating and possibly ventilating and cooling that operates at a lower level during off-hours. This eliminates the need to operate the HVAC system at its highest level when the building is not in use.

This type of two-level system can be designed to operate on a time-programmed cycle (daily, weekly, or monthly) with all the system components controlled to maintain safety criteria at a reduced volume of airflow. This system can also be designed to allow the option of using filtered, unconditioned outdoor air when ambient conditions satisfy the temperature and humidity requirements.

It is also possible to convert, for a cost, a one-level constant airflow system to a two-level programmed system. During the hoods' operating hours, the system is set up to operate at the highest level of airflow needed to protect the hood operator from the operation. During off-hours the system operates at a much lower level that is sufficient to contain fumes within the hood when the hood sash is closed and to force these fumes up the exhaust stacks. The conversion requires the use of building automation and monitor and control systems, which are costly to acquire and install, but pays for itself by substantially reducing operation costs.

Regardless of the type selected, the HVAC system has to be sized to provide all the air needed to satisfy ventilation requirements at their peak and to raise or lower the temperature to the levels required during peak winter or summer months.

Ventilation for Special-Purpose Rooms

Some areas in a laboratory may need only ventilation and not heating or cooling. A study should analyze the cost-efficiency of using conditioned air to ventilate these areas versus a mechanical ventilation system to provide air in sufficient separate quantities to satisfy equipment or occupant's requirements.

Special-purpose rooms or areas, where temperatures are controlled individually and may be substantially different from that of the rest of the building, should have either individual systems or be grouped and serviced through zone systems. The volume of air delivered to and exhausted from each space or section within the laboratory should always remain balanced.

Exhaust and Supply Ducts and Connections

The ducts and connections for the exhaust and supply of air must be equipped with balancing devices. Access to these devices must be easy to allow for periodic maintenance and adjustment. The air exhaust and supply ductwork and plenums should also be designed to allow easy access to all motors, bearings, control valves, steam traps, and other devices. The laboratory owner or manager should require the mechanical engineer designing the system to select motors, bearings, control valves, steam traps, and other devices that are easy to service and maintain. The contractor should also be required to place these devices in a location that can be easily reached with minimum disruption to the laboratory operations.

HVAC System Balancing

Regardless of the air exhaust and supply system used, the HVAC system should be designed with sufficient flexibility in its air balancing to allow all the hallways of the laboratory section of the building to be maintained at positive pressure. Each laboratory containing one or more fume hoods should be maintained at negative pressure with respect to the corridor. The entire laboratory section of the building needs to be at positive pressure with respect to the outside atmospheric pressure, but at a negative pressure—generally equal to 0.05–0.1 inches (1.3–2.54 mm) of water—with respect to the adjacent part of the building. Laboratory clean rooms, however, should be maintained at positive pressure with respect to the corridors servicing them. In a laboratory wing where general chemistry and biochemistry laboratory rooms are mixed with clean rooms, it is advisable to have the clean room(s) separated from the main corridor by a vestibule. This vestibule's air pressure should be negative to the corridor and to the clean rooms.

VENTILATION PROTECTIVE DEVICES

Low-hazard laboratory rooms, where there is no need to protect the operator from his operation, need only be under negative air pressure in relation to the corridor and to the outside of the building and have a certain number (generally 8–12) air changes per hour. However, when the level of hazard is moderate or high, these simple precautions are not sufficient. The rooms must have protective devices—the most common of which are fume hoods, biological safety cabinets, and glove boxes. These three devices affect the HVAC and other air supply systems servicing the laboratory, as is explained below.

Fume Hoods

Fume hoods and glove boxes can be defined as extensions of the exhaust duct system. They consist of a ventilated enclosure and are designed to allow the undertaking of tasks and operations that produce hazardous or noxious fumes, odors, or dusts. They contain fumes, dusts, and other contaminants, preventing their escape into the laboratory. Their purpose, therefore, is to separate the task and its byproducts from the operator undertaking it and to discharge the byproducts in a controlled and safe manner.

HOOD SIZES

The work space needed inside the hood determines the size of the hood. Hoods' depths vary from 34 to 48 inches (0.86 to 1.22 m). The width of the hood is therefore be determined by the working space required. Hood sizes are expressed by the outside width of the hood. The most common hood widths are 3′-0″, 4′-0″, 5′-0″, 6′-0″, and 8′-0″ (0.91, 1.22, 1.52, 1.83, 2.44 m). Custom-designed hoods may have widths up to 24′-0″ (7.32 m).

INTERIOR LINING OF HOODS AND EXHAUST SYSTEMS

The materials used to line a hood's interior and the interior surfaces of the exhaust system should be determined according to the substances, or combination of substances, that will be handled in the hood. Among the determinant factors are the concentration of such substances, their frequency of use, and the health and safety requirements related to their use.

Hood effluents are generally classified as organic or inorganic chemical gases, vapors, fumes, or smoke. Hood effluents can be classified qualitatively as fume acids, alkalis, solvents, or oils. Such effluents can damage the hood liner by corroding it, dissolving it, or melting it. Also important to consider are the materials used to decontaminate the hood after certain operations, as it is possible that those materials could affect the hood liners. Generally, the hood manufacturers are qualified to recommend the type of liner materials best suited for the operations.

HOOD SASHES

Sashes are transparent, windowlike barriers on the face of the hood that provide physical protection from splashes or minor reactions. Sashes are generally made

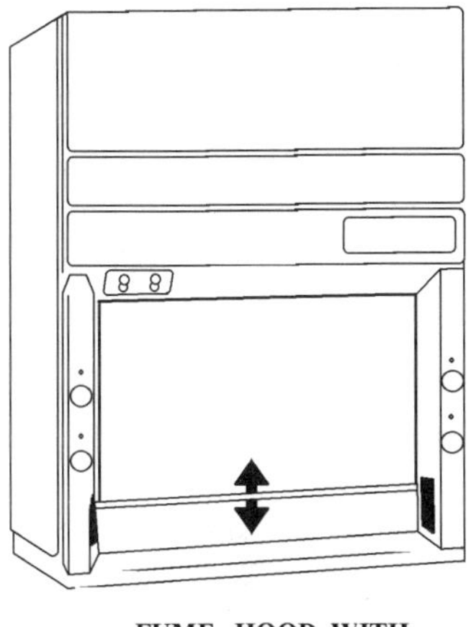

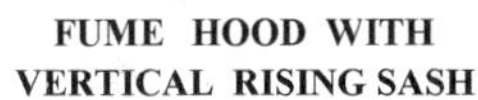

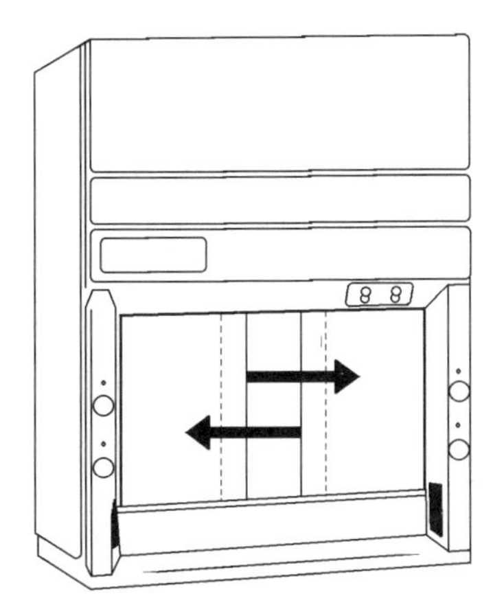

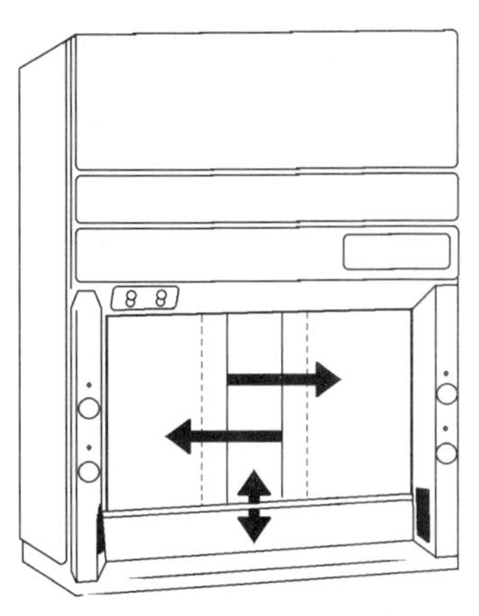

FIGURE 9-3 (left)
Fume hood with vertical rising sash

FIGURE 9-4 (center)
Fume hood with horizontal rising sash

FIGURE 9-5 (right)
Fume hood with combination sash

of safety glass, which is the most common and economical choice. Polycarbonate sashes should be used for operations involving hydrofluoric acid, as it will not fog or etch when exposed to hydrofluoric acid fumes. Sashes rise vertically, slide horizontally, or do some combination thereof. Figures 9-3 through 9-5 illustrate how the sashes move. Vertical sashes should be used when large apparatuses are to be moved or used in the hood. Horizontal sashes shield the hood operator while still allowing him or her to reach any area in the work space of the hood. Because the sash opening is smaller, the volume of air exhausted is also smaller.

OPTIMUM HOOD FUNCTIONING

The major function of the fume hoods is to protect the operator from contaminants. This is achieved by keeping the concentration of contaminants as low as possible in the zone in which the operator undertakes his tasks. This concentration should never exceed the acceptable threshold; appropriate conditions and ways of measuring the threshold are proposed in the *American Conference of Governmental Industrial Hygienist Handbook.*

In any case, the threshold should not be used alone in determining the type of hoods and air velocity required. In order to assure that the concentration of contaminants in the operator's breathing zone is below the threshold, it is essential to select the right type of hoods for the kind of tasks, to have these hoods working properly according to their specifications, and to satisfy related ventilation and balance requirements.

There are, however, other factors that affect the area of exposure of the operator and create air movement at the face of the hood. For full protection, the following factors have to be understood and considered in the design: the effect of the operator's movements on the airflow pattern at the face of the hood, the air movement and hood location in the room, and the turbulence within the hood's working space.

The movements of the operator standing in front of the hood affect the airflow patterns at the face of the hood. These movements create turbulence that, in turn, disrupts the regulated inflow of air toward the face of the hood. This can cause backflow of contaminated air from the hood to the operator's breathing

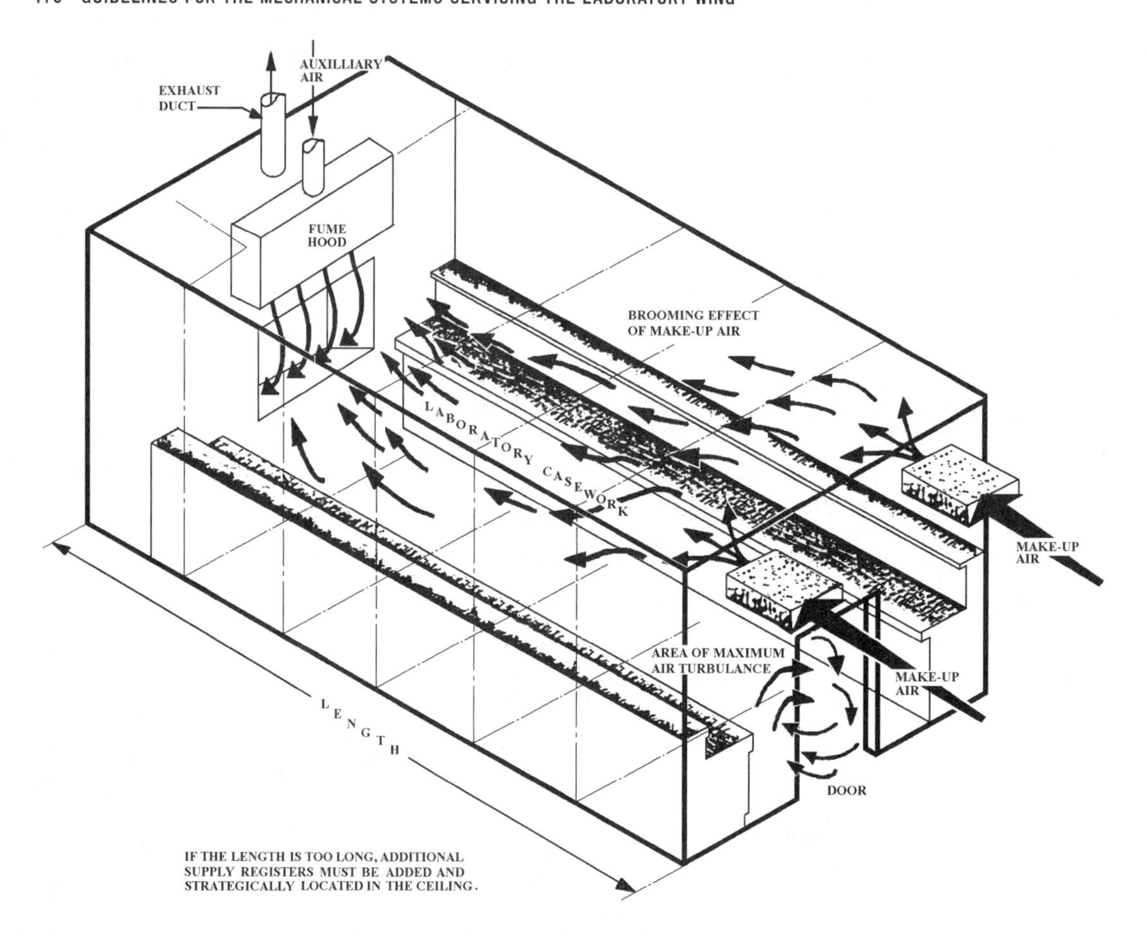

FIGURE 9-6
Air movement in a one-module laboratory room

zone. Correct balance of make-up air and auxiliary air regulated to flow equally with acceptable face velocity, throughout the entire face of the hood, will help fill all voids and minimize the low-pressure area in front of the operator and any backflow from the hood to his or her breathing zone.

The air movement in a room affects the performance of hoods. This is illustrated in figure 9-6. There are many factors that create movement of air in a room, the most important being the location of the air supply outlets, in relation to hoods and door locations, and the hood locations, in relation to doors, windows, and pedestrian traffic.

The air supplied to a room through air outlets can have a velocity several times higher than that at the face of the hood. When this occurs, the make-up air causes turbulence at the face of the hood. This can create voids at the hood face, which forces backflow or displacement of contaminated air to the operator's breathing area. If the velocity of the air supplied to the room is lower than 25 fpm (0.13 m/s), as measured at the hood face when the hood is not operating, no significant turbulence is created.

The opening and closing of doors and the movement of people walking also causes displacement of air, which can directly affect the air introduced into the hood or can indirectly affect movement of air at the face of the hood by interfering with the air introduced through the supply outlet. This is why the interior arrangement of laboratories is so important. Again, figure 9-6 shows that the best hood location is at the end of the room or bay, away from the door and windows. There essentially should be no other traffic in the room except that generated by the hood operator.

The supply air outlets need to be strategically placed away from the hoods and doors. Air should be induced in the space so that it has a brooming effect toward the hood. This air velocity should not exceed 25 fpm (0.13 m/s) at the face of the hood when the hood is not in operation. When a laboratory is arranged as in figure 9-6, hoods operate most efficiently with a face velocity of 80–100 cfm (0.41–0.64 m/s) with the sash in the full open position. Higher face velocities may create turbulence similar to that described in the next paragraph.

The last of the major factors that create air movement at the face of the hood and affect the operator's breathing zone comes from within the hood itself. Once the air enters the hood, it is drawn through the equipment and the contaminants toward the exhaust slots. The air within the hood is continuously in a turbulent state. The airflow within the hood should be just enough to contain the contaminants within the hood and direct them toward the exhaust outlets. If the airflows are greater than needed to provide directional containment, the turbulence becomes excessive, causes a greater rolling effect in the hood, and eventually causes backflow of the contaminated air into the room.

TYPES OF FUME HOODS

There are three types of fume hoods: conventional, bypass, and auxiliary air. Fume hoods use one of two kinds of exhaust systems—the constant air volume system or the variable air volume (VAV) system.

Constant Volume Conventional Hoods

The most common and least expensive type of fume hood is the constant volume conventional hood. This hood, shown in figure 9-7, is an enclosure with three sides and a sliding glass sash in front. This sash can be raised and lowered.

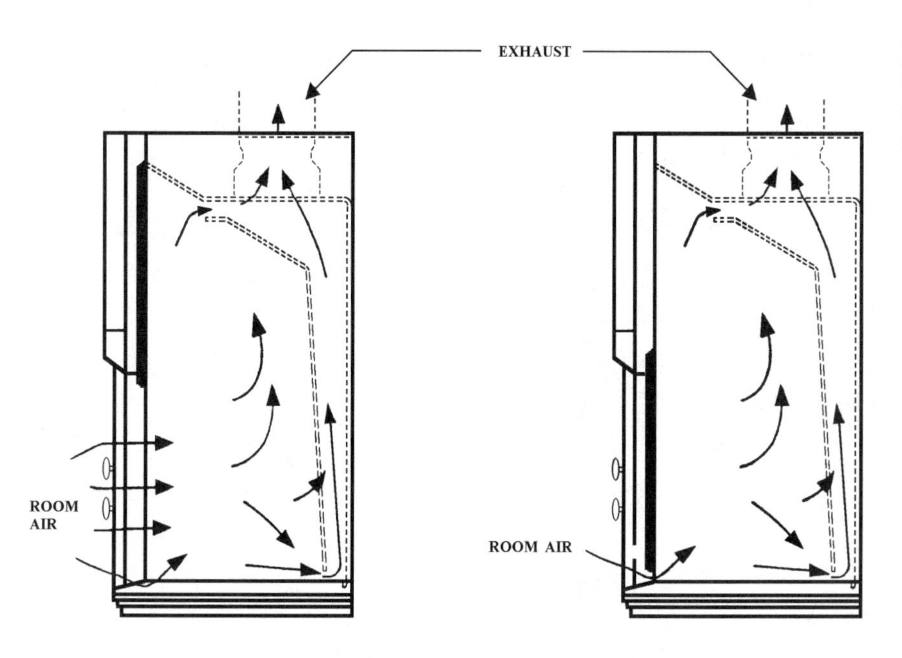

FIGURE 9-7
Conventional fume hood with sash open and almost closed

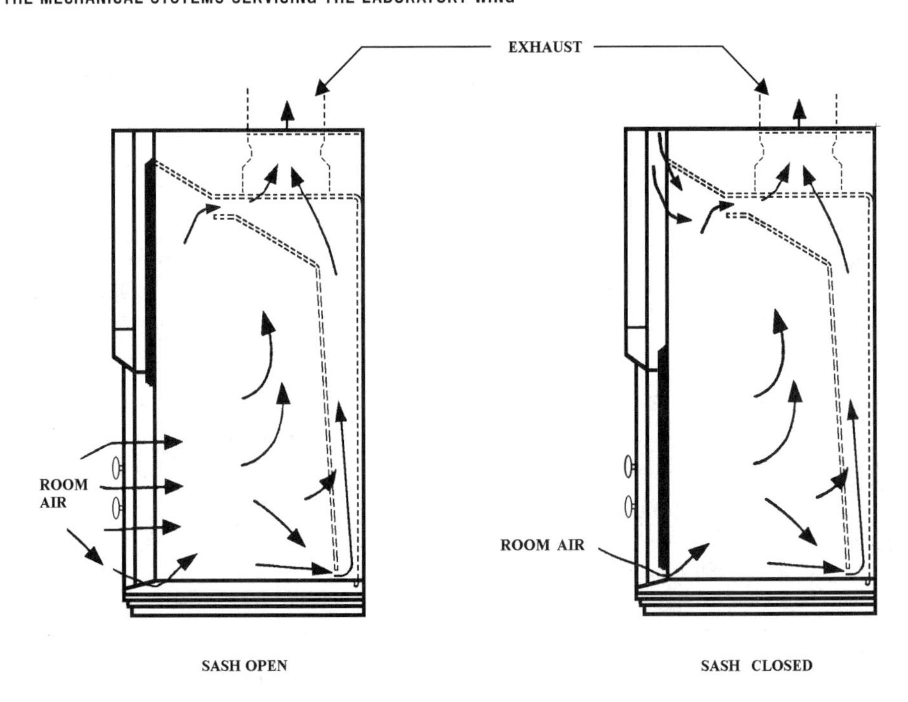

FIGURE 9-8
By-pass fume hood with sash open and closed

The hood has a baffle near its rear wall with adjustable slotted openings. A duct is connected to the space between the baffle and the rear wall, with a fan at its other end. This fan draws air from the room, through the sash opening, to the slots in the baffle, to the back plenum, to the exhaust duct, and to the outside of the building. The slotted openings can be adjusted to draw cold and hot fumes at a uniform face velocity. The constant volume conventional hood operates at a constant exhaust volume. The total volume of air exhausted enters the hood through the sash opening. This hood's performance and face velocity depend on the sash position. As the sash closes, air velocity increases, causing turbulence and possible damage to the operation taking place inside the fume hood.

Constant Volume Bypass Hood

A different and also very common version of the constant volume hood is the constant volume bypass hood (figure 9-8). This type of hood permits the room air to be drawn into the hood through a bypass as the sash is lowered or raised. It is designed so that as the sash is closed, the air entering the hood is redistributed; therefore, it prevents the face velocity from increasing or decreasing too much and assures continuous air removal from the laboratory.

Constant Volume Auxiliary Air Hoods

Another frequently used type of hood is the auxiliary air hood. This hood has many names—it is also called induced air hood, add air hood, make-up air hood, and balanced air hood. Shown in figure 9-9, this hood is designed to permit the introduction of outside air directly in front of the sash but outside the hood. This air is referred to as auxiliary air and is differentiated from the air drawn from the room. The air provided to the room by the main HVAC system is referred to as "make-up air." Make-up air is heated, air conditioned, humidified, or dehumidified like the air induced to the rest of the building. Make-up air is supplied to

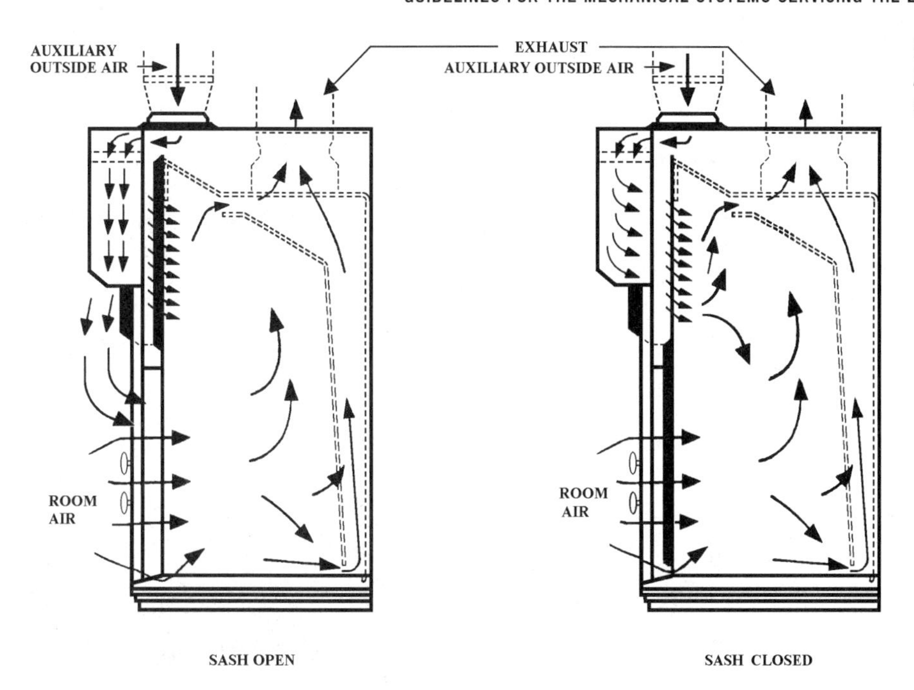

FIGURE 9-9
Auxiliary air fume hood with sash open and closed

a room in sufficient volume to satisfy the number of air changes required by the room and by the hood face velocity.

In some old models of auxiliary air hoods, the auxiliary air was directly introduced into the hood itself. These types of hoods allowed backflow of air and fumes into the room if the hood's exhaust system failed or was turned off. These hoods are dangerous and should not be used.

Auxiliary air hoods are used in laboratories needing more hoods than the number necessary to evacuate the air introduced for building ventilation. In other words, auxiliary air hoods are useful for air-starved laboratories. These are the laboratory rooms where the make-up air volume is not sufficient to satisfy fume hood's need for air. In most instances, the volume of air needed to ventilate the laboratory building is introduced by the main air handlers and is either cooled or heated. Auxiliary air is not generally heated or cooled in temperate climates but may need to be dehumidified. Thus, auxiliary air hoods often conserve energy. The level of conservation depends on how much the auxiliary air must be tempered. Naturally, the conservation will be minimal if the air must be brought to a temperature close to that of the room. On the other hand, temperature extremes caused by untempered air can adversely affect the hood's containment ability and cause an uncomfortable environment for the hood operator.

Auxiliary air supplied to hoods should be provided by an air supply system that is different and separate from that of the HVAC system. This allows each system to provide air at selected temperatures and levels of humidity. An auxiliary air supply system may be costly to install and difficult and costly to balance and maintain. Thus, it is necessary to make a thorough analysis of costs and benefits—initial as well as for operation and maintenance—and compare it with the costs and benefits of a system using standard hoods that exhaust only air supplied by the HVAC system.

When one or more hoods are being operated in a room—in the rare case when an operator is allowed to turn the hoods on and off—each hood should automatically activate a device that provides its bonnet with the required volume of auxiliary air. It should also activate a device that increases the volume of air introduced into the room by an amount equal to the volume exhausted by the operating hood or hoods. This extra level of complexity is not present when hoods are integral to the ventilation system.

Auxiliary air systems also must be carefully balanced to prevent turbulence at the hood's face and the air should be clean, dry, and tempered enough not to negatively affect the operations taking place in the fume hood. In an auxiliary air hood, the air introduced from outside the hood in front of the sash may interfere with the smooth laminar flow through the sash opening. This is particularly true if the air is introduced around the periphery of the sash opening immediately in front of the sash and at right angle to the inward airflow. If the total volume of air exhausted diminishes for any reason—due to a dirty exhaust system or a faulty damper, for instance—and the auxiliary air continues to be supplied, the smooth flow of air into the hood will be disrupted. This produces eddy currents and leaks and may push the fumes out of the hood. Most of these problems, however, can be compensated for in the design.

When an auxiliary air system is used in addition to the main HVAC system, the two systems need to be coordinated so that the make-up air is provided by the HVAC system to each room containing hoods. This make-up air volume should be equal to the volume of air exhausted from the room, either through the hood (when the hoods are in operation) or by other means. Further, when the hoods are not in operation, no auxiliary air should be provided to their bonnets. However, the HVAC system should continue providing these rooms with enough air to allow for the required number of air changes and to maintain the room temperature and humidity.

Variable air volume hoods

The variable air volume (VAV) hood is either a conventional or a modified bypass hood that allows for variations in the volume of air exhausted and simultaneously maintains its face velocity within a preset range. This is done by a damper that opens and closes as the sash changes position. Modified bypass hoods maintain an airflow sufficient to contain and dilute fumes even when the sash is lowered. The VAV hoods (conventional or modified bypass) offer the advantage of a consistent airflow performance regardless of the sash position. They can also be equipped with alarms to alert the hood operator should an unsafe air condition occur. The VAV hoods are either built as such or are conventional or bypass hoods with an accessory that makes them act as such. VAV hoods can also be designed to interface with the laboratory wing's HVAC system so that the exhaust can be monitored and the supply can be controlled from a central location. The initial cost of these hoods is high but in the long run they save energy. A schematic drawing of a modified bypass VAV hood is shown in figure 9-10.

Perchloric acid hoods

The perchloric acid hood is a standard or auxiliary air hood in which perchloric acid or other hazardous oxidants are used. This type of hood is built with

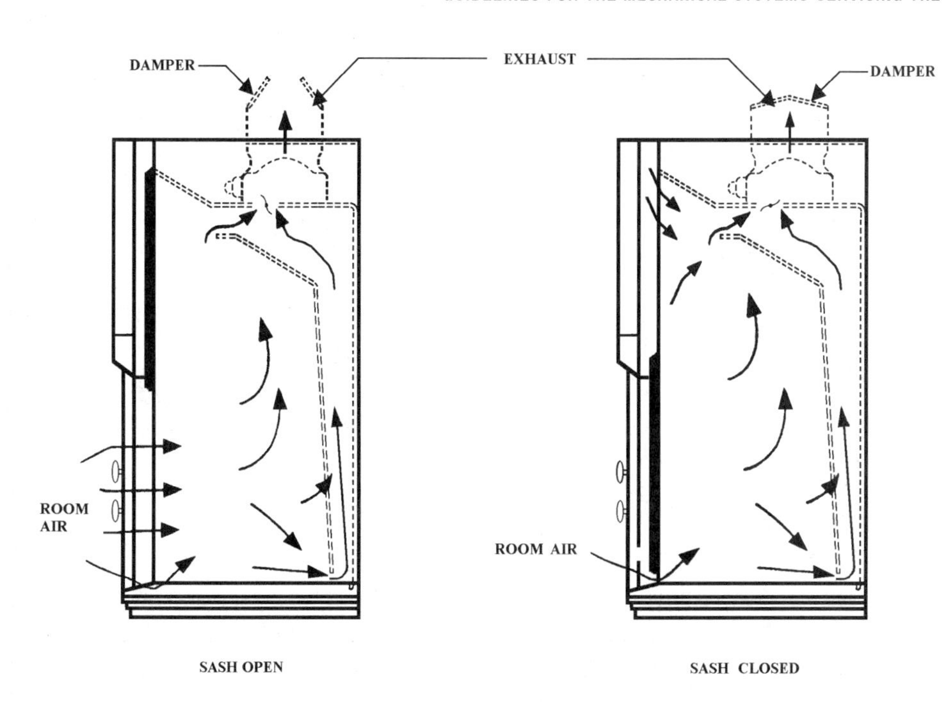

FIGURE 9-10
Variable air volume fume hood with damper control and modified by-pass

water sprayers in the entire length of the exhaust duct and in the space between the baffle and the back wall. These sprays are periodically turned on to wash away any acid crystals or organic matters that may have accumulated. The inside surface of this hood must be hosed down or hand washed to prevent build-up of acid crystals or oxidates from producing violent explosions. This type of hood should never be manifolded with other hoods. A drawing that depicts this hood's wash-down system is shown in figure 9-11.

Radioisotope hoods

The radioisotope hood is another variation of the constant volume hood. This hood is built to facilitate cleaning. It must have integral work surfaces and coved angles to minimize contamination build-up. This hood's liner material must be impermeable to radioactive materials. It should also have a base and supporting cabinetry capable of carrying very heavy loads, since lead shielding bricks may be placed in it to shield radioactive material. This hood's installation should facilitate the use of high-efficiency particulate air (HEPA) filters to collect the exhausted radioactive particles. This type of hood should never be manifolded with other hoods.

Distillation hoods

A distillation hood can be a constant volume conventional hood, a constant volume bypass hood, an auxiliary air hood, or a VAV hood. What differentiates it from regular fume hoods is its additional interior height, which accommodates large apparatuses. This type of hood is generally mounted on a platform instead of a base cabinet or a bench. A drawing of this hood is shown in figure 9-12.

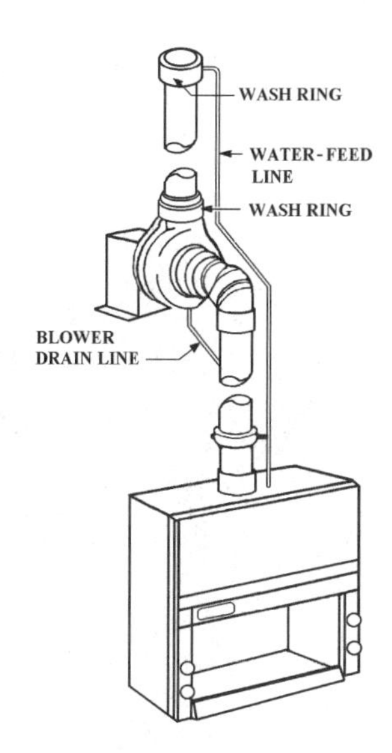

FIGURE 9-11
Perchloric acid fume hood with wash down system connected to duct work

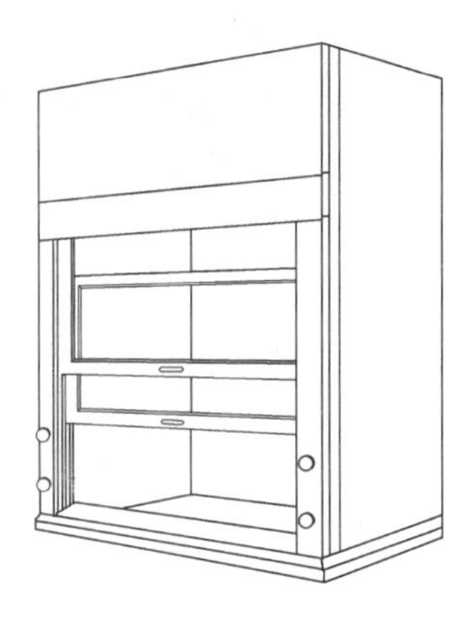

FIGURE 9-12
Schematic drawing of a distillation fume hood

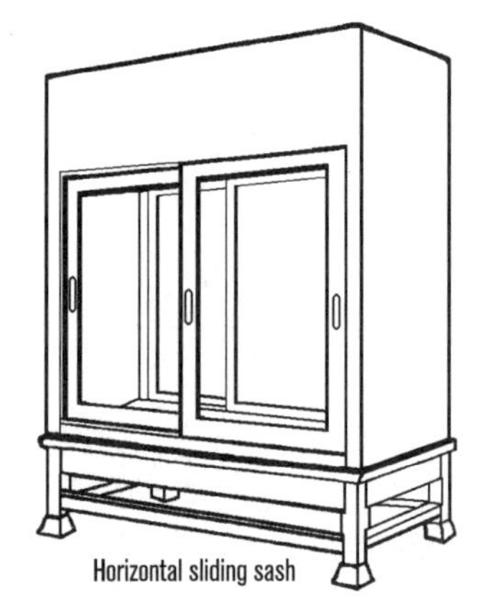

FIGURE 9-13
California fume hoods with sash open entry on front and back.

California hoods

A California hood is a distillation hood that also can be any of the previously mentioned types of hoods. What differentiates it from a regular distillation hood is that it has sashes on both its front and back. This permits access from both sides. These sashes can be vertical or horizontal. Two schematic drawings of this type of hood are shown in figure 9-13.

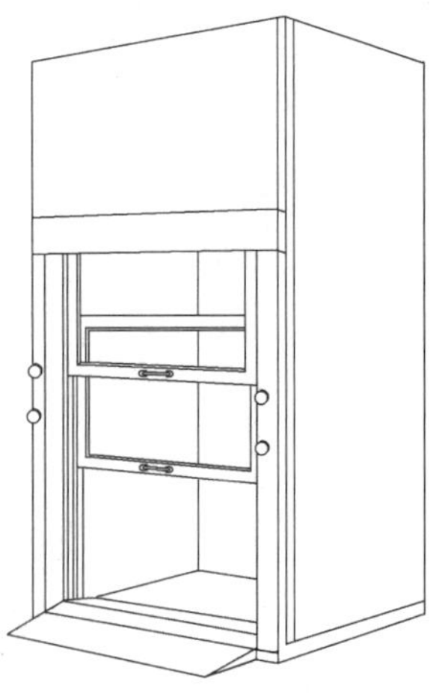

FIGURE 9-14
Schematic drawing of a walk-in fume hood

Walk-in hoods

A walk-in hood is just like a regular distillation hood, except that it is mounted on the floor, which permits the rolling of heavy or bulky equipment inside the hood. It should be noted that the hood operator should never step or stand inside the hood while the hood is being used and fumes are generated. A schematic drawing of this type of hood is shown in figure 9-14.

Ductless hoods

Ductless hoods are not connected to an exhaust system. In these hoods vapors and fumes are trapped by special filters before air is recirculated into the room. These filters are made of specially treated charcoal media that absorb chemical fumes, including the fumes generated by some organic solvents and acids. At the same time that chemicals are being selectively trapped by these filters, the volume of air resulting from the hood's low face velocity may not be sufficient to effectively capture all contaminants, and some chemicals that should be trapped may not be trapped by the charcoal filter. This type of hood should be carefully evaluated for its adequacy for the operations. The ability of its filters to trap the vapors and fumes generated by these operations must also be confirmed. These hoods should never be used in laboratories where toxic or other types of hazardous chemicals are used. A schematic drawing of this hood is shown in figure 9-15.

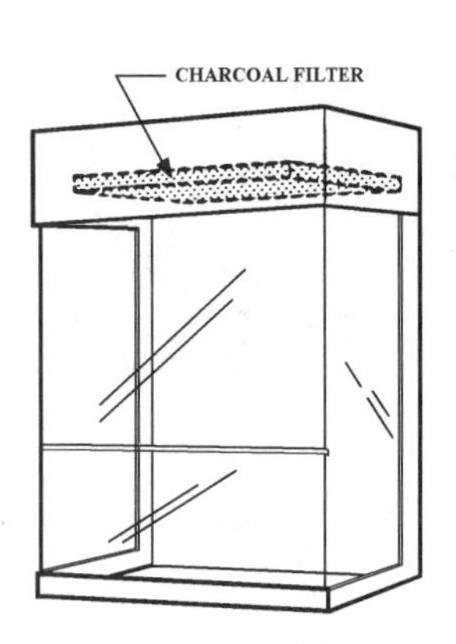

FIGURE 9-15
Schematic drawing showing a ductless fume hood with sash built-in charcoal filter

Hoods in compliance with the American with Disabilities Act

Most hood manufacturers build and equip most hoods with features that meet the requirements of the American with Disabilities Act (ADA). The height of the work surface and the clearance under the hood allow a person in a wheelchair to work comfortably. The switches, controls, and all written information are located where they can be seen and comfortably reached by a person seated in a wheelchair. The alarm should be audible and visual. Audible alarms must have an intensity and frequency that can be heard by a person with partial hearing loss. A schematic drawing of this type of hood is shown in figure 9-16.

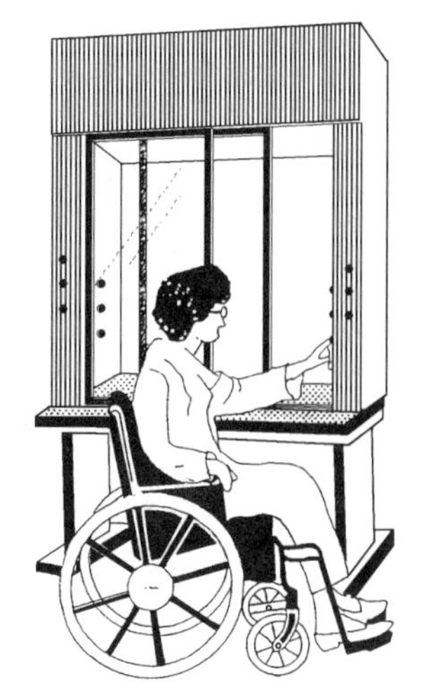

FIGURE 9-16
Schematic drawing showing a fume hood in compliance with the American with Disabilities Act

Canopy hoods

These hoods are used to draw heat, moisture, and nonhazardous fumes from localized areas. They are generally placed over large or bulky apparatuses such as ovens and autoclaves to dispose of heat and steam. To be effective, canopy hoods should be placed about 12 inches (0.30 m) above the equipment they service. This type of hood should never be used to remove contaminants and other hazardous substances. A schematic drawing of this type of hood is shown in figure 9-17.

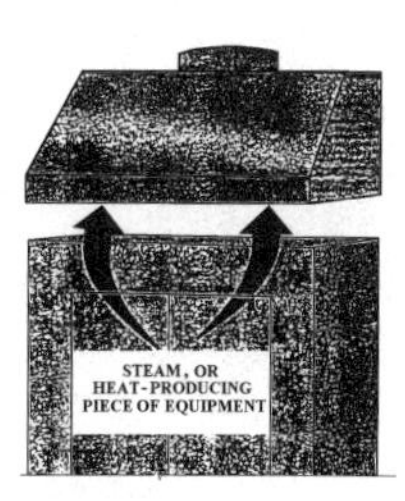

FIGURE 9-17
Schematic drawing showing a canopy hood over a steam or heat producing piece of equipment

Slot hoods

Slot hoods draw air into slots built into one or more of the side walls framing a working surface such as a table or a sink. The blower is usually mounted below or behind the working surface and the air is pulled out and exhausted to the outside. Slot hoods are generally used in conjunction with sinks used for glass washing and preparation. They should never be used to remove contaminants and other hazardous substances. Side baffles improve the performance of these type of hoods. A schematic drawing of this type of hood is shown in figure 9-18.

Downdraft hoods

Downdraft hoods draw air down through a perforated or mesh work surface and then exhaust it. The blower of downdraft hoods is generally placed below the hood's work surface. These hoods are generally used for operations using materials heavier than air, such as dust and powders.

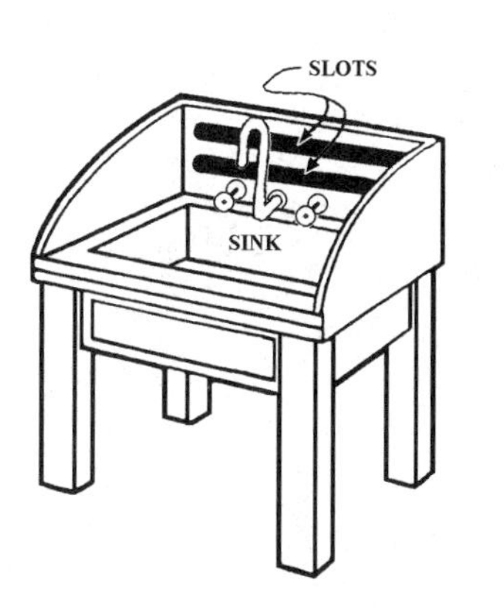

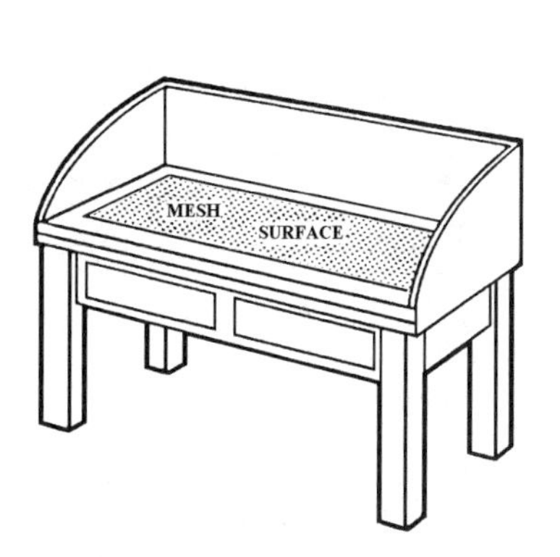

FIGURE 9-18 (left)
Schematic drawing showing a slot hood over a glass wash sink

FIGURE 9-19 (right)
Schematic drawing showing a downdraft hood

Often these materials are recovered and reused. Downdraft hoods should not be used to remove contaminants and other hazardous substances. A schematic drawing of this hood is shown in figure 9-19.

Elephant trunks

Elephant trunks, also known as snorkels, are flexible ducts connected to the laboratory's exhaust system that capture discharges from gas chromatographs and other instruments that produce vapors or odors. They cannot capture vapors, contaminants, or other discharges that are farther away than one-half of the snorkel's diameter. To be effective, the snorkel must be placed on top of the discharge, with the end of the discharge protruding inside it.

Biological Safety Cabinets

These cabinets are often referred to as laminar flow biohazard hoods although they are not really hoods. The cabinets are containment devices equipped with HEPA filters. They are designed to protect personnel, and sometimes the product and environment, from biohazardous materials. Biohazard is defined as an infectious agent or part of an agent that presents a real hazard to people, animals, and/or plants through infection or through disruption of the environment. There are three classes of biological safety cabinets. Each type of cabinet responds to specific biosafety levels and requirements. (See the introduction for the definition of the four biosafety levels.)

The selection of the right biological safety cabinet depends on:

- The kind of protection required (this may be the protection of the product, the protection of the product and the operator, or the protection of the product, the operator, and the environment)
- The kind of operations that will take place in the cabinet
- The type and quantity of hazardous and toxic chemicals that will be used in the cabinets
- The type of exhaust system that will be needed

CLASS I BIOLOGICAL SAFETY CABINETS

Class I biological safety cabinets are ventilated and designed to protect the operator and the environment. The biological safety cabinet does not protect the product from contamination. For the protection of the operator, it uses one-pass air, with the airflow path moving from the room, behind the operator, to the cabinet. This cabinet has a HEPA filter at the exhaust outlet to filter the exhausted air and may or may not be connected to the room's exhaust system. This type of cabinet is suitable for use with agents that require biosafety level I, II, or III containment. The face velocity of this type of cabinet varies between 80 and 100 fpm (0.41 and 0.64 m/s). A schematic drawing of this type of cabinet is shown in figure 9-20.

CLASS II BIOLOGICAL SAFETY CABINETS

These are ventilated cabinets designed to protect their operators, the products used in them, and the environment. For the protection of the operator, they use

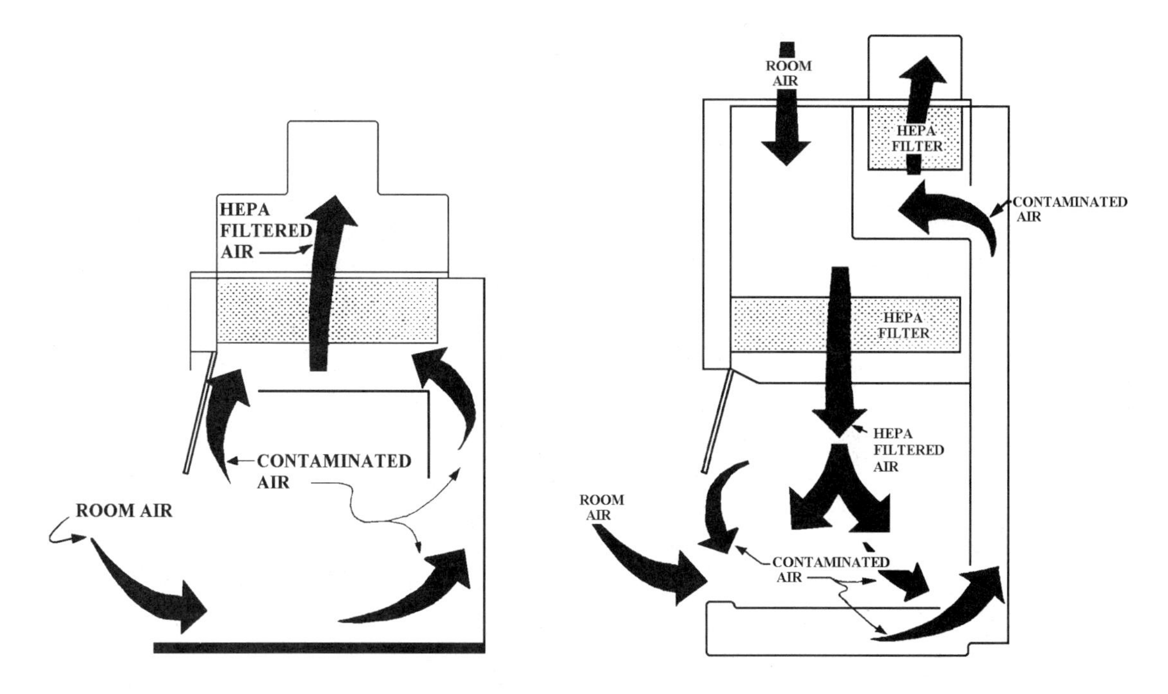

FIGURE 9-20 (left)
Schematic drawing showing a class I biological safety cabinet diagram

FIGURE 9-21 (right)
Schematic drawing showing a class II type A biological safety cabinet diagram

one-pass air, with the airflow path moving from the room, behind the operator, to the cabinet. This cabinet has a downward, HEPA filtered, laminar airflow (for protection of the product), and a HEPA filter at the exhaust outlet to filter the exhausted air (for protection of the environment). This type of cabinet is suitable for use with agents that require biosafety level I, II, or III containment. Class II biological safety cabinets are further subdivided into type A and type B cabinets. Type A cabinets may have contaminated plenums under positive pressure that are exposed to the room. Their exhausted air may be discharged into the laboratory room or through the exhaust system to the outside of the building. The minimum face velocity of type A cabinets is 80 fpm (0.41m/s). Figure 9-21 shows a class II type A cabinet, and figure 9-22 shows a class II type A cabinet with a canopy-type connection. The contaminated plenums of class II type B cabinets are under positive pressure but are surrounded by ducts under negative pressure. This type of cabinet must have a dedicated exhaust system, with a remote blower, that connects to an alarm system. The minimum face velocity of class II type B cabinets is 80 fpm (0.41 m/s).

Class II, type B cabinets are further subdivided into type B1, B2, and B3 cabinets. Class II, type B1 cabinets, shown in figure 9-23, are suitable when the operations conducted in them contain agents treated with minute quantities of toxic chemicals and trace amounts of radionucleides, when they are required as an adjunct to microbiological studies. For these cabinets to be suitable, the work must be done in the directly exhausted portion of the cabinet, or the chemicals or radionucleides should not interfere with the work when recirculated in the downflow air.

Class II, type B2 cabinets, shown in figure 9-24, are often referred to as total exhaust cabinets. They may be used with biological agents treated with toxic

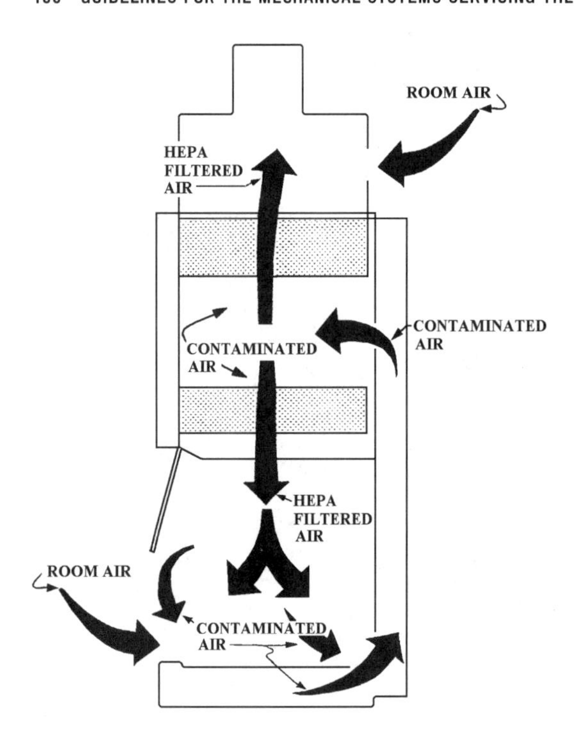

FIGURE 9-22
Schematic drawing showing a class II type A biological safety cabinet diagram with a canopy-type connection

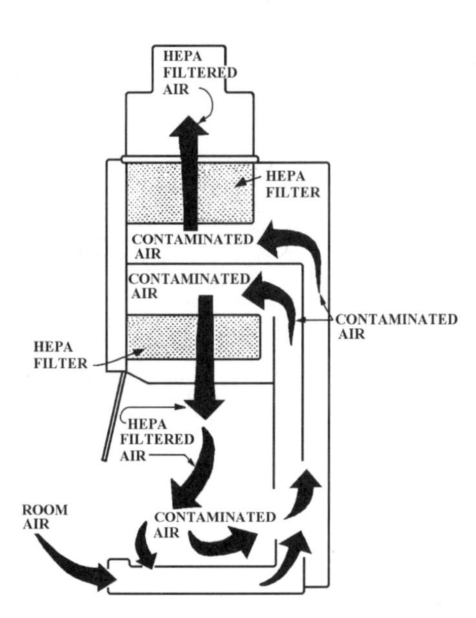

FIGURE 9-23
Schematic drawing showing a class II type B-1 biological safety cabinet diagram

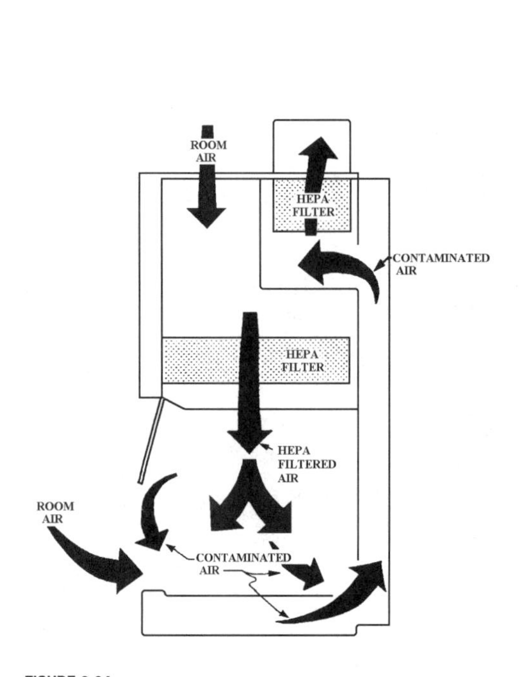

FIGURE 9-24
Schematic drawing showing a class II type B-2 biological safety cabinet diagram

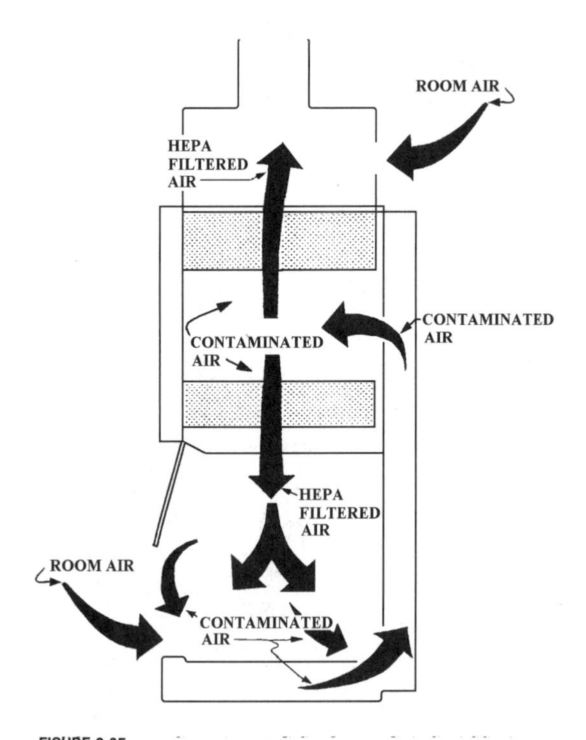

FIGURE 9-25
Schematic drawing showing a class II type B-3 biological safety cabinet diagram

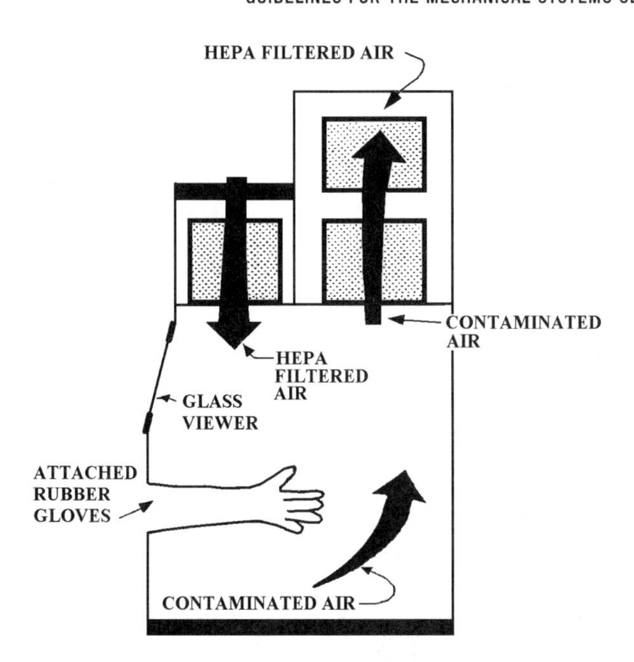

FIGURE 9-26
Schematic drawing showing a class III biological safety cabinet diagram

chemicals and radionucleides, when they are required as an adjunct to microbiological studies.

Class II, type B3 cabinets, shown in figure 9-25, are often referred to as convertible cabinets. They are suitable for work with biological agents treated with minute quantities of toxic chemicals and trace quantities of radionuclides that will not interfere with the work conducted in the cabinet if they are recirculated in the downflow air.

CLASS III BIOLOGICAL SAFETY CABINETS

Class III cabinets, shown in figure 9-26, are totally enclosed, ventilated cabinets of gastight construction. The operators of these cabinets use rubber gloves tightly connected to the cabinet body. These cabinets are maintained under a negative pressure of at least 1/2 inch (0.01 m) water gauge. Air is supplied to the cabinets through HEPA filters. The exhausted air is filtrated through two HEPA filters, one after the other, or once through HEPA filters and then through incineration. Class III cabinets are suitable for use with agents that require biosafety level I, II, III, or IV containment.

Glove Boxes and Exhaust Boxes

Glove boxes and exhaust boxes may be referred to as full-containment fume hoods; however, they are generally smaller than standard hoods. They are completely sealed chambers with a viewing window for observing ongoing operations inside them. They are also equipped with small openings with sealed gloves to allow the performance of tasks inside them. The access to a glove box is through an air-locked space with interlocked doors. This allows the introduction and removal of hazardous samples without backflow. Because they provide a physical barrier between the operator and the substances inside, glove boxes are used for the most hazardous operations—those requiring the greatest protection against inhalation.

Glove boxes require much less air than the standard fume hoods and exhaust at a much slower rate. Air is introduced into these boxes either directly or through an air filter. The air inlet or the filter at the inlet has an airtight plug to stop airflow when the box is not in operation. The chambers of glove boxes in which hazardous materials such as radioisotopes and carcinogens are handled must have HEPA filtered air that is then exhausted through a duct system to the outside. Certain types of glove boxes may be used for containing atmosphere-sensitive materials; those boxes may or may not be ducted to the outside.

CONTROLLED ATMOSPHERE GLOVE BOXES

Controlled atmosphere glove boxes are also known as dry boxes. These enclosures maintain a leak-free environment so that the operations performed in them remain under controlled conditions. Operations involving biohazardous materials should not be carried out in this type of glove box, as the box has no means of capturing aerosol fumes generated within the work area. However, these boxes are good for operations that must be performed in an atmosphere of inert gas or under controlled pressure, as well as for operations that must be performed in an atmosphere without oxygen or moisture. In most instances, these enclosures are used for oxygen-sensitive organic, inorganic, organometallic, and nonhazardous biochemical materials. A schematic drawing of a controlled glove box is shown in figure 9-27.

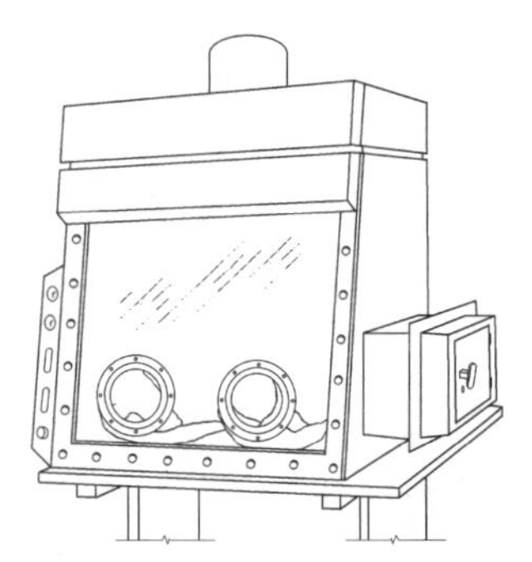

FIGURE 9-27
Schematic drawing showing a controlled atmosphere glove box diagram

VENTILATED GLOVE BOXES

Ventilated glove boxes are enclosures that are fitted with special filters at the supply and exhaust ends to protect the product inside the box from particulate contamination. The glove box operator is protected from hazardous materials used inside the box by the physical barrier. They are generally used to weigh reactive solids, to load capillary tubes for x-ray diffusion, and to transfer low levels of radioactive materials as well as carcinogenic and toxic materials. A class III biohazard cabinet is a type of ventilated glove box specially designed to sterilize and disinfect materials or products before they enter or exit the work area of the enclosure. Other ventilated glove boxes should not be used with biohazardous materials since there is no way to disinfect these materials before they are removed from the work area.

Clean Benches

Clean benches are another form of laboratory enclosure. They are generally used in clean rooms or when operations need to be conducted in an environment cleaner than the laboratory room's general environment. Clean benches use a blower to force room air through a HEPA filter and from there over a work surface. The filtered air is directed over

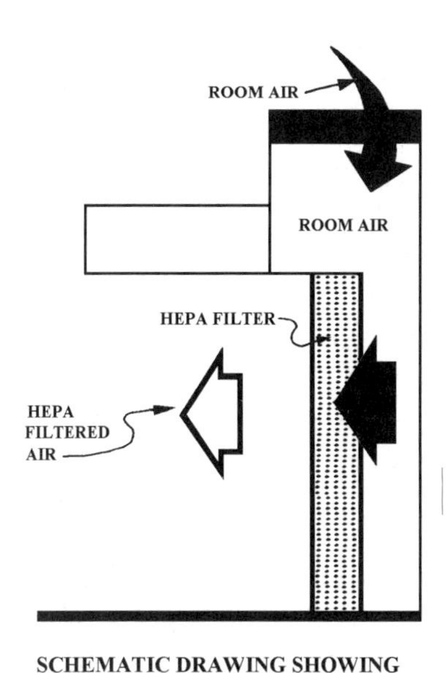

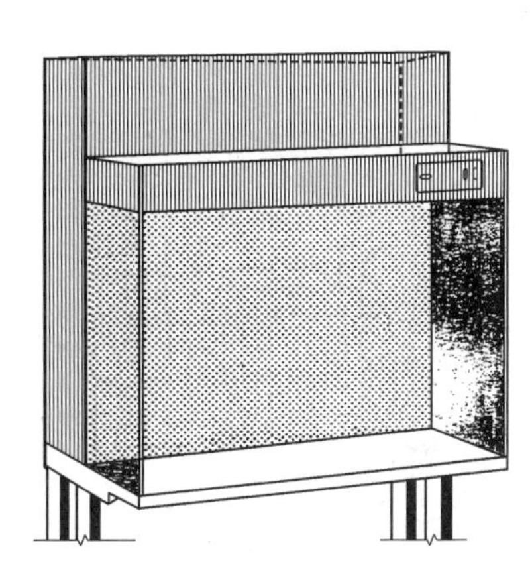

FIGURE 9-28 (left)
Schematic drawing showing a horizontal clean bench diagram

FIGURE 9-29 (right)
Schematic drawing showing a perspective of a horizontal clean bench

the work area either vertically (vertical laminar flow) or horizontally (horizontal laminar flow). The laminar flow of clean air protects materials in the work area from particulates and cross-contamination. Clean benches were developed as part of the clean room technology; therefore they provide only product protection.

Clean benches are often used in the electronics and pharmaceutical industries, in research laboratories for tissue culture and media preparation, and in hospitals and pharmacies for filling syringes and parenteral admixture formulation. Clean benches do not protect the operator from aerosols or other products generated by the work performed; therefore, work with toxic or biohazardous materials should not be performed in a clean bench. Clean benches in hospitals should not be used to prepare intravenous antineoplastic drugs. Figure 9-28 shows a schematic drawing of a horizontal clean bench, and figure 9-29 shows a perspective of the same clean bench.

FUME HOOD EXHAUSTS

Manifolding Fume Hood Exhausts

A manifolded exhaust system is one in which several fume hoods are exhausted together by means of one or more blowers. Hoods' exhausts can be manifolded safely under certain conditions and if certain precautions are taken. The designer should determine the size and location of the manifolded duct system and of its exhaust blowers, based on the number of hoods simultaneously used. The designer must also incorporate an automated monitoring and regulating system so that a change in the number of hoods used would not affect their face velocity. The design should also include devices such as dampers to prevent or minimize the risk of fire spreading through the exhaust duct system.

In addition, the dilution factor must be large enough to assure that fumes from different hoods will not mix and react in the manifolded areas. One or more blowers of an adequate size placed as back-up on a single manifold helps prevent backflow should a blower fail to function.

Heat Recovery From Fume Hoods' Exhaust Air

In laboratories where one-pass air is used, a very large volume of air, either cooled or heated, passes one time through the laboratory rooms and is then exhausted. Because the heating or cooling of the air consumes much energy, the temptation has been to recapture the heat or cold from the air prior to exhausting it. There are two types of energy recovery systems. One is the indirect type, where a separate heat transfer medium is moved between two points. A run-around coil system is an example of an indirect energy recovery system. The other is the direct type, in which two air streams are located on opposite sides of the heat transfer surface. Energy recovery wheels and air-to-air heat exchangers are examples of direct energy recovery systems. These systems use heat exchangers to transfer energy between the exhaust and supply airstreams. The type of heat exchanger defines the type of system.

The most efficient system is the direct type. An energy recovery wheel recaptures energy from the air exhausted and directly provides it to the supply air, and vice versa. Both airstreams are brought together at the wheel to allow the energy exchange to occur. An air-to-air heat exchanger consists of static plates that allow energy to transfer from one airstream to another. This type also requires the two airstreams to be ducted to the same location.

In the run-around coil system, water is the heat transfer medium. The system consists of a water coil in both the exhaust stream and the supply stream. The water is pumped between both the exhaust stream and the supply stream coils. Glycol can be mixed with water to prevent freezing. This system is not as efficient because there are two steps of heat transfer, but it allows the two airstreams to be located remotely from each other.

Two concerns that must be addressed in any energy recovery system are the corrosion of the heat energy device in the exhaust airstream, and the cross-contamination of airstreams. The feasibility of protecting the exhaust heat transfer surface must be evaluated, taking into account the type of chemicals expected in the airstream and the cost of the protection. Most types of protection also reduce the efficiency of the system. Cross-contamination between the supply airstream and the exhaust airstream is a concern with the direct type of energy recovery. The air-to-air heat exchanger has the advantage of not having a seal (like the energy recovery wheel does), which can corrode and leak. Energy recovery systems are generally only applied to manifolded exhaust systems, since the initial cost of energy recovery equipment for individually exhausted fume hoods would be prohibitive.

Treatment for Fume Hood Exhaust Air

In most instances it is not necessary to treat fume hood exhaust, as the effluent quantities and concentrations are relatively low. However, certain materials used in experiments would produce hazardous fumes and particles if exhausted without prior treatment. If possible, when that is the case the experiment should be designed so that the toxic materials and other harmful effluents are collected in traps or scrubbed rather than being exhausted. Incineration is another way to destroy combustible compounds. However, this is a difficult and demanding task that cannot normally be conducted in a regular fume hood.

If the two above methods of disposing the toxic materials and other harmful effluents cannot be used, it becomes necessary to use HEPA filters to remove highly toxic particulates and liquid scrubbers to remove particulates, vapors, and gases. Inert and/or chemical adsorbents can also be used to filter certain vapors and gases. When any of these three systems is used, the fume hood exhaust fan where such filtration takes place must be capable of working with the selected system.

Scrubbing of Fume Hood Exhaust Air

Fume hood scrubbers are generally used out of concern for the environment. They are also used to avoid or reduce duct corrosion and hazardous stack emissions. Few, if any, government regulations require the use of fume hood scrubbers. Because of this lack of guidance, it is important that designers educate themselves about the selection criteria for fume hood scrubbers and how they work.

To start with, one should know what a scrubber is. A scrubber is a device that can be added to a fume hood exhaust duct or to the exhaust system of up to three fume hoods, with the aim of eliminating or substantially reducing the release of soluble acid, base, and organic fume contaminants (as well as of particulates from the exhaust duct or system) into the airstream.

Fumes generated from the operations in the hood are carried through the exhaust stack from the hood to the scrubber. Fumes enter the scrubber, are exposed to a liquid spray dripping from the packed bed, and then move through the packed bed and the liquid spray section. As the fumes flow upward (into the packed bed, the liquid spray section and the mist eliminator) and out to the exhaust system, the scrubbing liquid flows downward, from a spray manifold, into the bottom of the scrubber in what is called the scrubbing liquid sump. From the sump, the scrubbing liquid is pumped back through a filter to the spray manifold at the top of the scrubber. Figure 9-30 is a schematic of a typical fume hood scrubber. It clearly shows that the fumes and the scrubbing liquid paths cross each other for a contact that is efficient and as complete as possible. Water-soluble fumes, aerosols, and vapors are dissolved in the scrubbing liquid, and particulates are captured in it. Scrubbers are particularly effective with water-soluble acids and bases, capturing as much as 95 percent of these chemicals.

HEPA Filters

HEPA filters are generally used for environmental reasons. They let gases pass freely but retain airborne particles and microorganisms. They do so by use of five distinct mechanisms: sedimentation, electrostatic attraction, interception, inertial impaction, and diffusion. Sedimentation takes place when the particles settle into

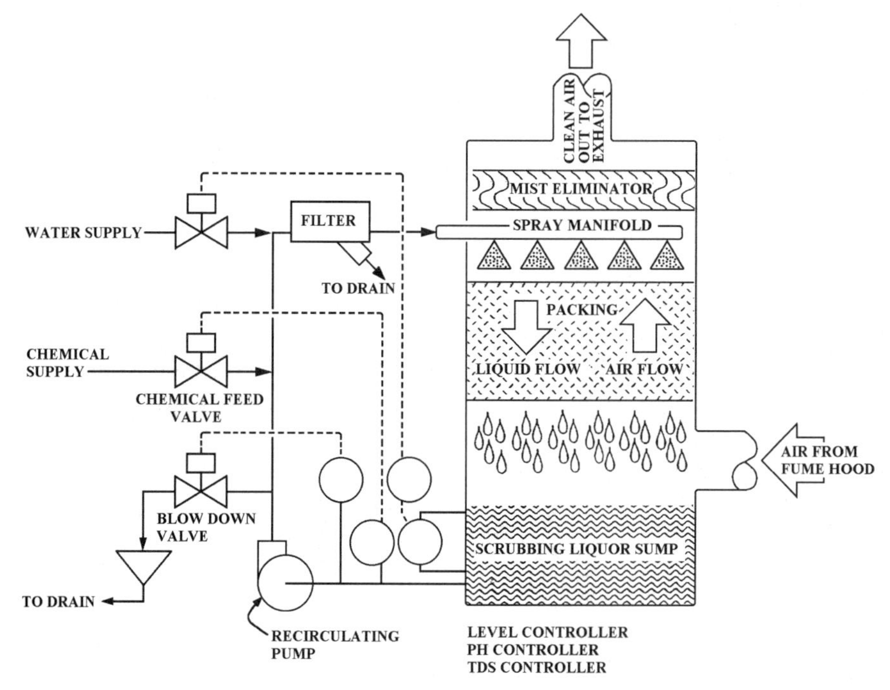

FIGURE 9-30
Schematic drawing showing a typical fume hood scrubber

the filter fiber because of the gravitational force. Electrostatic attraction occurs when particles are attracted to the filter fibers because of their opposite electrical charges. Interception occurs when the airstream particles hit the filter fibers and are retained. Inertial impaction occurs when the large particles leave the airstream to directly impact on the filter fiber. Diffusion occurs only with very small particles.

HEPA filters typically trap 99.97 percent of all particulates 0.3 micrometers or larger in diameter. It is recommended to select a HEPA filter that can have its filter changed without exposing the maintenance person or the environment to the particulates or other agents retained by it. The "bag-in, bag-out" method of replacing the filters is the best way to prevent exposure. This method separates the contaminated filter and housing from the maintenance person with a barrier bag. When using this method the maintenance person must follow special procedures, which are dictated by the HEPA filter manufacturer. When performed adequately, this method protects the environment and the operator removing the filters.

Gas-Phase Filtration

There are two types of gas-phase filtration: the inert adsorbent system and the chemically active adsorbent system. Each of the two systems collect gases and vapors or dissolved substances in condensed form on the surface of their components. The inert filters use activated carbon, activated alumina, or a patented formula called "molecular sieves." These substances come in bulk form, as cartridges or as panels. When in bulk form, they are placed in a deep bed, filled in cartridges or housed in panels similar to those used for HEPA filters. Regardless of the form in which they come, the beads that constitute them are porous and have extremely large surface areas where gases and vapor are trapped or adsorbed as they pass through. The chemically active adsorbents are inert adsorbents impregnated with a strong oxidizer such as potassium permanganate, also known as purple media. This strong oxidizer reacts with and destroys the organic vapors. There are a number of other oxidizers used to adsorb specific compounds, but potassium permanganate is the most popular.

The processes of the inert filters and the chemically active filters are pretty much the same. As the air containing gases and vapors passes through the adsorbent bed, the gases and vapors are removed in a section of the bed. As the bed fills with gases and vapors, the contaminants may break through the end of the bed, except if the adsorbent is continuously regenerated or replaced. Should a breakthrough occur, gases and vapors will pass through the bed steadily, at a higher and higher concentration, until the upstream and downstream levels equate.

Materials for Fume Hood Exhaust Ducts

Exhaust ductwork materials are generally exposed to chemical attacks and to very high temperatures. Therefore, corrosion, dissolution, and melting can occur if the design criteria do not account for them.

The selection of exhaust duct material therefore results from careful analysis during the design phase. Since no material is suitable for all chemicals, the selection process should include the following steps:

1. Determine which fumes will be exhausted through the exhaust ductwork.
2. Obtain and analyze information on the chemicals used, their level of concentration, their temperature, their dew point, and the lowest possible out-of-duct temperature they can stand.
3. Determine the condensation possibilities.
4. Select the material based on its resistance to the corrosive agents to which it will be exposed, to its flame-spread rating, as well as to its cost, weight, and flow resistance.
5. Check with the manufacturer about the duct material's specifications, as they relate to the above-mentioned factors.
6. Consider the ease of construction and installation, as well as the normally required repairs, maintenance, and replacement.

Several codes have specific requirements for laboratory exhaust ductwork. NFPA-45 requires that ducts from laboratory hoods and local exhaust systems be constructed entirely of noncombustible materials, except when the material has a flame-spread index of 25 or less and is sufficiently strong and rigid to provide protection against mechanical damage. ANSI/AIHA requires the material of the exhaust ductwork to be fire- and corrosive-resistant and to be provided with a protective coat, if needed. Building Officials and Code Administrators (BOCA) mechanical codes require that exhaust ductwork be constructed of G90 galvanized sheet steel, with the thickness determined by the duct size. The use of other materials is permitted when the fumes to be exhausted are detrimental to galvanized steel. These materials, however, must be entirely noncombustible or have a flame-spread index of 25 or less. They should also be sufficiently strong and rigid to provide protection against mechanical damage. The Uniform Mechanical Code classifies the materials that can be used for laboratory exhaust ductwork in five categories and imposes limits on the use of galvanized steel and aluminum.

Flexible connections should be avoided even when their use is permitted. If it is absolutely necessary to use them, they should not be placed in horizontal runs or in such a way as to create or contain pockets in which conveyed materials could collect. They should be placed so that they can be completely visually inspected.

The most commonly used materials for fume hood exhaust ducts are galvanized steel, 304 and 316 stainless steel, plastic coated steel, PVC and FRP plastics, and borosilicate glass. Galvanized steel exhaust ducts are the least expensive of these. However, ducts made of this material have little chemical resistance to many of the fumes generated in the fume hoods. Thus, galvanized steel ducts may need to be replaced frequently. However, if these ducts are used with scrubbers, so the duct's walls are continuously cleaned (in the same manner as in the perchloric acid hood exhaust ducts), these ducts can provide a low-cost way to exhaust the hood's fumes.

Exhaust ducts made of 304 and 316 stainless steel are about one and a half times as expensive as the ducts made of galvanized steel. Stainless steel ducts are also attacked by common acids and other chemicals. Such chemicals, depending on their concentration and potency, eat away the stainless steel duct's wall in a relatively short span of time. One solution in a manifolded system is to upgrade the chemical resistance of the material of the ducts directly connected to the hoods, as this is the location where chemical attacks are most likely to occur. Then use another material from the point where the corrosive fume hood exhausts are mixed, combined, and diluted with less corrosive fumes generated by other fume hoods.

Fiberglass reinforced plastics are about three times as expensive as the ducts made of galvanized steel. They are generally used in a salty atmosphere to exhaust fumes from hoods where most operations use acids. This is because they are relatively resistant to most acids. These ducts, however, are not always approved by fire marshals, since fiberglass will sustain structural damage in a fire. These ducts must be coated if they are exposed to the sun, as ultraviolet rays degrade the duct's surface.

PVC exhaust ducts are used inside the building to exhaust fumes from hoods where most operations use acids. This is because they are relatively resistant to most acids. Like the fiberglass ducts, these ducts are not always approved by fire marshals because they will sustain serious structural damage in a fire. If exposed to fire, they can burn and generate thick toxic fumes. This type of duct also should not be exposed to the sun unless it is coated, as ultraviolet rays degrade the duct's surface.

Borosilicate glass exhaust ducts have the highest chemical resistance of any material on the market. Installed, they cost about the same price as stainless steel. They are virtually leakproof and any blockage can easily be spotted. Borosilicate glass exhaust ducts' thermal expansion and fire resistance are among the very best; however, their mechanical strength is low.

BUILDING COMMISSIONING

Commissioning is a process intended to insure that the building systems are functioning as designed, that the functioning of the different systems is documented for the owner, and that the maintenance schedule and requirements are provided so that the building is maintainable after the contractor is gone. Often such requirements are not completed at the end of the project, as the contractor is in a hurry to move to other projects and the owner is anxious to occupy the building.

The commissioning process should be addressed during the programming phase when the program of requirements is prepared. The commissioning process should be discussed with the owner in order to develop design criteria. Questions such as whether the owner prefers a separate contract for commissioning services apart from the contractor should be raised. Some owners may prefer separate, independent testing and balancing services, as well as separate commissioning services. The requirements of the commissioning process must be clearly spelled out so that when the A/E develops the design concept for the project in the form of system descriptions, they include the commissioning process criteria. When the design documents and specifications for the project are developed and reviewed, they must be in compliance with the commissioning criteria. These criteria should normally be in conformance with American Society of Heating, Refrigerating, and Air Conditioning Engineers (ASHRAE) guidelines 1-1989, entitled "Guidelines for Commissioning of HVAC Systems," as well as with other industry guidelines and good practice. Always use the most recent edition of the ASHRAE guidelines.

If commissioning requirements are not clearly identified in the bid package as a separate line item in the schedule of values, contractors may underprice it, and the true cost of commissioning would not be included in the project bid. This results in an incomplete process. The commissioning of the mechanical work on large projects can be as high as 1 percent to 2 percent of the system cost. Requiring the identification of the commissioning costs in the bid package forces the contractors to address this cost and prevents any cost omission or insufficiency. There should be one commissioning authority responsible for assuring that all the required commissioning activities take place at the end of the project, before the end of the contract. The responsibility for commissioning can be bestowed on any member of the design team. Depending on the project' complexity, size, critical nature, and the owner's preference, this responsibility can be given to the owner's representative, the contractor, or a member of the design team. Whoever assumes this responsibility becomes the commissioning authority.

The project's specific commissioning procedures should be developed by the commissioning authority, shared with and agreed upon by the A/E, the owner, and the contractor. Contractor submissions should be reviewed and approved first by the A/E team, then by the commissioning authority. The performance testing of all systems, known as "start-up and checkout," is a most important part of the commissioning process. In this phase, the mechanical, controls, and test-and-balances subcontractors should work closely with one another and with the A/E representatives. The commissioning authority should, however, orchestrate this effort and determine the success or failure of the process. The final documentation of the as-installed conditions and the equipment and maintenance information is the last step in providing a fully commissioned building. Should the closeout occur in the spring or fall, when the cooling and heating needs are minimal, it may take several months before system deficiencies are discovered, and the system will have to be corrected during a time when it is most needed.

Chapter 10

GUIDELINES FOR DESIGN OF ELECTRICAL SYSTEMS SERVICING THE LABORATORY WING

THE electrical requirements for a laboratory facility are complex and vital to the safe and efficient conduct of the operations. A designer can choose among several solutions that may respond to these requirements. Experience will guide the designer in selecting the solution best suited for a given job within the framework of the design criteria. Flexibility, reliability, and safety are paramount among the criteria used for the electrical design of laboratory requirements.

FLEXIBILITY

Every wiring or lighting system in a laboratory facility should incorporate sufficient flexibility of design of branch circuitry, feeders, and switchboard to accommodate all probable patterns, arrangements, and locations of electric loads. In addition, the system must be modular and its modularity must follow that of the laboratory rooms and of the mechanical systems servicing these rooms. Further, the system must be designed for expansion, accommodating the management's projection for expansion for the next 10 to 15 years. It may be wise, however, that at a minimum such expansion projection be of about 25 percent of the actual design loads, as laboratories usually experience an expansion of electric load demand.

RELIABILITY

The reliability of the incoming utility line is beyond the control of any laboratory management. However, feeding may be available from more than one substation. When it is important for a laboratory to continuously have power, it becomes necessary to have another source of power for emergency use. Further, certain operations may require that their tasks be uninterrupted, thus requiring uninterrupted service power. The quality of equipment and of the wiring system also affects the reliability of the system—for instance, economy-grade equipment can cause damage. Even if the finest equipment is used, equipment failure is always possible. Therefore, the system must be designed so that equipment failure is readily detected and corrected. When no disruption can be tolerated, duplicate equipment facilities must be provided.

SAFETY

Adherence to code requirements assures an initially safe electrical installation, but this is not enough. The designer needs to be alert to and provide for electric hazards caused by the misuse or abuse of equipment or by equipment failure. In most instances, the level of reliability and safety, as well as flexibility, requires the use of three types of power in laboratory designs:

- Regular service power
- Emergency service power
- Uninterrupted service power

Each of these should be designed to be received, transformed, circuited, and provided according to needs and code requirements. Sufficient and ample regular service power should meet all the present and expanded needs of a laboratory facility. Service is generally supplied to a laboratory by a high-voltage primary feeder to a transformer vault where main switches, transformers, and fuses or circuit breakers are located. Transformer vaults must be sized and located so that large equipment can be removed and installed without problems. With regard to the number of transformers provided, the "rule of three" applies. That is, provide three transformers sized so that two will accommodate the entire load. This way, if one transformer fails or is taken out of service for any reason, it will not be missed. And, if one fails while another is out of service, critical loads can still be served. The equipment itself, as well as the space where such equipment will be located, must be ventilated mechanically. Switchgear should then be located in a separate room so as not to be subject to the high ventilation rate of the transformer vault.

Secondary service conductors supply power from the transformer vault to the main switchboards servicing different sections of the laboratory wing(s). The main switchboard should contain the main switches and circuit breakers servicing the laboratory section. From this switchboard, main feeders can provide power to several distribution panels.

It is good design practice to have each laboratory room serviced by a distribution panel. This panel should be located in the corridor near the room's entrance door. Each distribution panel should contain an emergency cut-off switch and circuit breakers for all the circuits (except for lighting) servicing the laboratory. Circuits servicing laboratory space should not have more than two 15-amp convenience outlets per circuit. These outlets should not be adjacent to each other and should have between them an outlet fed by a different circuit.

All outlets must be properly grounded according to code. In addition, ground fault protection should be provided when instruments or equipment are used in wet areas or near sources of water. The provision of such protection should be carefully designed, as some instruments and equipment can continuously trigger the interrupter of the ground fault protection device, thus rendering them inoperative. When the task or operation undertaken in a laboratory involves the use of equipment requiring grounding, a ground bus should be provided. How this bus is connected to ground is often determined by the specific type of operation involved. In some cases, no earth connection is necessary but in those cases it will be necessary to have all equipment at the same potential. If this is the case, provision for future ground connection should also be included.

Other times it may be necessary to have a completely separate grounding system for the ground bus. In these cases, the bus should be connected to a separate ground rod. Although an isolated ground conductor is preferable to the final grounding point, separation of the system ground and of the equipment ground from the laboratory ground connection may not always be desirable because of the possibility of circulating currents between grounds. In some instances a separate insulated ground wire to the transformer neutral grounding system may be sufficient. Emergency service power should be available to a laboratory in which tasks and experimentation require constant, mechanically controlled conditions, such as temperature and humidity control, ventilation, or exhaust. Emergency power is generally provided by a fuel engine–driven generator that automatically starts in the event of a complete power failure. The system should be designed so that the electrical load required for emergency power is automatically transferred to the emergency power generator. Further, the system should be sized to carry the entire calculated loads plus an additional 25 percent of those loads for future expansion.

The system also should be designed so that in the event of lighting circuit failure, the emergency lights in the areas served by the failing circuits are automatically activated and powered by the emergency generator. The emergency power system should be sized so that it can automatically provide for all waste disposal and neutralizing systems, all balance room functions, all perchloric acid hoods, and the exhaust fans of certain other hoods. It should also energize all incubators, refrigerators, freezers, certain drying ovens, instrumentation, and other items, as well as the fire detection and alarm system and the automatic sprinkler system.

Uninterrupted service power is sometimes required when short-term (less than 15-second) power shortages would affect the equipment and result in loss of data or irreparable damage. Uninterrupted service is achieved by use of batteries. The batteries provide power to the equipment and are continuously energized by regular service power. They may also be connected to the source of emergency service power, which supplies them in case of failure of the normal power system. Many new, computerized laboratory instruments require the use of uninterrupted power, as a shortage would result not only in loss of data but also would require recalibration of the equipment, a process often long and tedious.

Chapter 11

GUIDELINES FOR THE DESIGN OF SPECIAL LABORATORY ROOMS: PARTIAL- AND FULL-CONTAINMENT LABORATORIES, BIOSAFETY LABORATORIES, AND CLEAN ROOMS

THIS chapter explains three types of special laboratory rooms, as these rooms must respond to very definite and rigid requirements: the partial- and full-containment laboratories, biosafety laboratories, and the clean rooms. The guidelines provided for each of these types of rooms will educate professionals about the uses of these rooms and about how intricate their requirements are. Further, professionals will be able to familiarize themselves with the design and operation requirements for each of these types of rooms.

The guidelines provided are general in nature and will probably need to be supplemented, as each type of room may have additional requirements dictated, for example, by the experimentation for which the room is being designed. Additional safety requirements related to these operations may also be needed, and scientists' wishes may have to be considered. Further, the management may dictate certain quality control and other requirements, in accordance with the policy of the entity conducting operations.

In many situations, it is necessary to combine the requirements of two types of special laboratory rooms. The design criteria for all of these possible combinations is highly complex and requires the involvement of many professionals with different areas of expertise to safely and efficiently resolve the design challenges presented.

Because of the complex and conflicting requirements, no professional should attempt to design any of these types of laboratory rooms without being very experienced with such design and very knowledgeable about the codes and standards governing them. Further, even very experienced professionals must seek the advice of industrial hygienists, certified safety officers, fire engineers, health physicists (if radiation work is involved), and other pertinent professions.

CONTAINMENT LABORATORIES

Containment laboratories are laboratories where highly hazardous, concentrated, potent chemicals and other materials, of known or unknown nature, are handled. Containment laboratories therefore require special consideration in design and construction, as well as in the way in which they are used. These special considerations are necessary in order to safeguard the health and safety of the handler of such materials and to protect the immediate environment.

The definition of what constitutes a highly hazardous and potent chemical should be provided by the user's chemists, toxicologists, industrial hygienists, or other qualified experts. This category of chemicals generally consists of analytes: known samples or compounds, mixtures of chemicals for which little is known about their toxicity, unknown samples of constituents, or other chemicals needing to be handled for industrial or other reasons. These should be analyzed or worked on in such a way that any fumes, effluents, or solid particles emitted always remain fully contained and are disposed of under restrictive conditions. The disposal of these fumes, effluents, and particles should take place in such a way as to prevent contamination of the operator and the immediate environment, inside and outside the building.

The design of containment laboratories should be governed by common sense and prudent practice when no required codes or standards exist. The guidelines provided in this book are the result of my experience with these laboratories and should by no means replace codes or standard requirements.

Containment laboratories fall into one of the two following categories: full-containment laboratory suites and partial-containment laboratory suites. A full-containment laboratory suite consists of a set of laboratory rooms in which potent chemicals and other very hazardous materials of known or unknown nature are constantly and routinely handled. The handling of these hazardous chemicals and materials may include analyzing them, mixing them into compounds, separating them from each other, or dealing with them in any other manner. A full-containment laboratory may consist of one containment chamber and one or more adjacent laboratories, or laboratory support rooms. It may also be an entire facility with many containment chambers and many laboratory support rooms. This facility may be part of a laboratory wing or of a separate structure altogether.

The containment chamber is that part of the containment laboratory where the most hazardous work is performed. It has very restrictive requirements, a description of which will follow. This chamber has, as a main feature, at least one glove box and an adjacent fume hood connected to it by an interlocked pass-through chamber. A particularity of a full-containment laboratory suite is that any person, piece of equipment, fume, effluent, particle, or anything that leaves the suite must first be decontaminated or totally encapsulated and disposed of in accordance with codes and regulations.

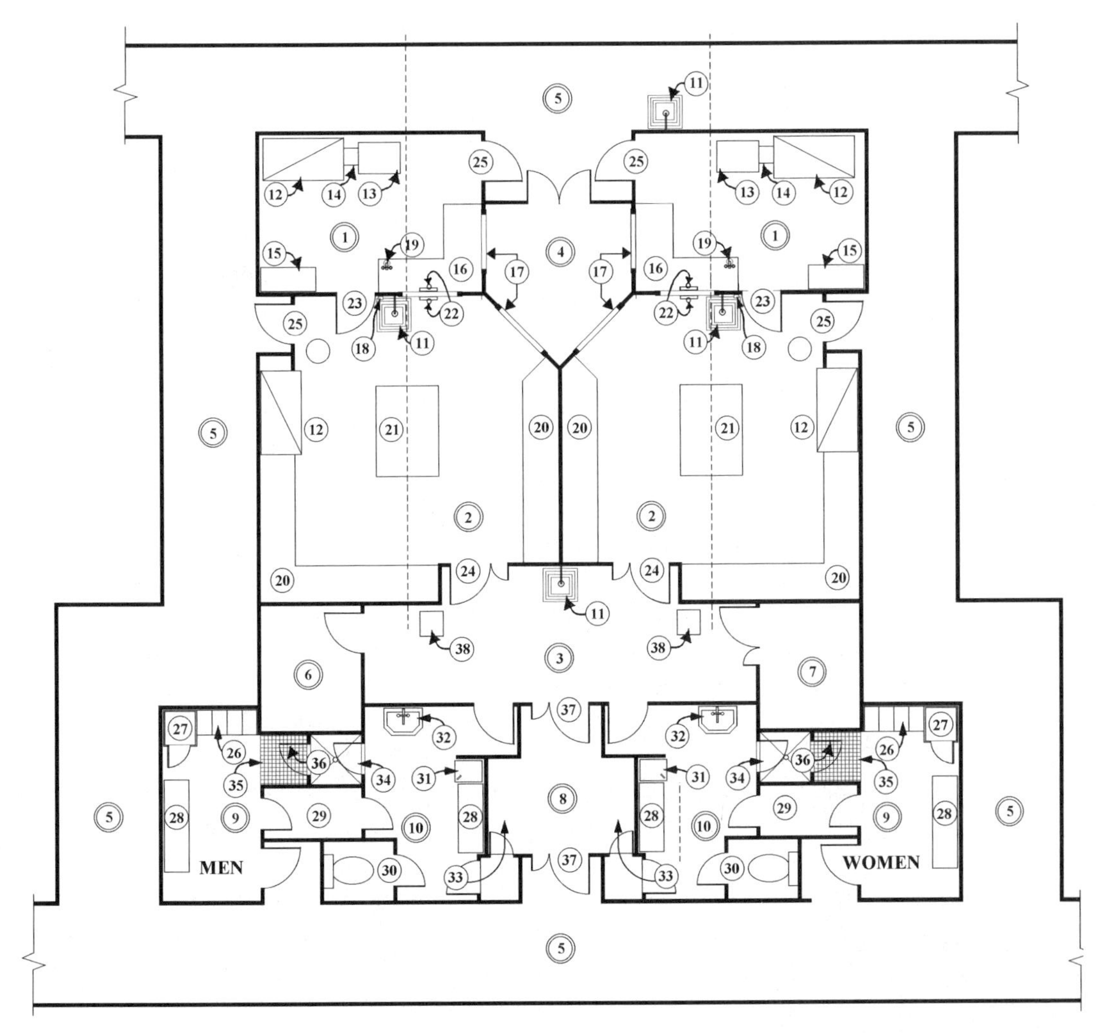

FIGURE 11-1
Schematic plan for a full containment laboratory suite

1. Containment chamber 2. Containment laboratory support room 3. Containment suite corridor 4. Observation room 5. Laboratory corridor 6. Cleaning equipment storage room 7. Laboratory equipment storage room 8. Decontamination air lock vestibule/soiled garment alcove 9. Street clothing change room 10. Laboratory clothing change room 11. Emergency shower/eye and body wash 12. Fume hood (must have 100 fpm face velocity) 13. Glove box 14. Air lock chamber between fume hood and glove box 15. Chemical storage cabinet vented to the outside 16. Counter top with no cabinets under it 17. Sealed view window 18. Switches for lights, fume hood, and glove box in the containment chamber 19. Eye wash 20. Counter top with cabinet under it 21. Center isle with cabinets under it. May or may not be needed 22. Intercom system 23. Gasketed door interconnected with laboratory support room 24. Door interconnected with containment chamber doors 25. Alarmed emergency exit door 26. Clothing locker 27. Disposable garment closet 28. Bench 29. Air lock vestibule leading to the laboratory clothing change room 30. Toilet 31. Water cooler 32. Lavatory 33. Pass through to soiled garment special bag 34. Shower stall 35. Decontaminating floor pad 36. Door interconnected with shower stall and air lock vestibule doors to the laboratory change room 37. These two sets of doors must be interconnected 38. Alternate receiver of contaminated protective wear

Individuals entering and/or leaving a full-containment laboratory suite are required to take special precautions in a set of antechambers specially designed for that purpose. They first enter in the street clothing change room, where they remove their street clothes and store them in lockers located either in this room or in a part of the room designated as the clothing locker/clean garment alcove. They then put on one of the special disposable garments provided before entering the full-containment laboratory suite through an air-locked vestibule.

Individuals leaving the full-containment laboratory suite take off the disposable garments in the clothing change room and put the garments in a special bag, through a pass-through. This bag is generally placed in a soiled garment alcove located in the decontamination air-locked vestibule. This vestibule must be next to the clothing change room, but within the area of the containment laboratory suite. Then they must shower at one of the shower stalls before entering the street clothing change room, where they initially left their street clothes. They then dress and leave the containment-related antechambers. Any instrument or piece of equipment may also have to be decontaminated in the decontamination air-locked vestibule prior to being removed from the full-containment laboratory suite.

Figure 11-1 shows a schematic plan of a full-containment laboratory suite and of the antechambers servicing this suite. A description of each of these antechambers and their requirements is given later in this chapter.

In a partial-containment laboratory suite, small quantities of potent chemicals or other hazardous materials of known or unknown nature are only occasionally handled. It may be useful to have a partial-containment laboratory suite next to a sample receiving room where some samples are of unknown nature and need to be opened and identified under containment prior to being processed. A partial-containment laboratory suite consists of a containment chamber with one or more glove boxes with adjacent and connected fume hoods and of an adjacent laboratory support room. A full description of a partial-containment laboratory suite is provided later in this chapter. Figure 11-2 shows a schematic plan of a partial-containment laboratory suite.

The handling of analytes (this word refers to samples, compounds, or other matter that is handled) is very much the same in full-containment and partial-containment laboratory suites. The analyte should be taken to the containment chamber in a double-sealed container. This container must be opened in a glove box, where the analyte undergoes a preliminary evaluation if it is of unknown nature. There are two possibilities for handling this analyte. One is to do whatever is needed in the glove box. The other is to dilute part of the analyte to a less potent level in the glove box and then move the diluted analyte through a pass-through to an adjacent fume hood. There, the diluted analyte can be handled by the operator as required, or moved to the adjacent laboratory support room for further handling.

At no time during this process should fumes or particles be released, except from the glove box or fume hood to the exhaust system servicing them. Hazardous effluents should be contained in a double-walled safety container and disposed off as required by codes and regulations. Further, any element that comes into contact with the analyte must be bagged and disposed of in accordance with codes and applicable standards.

Containment Laboratory Suite

LOCATION

All containment laboratory suites (partial or full), must be located within the laboratory section of the building and must adhere to codes regulating their sep-

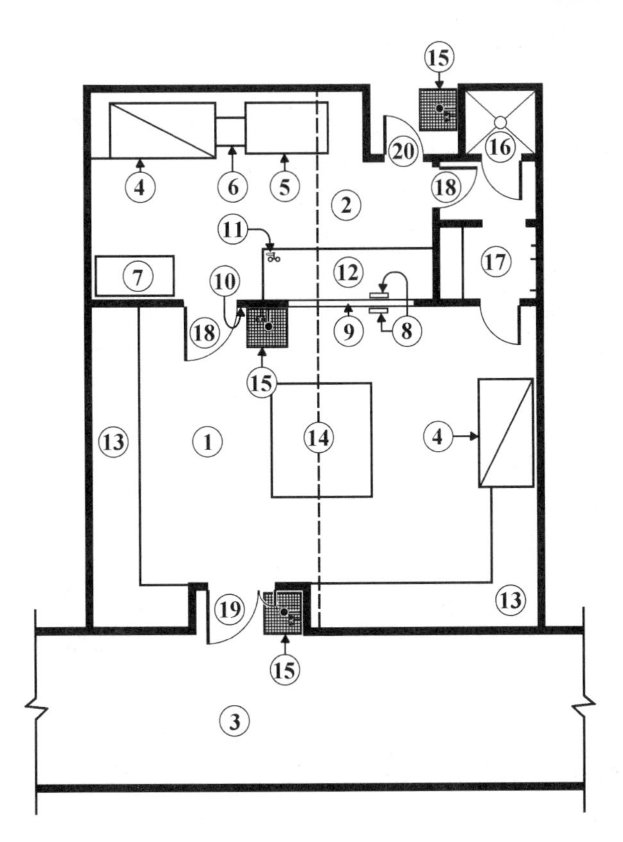

FIGURE 11-2
Schematic plan for a partial-containment laboratory suite

1. Partial containment laboratory support room 2. Containment chamber 3. Laboratory corridor 4. Fume hood (must have 100 fmp face velocity) 5. Glove box 6. Air lock chamber between fume hood and glove box 7. Chemical storage cabinet vented to the outside with exterior fan, motor, and fire damper 8. Intercom system 9. Sealed view window 10. Switches for lights, fume hoods, and glove boxes 11. Eye wash 12. Counter top with no cabinets under it 13. Counter top with cabinets under it 14. Center isle with no cabinets under it 15. Emergency shower and body and eye wash 16. Shower stall 17. Dressing room with bench, shelf, and clothes hook 18. Gasketed door interconnected with the two other chamber doors and with the containment laboratory support room 19. Door interconnected with the containment chamber door 20. Alarmed and gasketed emergency exit door

aration from office, laboratory, and nonlaboratory areas. Some full-containment laboratory suites are completely independent facilities. Others are integrated into a building, as long as the containment laboratory suite and the surrounding parts of the building meet all the adjacency and other requirements set by codes and standards. A partial-containment laboratory suite's containment chamber must be located next to and be part of the partial-containment laboratory support room. This is the room where the analyte, once diluted to a less potent level, is treated as needed before being passed to other locations in the laboratory section for further work. The laboratory support room must be directly connected to the containment chamber by a labeled "gasketed" door with a viewing window. One of the requirements for a partial-containment laboratory suite is that a shower and a clothing change room with lockers must be located as close as possible, but not necessarily adjacent to, the partial-containment laboratory.

FURNISHING

A containment chamber located in a partial- or full-containment laboratory suite must include the following furnishings:

- At least one glove box with an air-locked compartment connected to a fume hood
- At least one fume hood connected to the glove box with a pass-through interlocked box
- A solvent (flammable liquid) chemical storage cabinet with exterior fan and motor, as prescribed in NFPA-45
- An acid/base chemical storage cabinet with exterior acid-resistant fan and motor
- A countertop area on one side of the glove box and on one side of the fume hood and a countertop area across the passage facing the glove box and the fume hood

The containment laboratory support room should be furnished as a regular laboratory. Its furnishings, including fume hoods, will be as required by the operations conducted in the room. A typical arrangement of the containment chamber and of the partial-containment laboratory support room is shown in figure 11-2.

ARCHITECTURAL REQUIREMENTS

Walls

The containment chamber's walls must be constructed and finished with hard nonporous materials free of holes, cracks, and crevices. Wall finishings must be smooth, and the corner joints between walls, between walls and the floor, and between walls and the ceiling must be coved. Walls should be painted with a glossy washable paint. Fixtures must be recessed and flush with the wall surface to avoid accumulation of dirt. Electrical and other types of outlets should be flush with the wall to eliminate the lodging of pollutants in cracks, corners, and crevices. It is prudent to construct and finish the support room's walls like those of the containment chamber.

Floors

The containment chamber's floors must be covered with a tough, durable, and seamless material capable of withstanding the wear imposed by the operations. This material should be nonporous and easy to clean. The junctions between the floor and the walls and between the floor and the fixed cabinets and equipment should be coved.

The containment laboratory support room's floors do not need to be seamless and coved like the floor of the containment chamber. However, the flooring material should be resistant, durable, and capable of withstanding the wear imposed by the operations. This material should be nonporous and easy to clean.

Ceiling

The containment chamber's ceilings should be constructed and finished with hard, nonporous materials. They should be free of holes, cracks, and crevices. The junctions between the ceiling and the walls should be coved. The containment laboratory support room's ceiling should preferably be constructed and finished with hard, nonporous materials. This ceiling should be free of holes, cracks, and crevices. However, the junctions between the ceiling and the walls do not need to be coved.

Casework

Casework and other types of furniture for containment chambers and containment laboratory support rooms should be constructed of smooth, nonporous materials that are easy to clean. The junctions between the fixed cabinets and between the fixed cabinets and other equipment must also be smooth and free of cracks. The junctions between the fixed cabinets and the floor in the containment chambers must be coved.

Doors

Containment chambers as well as containment laboratory support rooms must have at least two doors, one for regular access to the room and the other for emergency exit. These doors should be located so as to always allow an escape on either side of the room should an emergency situation occur. Doors to and from a containment chamber must be gasketed, equipped with panic hardware, and airtight to keep the containment chamber under negative pressurization. They also must be equipped with flush view windows. Doors at opposite end of an air lock, if an air-locked vestibule is used, should be interlocked so that only one door can be opened at a time. At least one door accessing both the containment chamber and the containment laboratory support room must be wide enough to allow bulky equipment to be moved in and out of the room.

Windows

Windows (used either for light or as view windows) should be fixed and flush with the inside walls of the containment chamber. Windows in containment laboratory support rooms must also be fixed and preferably flush with the inside walls of the room. Their frames must be constructed to be smooth and without cracks. NFPA requirements may limit the size and construction components of windows installed in fire walls.

MECHANICAL REQUIREMENTS

The containment chamber's air pressure must be negative to that of the containment laboratory support rooms and of adjacent spaces or corridors, so that any leakage that occurs is from the adjacent rooms or areas to the containment chambers rather than from the containment chambers to the adjacent areas. Likewise, the containment laboratory support room's air pressure also must be negative to that of adjacent spaces and corridors (except for the containment chambers). The minimum positive pressure differential between a containment chamber and adjacent rooms or areas should be 0.05 inches (1.27 mm) of water, when all the doors are closed. When the doors are open, the capacity of the blower providing air to the containment chamber's adjacent rooms or areas should be such as to maintain a sustained outflow of air sufficient to prevent contaminates from migrating from the containment chamber.

The temperature of all containment chambers and containment laboratory support rooms should normally be maintained around 72°F (22.2°C). However, this temperature may vary depending on the requirements of the product being handled. In most instances, the temperature should be within a range of 67–77°F (19.4–25°C), plus or minus 5°F (2.8°C) in rooms where a constant temperature is not critical, and plus or minus only 0.5°F (0.28°C) in rooms where a constant temperature is critical.

Relative humidity in containment chambers and containment laboratory support rooms should be controlled. The range depends on the requirements of the product handled. The recommended maximum relative humidity should not be higher than 45 percent, plus or minus 10 percent for general applications, and plus or minus 5 percent for humidity-sensitive applications.

The containment laboratory suite, including all its chambers and rooms, must have its own, separate equipment for the supply and exhaust of air.

A separate air handler should be used to supply air to the containment laboratory suites. Controls should be in place to immediately stop the air supply should the exhaust fan(s) of the fume hoods, or of any other exhaust equipment, cease to function properly. When it is not possible to have a separate air handler to supply air to the containment laboratory suite, these rooms can use the same air handlers that supply air to the other rooms of the laboratory wing. When this is the case, the ducts supplying air to the containment laboratory rooms must, at each register in each of the containment laboratory rooms, be equipped with an airtight isolation damper with chemical-resistant neoprene edges. These damper controls immediately close the register's damper if air ceases to be supplied through the register and if the exhaust fan(s) of the fume hoods, or of any other exhaust equipment, cease to properly function.

With constant air volume systems, the air exhausted from a containment laboratory is, in most instances, exhausted through fume hoods. Containment laboratories should use only constant volume fume hoods. Occasionally in VAV rooms or in rooms with a heat load driving the need for ventilation, a separate exhaust duct is provided in addition to the fume hood.

All air exhausted from containment chambers must be filtered. Air filtration to remove particles, fumes, vapors, and gases is required. HEPA filters should be used to remove particles. It is good practice to use prefilters to prolong the life of the HEPA filters. The efficiency of the prefilters should be tailored to the expected level of contamination and to the desired expansion of the HEPA filter's life. The equipment, HEPA filters, and prefilters must be of the kind that allows for the bag-in, bag-out method, so that the particulates collected by the filters always remain contained until disposed of according to codes and regulations. The design of the filtering equipment should be such as to provide sufficient space between the filters and their supply ducts to allow a constant pressure across the filters' surfaces. Further, adequate space must be provided on the roof and around the filtering equipment to safely remove the bagged filters.

Air filtration is also required to remove certain vapors or gases produced by operations conducted in containment chambers. Such filtration must take place after the air is filtered by the HEPA filter. Depending on the vapor or gases to be removed, liquid scrubbers, adsorption train, carbon filters, or a combination of filters may be used. Figure 11-3 shows a schematic section through an exhaust

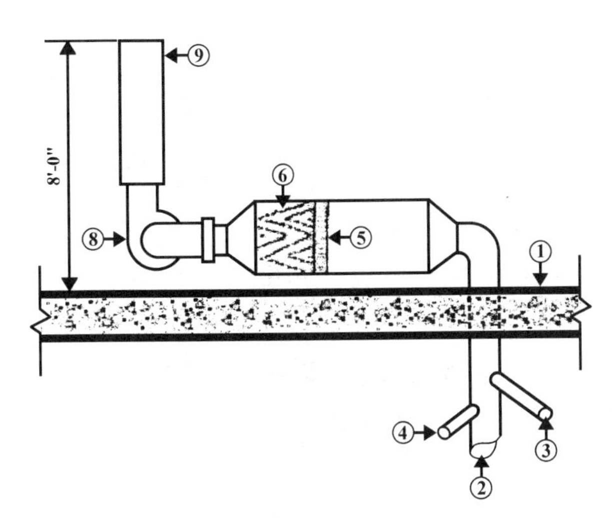

FIGURE 11-3
Schematic section through a containment chamber's exhaust duct system

1. Roof slab shown without finishing
2. Exhaust duct from fume hood
3. Exhaust duct from glove box
4. Flow switch with alarm
5. Solvent-resistant HEPA filter
6. Carbon filter (activated or impregnated)
7. Balancing damper
8. Centrifugal fan with backward curve
9. Zero-pressure weather cap

duct system out of a containment chamber to the filtering equipment. The air exhausted from containment chambers should not be manifolded prior to its filtration, except if the following conditions are met:

- The operations conducted in two or more adjacent containment chambers are always the same or are equally hazardous and use compatible materials
- Codes and standards permit their fumes to mix
- The chemicals used, even of a high potency, do not interact with each other

Generally, the air exhausted from the containment laboratory support rooms does not need to be filtered.

SERVICES TO THE CONTAINMENT CHAMBERS

Services and utilities may be required in containment chambers. The operations that will be performed in the room should determine which service or utility is needed. The necessary service and utility lines should be located outside the containment chamber and brought in through the walls as close as possible to the location where they are needed. The number of services and utilities placed in a containment chamber should be kept to a minimum. This is to reduce the possibility of contamination. No drains or water service should be permitted in the containment chamber.

Two emergency showers with hosed eye washes must be located outside the containment chamber, one next to the door connecting this chamber to the containment laboratory support room, and the other next to the emergency exit of the containment chamber.

All service and utility lines, including electrical lines and light fixtures, may have to be explosion-proof if the nature of the work and/or the type of chemicals handled in the chamber are explosive in nature. Even when this is not the case, it is good practice to have the switches for the light, the fume hood, and the glove box located outside the containment chamber's main door.

An intercom system should be provided between the containment chambers and the adjacent containment laboratory support rooms. The intercom should be located on both sides of the vision panel overseeing the containment chambers.

LIGHTING EQUIPMENT IN THE CONTAINMENT CHAMBER

The containment chamber's lighting equipment must be designed to provide shadowless, glare-free, uniform lighting at the intensity required by the operations that will be performed in the chamber. In most instances an intensity of 100–125 foot-candles is required. This light intensity should be maintained over the benches and at all the work areas. Light fixtures should be flush with and sealed in the ceiling to prevent cracks, crevices, or ledges that accumulate dirt. Emergency lighting must be provided in the containment chambers, as well as in all the rooms of the containment laboratory suite.

Containment Laboratory Antechambers

As was previously mentioned, all fumes, effluents, or solid particles resulting from the handling of the chemicals in the containment chamber, as well as any instrument used, must be decontaminated prior to entering any other space or must remain fully contained and then be disposed of according to codes and regulations. Likewise, all supplies and residues, such as tabs and filters, that come into contact with potent analytes must be bagged and properly disposed of.

As also mentioned earlier, those entering and/or leaving a full-containment laboratory are required to take special precautions in a set of antechambers specially designed for that purpose. They must remove their street clothes, put on one of the special disposable garments, and, when leaving, take off the special disposable garments, shower, and get dressed in their street clothes. In special circumstances, the protective suit may be decontaminated, cleaned, and reused.

Quite a few antechambers are required to perform these tasks. The following is a description of each of these antechambers and of their requirements.

LOCATION OF THE ANTECHAMBERS IN RELATION TO EACH OTHER

A street clothing change room is where the persons authorized to enter the full-containment laboratory rooms remove their street clothing and put on containment laboratory rooms' approved garments. There should be separate street clothing change rooms for men and for women. A clothing locker area is generally located within this room. A clothing locker alcove (also one for men and one for women) may also be used and is located adjacent to the street clothing change room. This alcove may be instead of, or in addition to, the clothing locker in the clothing change room.

Once individuals remove their street clothes and put on the disposable garments, they enter the full-containment laboratory area through the air-locked vestibule leading to the laboratory clothing change room. The laboratory clothing change room is where those who have completed their laboratory work in the contained area of the suite remove their disposable laboratory garments prior to showering.

One or more shower stalls must be located between each street clothing change room and the laboratory clothing change room. The location of the shower stalls should be such that the persons authorized to enter the containment laboratory rooms must cross one of these stalls and shower, as required, before reentering the street clothing change room. A soiled garment alcove, generally located within the decontamination vestibule and next to the shower stalls, should be designed to allow one to drop the soiled garment into a soiled garment container, itself contained, prior to showering. Figure 11-1 shows the location of the street clothing change rooms and their relationship to the connecting rooms.

In a partial-containment laboratory suite, a small street clothing change room or cabin should be placed adjacent to the shower stall. Figure 11-2 shows a possible location for this room and stall.

There are other types of antechambers related to the full-containment laboratory suites. Among them are the decontamination vestibule and the observation room. The decontamination vestibule is where an instrument or a piece of equipment is decontaminated prior to being removed from the containment laboratory suite. The observation room is where one may observe the work performed in a room or rooms that are part of the full-containment laboratory suite. The equipment storage room and the cleaning equipment storage room may also be part of the containment laboratory suite.

STREET CLOTHING CHANGE ROOM AND CLOTHING LOCKER ALCOVE

Furnishing

This room should be furnished with benches made of stainless steel, plastic, or any nonporous, washable material. It should have stainless steel shelves and mirrors on all walls. Lockers also made of stainless steel, plastic, or any nonporous, washable material should be used in the clothing locker alcove.

Architectural requirements

The walls, floor, and ceiling must be finished with a nonporous, washable material. Doors between the shower stalls and the laboratory clothing change room, as well as the door of the vestibule leading to the laboratory clothing change room, must be interlocked with the door between the laboratory clothing change room and the containment laboratory rooms, so that when one set of doors is open the other cannot be unlocked. Windows, if used, must have fixed panels.

Mechanical requirements

Recirculated air can be used in this room and its alcove. The air pressure may be the same as that of the corridor, but it must be positive to that of the shower room to which it is connected.

Electrical requirements

No special electrical requirements for this room.

SHOWER STALLS

Location

The location of these stalls (in relation to the street clothing change room and to the containment laboratory suite rooms) must be such that a person leaving the full-containment laboratory suite must go through one of the shower stalls to have access to the street clothing change room. Naturally, the number of shower stalls must be sufficient to accommodate the number of users.

In a partial-containment laboratory, the shower stall does not have to be situated like this. A single shower stall connected to a dressing cabin, which could be locked from the inside, may be used if the partial-containment laboratory suite is small, only occasionally used, and used by only one individual at a time. The shower stall and dressing cabin may be separated from but close by the partial-containment laboratory; in any case, they should be easily accessible from it.

Furnishing

Each stall should have a bench made of stainless steel or of a nonporous, washable material and be crack- and crevice-free. A slightly depressed floor area of about 3 by 3 feet (0.91 by 0.91 m) with a decontaminating pad should be located outside each shower stall on the side of the street clothing changing room. This area should have a towel hook. After showering, a person must step on the pad on the way to the street clothing changing room. The shower stall and the decontamination pad cabin must have floors and walls finished in nonporous ceramic or in other nonporous, washable materials. The shower-stall doors should be interlocked, built of nonporous, washable materials, and use stainless steel or nickeled hardware.

Architectural Requirements

The walls and floor should be finished with ceramic tiles with smooth joints or with similar nonporous, washable material. Ceilings must be finished with a nonporous material and painted with a glossy washable paint. Doors connecting the shower stalls with the street clothing change room and with the laboratory clothing change room must be interlocked so that when one door is open, the other cannot be opened. Windows, if used, must have fixed panels.

Mechanical Requirements

One-pass air must be used in all the stalls. The stalls' air pressure must be negative to that of the street clothing change room, but must be positive to that of the laboratory clothing change room to which they are connected.

Electrical Requirements

No special electrical requirements for the shower stalls.

LABORATORY CLOTHING CHANGE ROOM

Furnishing

This room should be furnished with benches and lockers made of stainless steel, plastic, or any nonporous, washable material. It should have stainless steel shelves and mirrors.

Architectural Requirements

The walls, floor, and ceiling must be finished with a nonporous, washable material. Doors between the shower stalls and the laboratory clothing change room, as well as the door between the vestibule and the laboratory clothing change room, must be interlocked with the door connecting the stalls and vestibule to the laboratory clothing change room so that when one set of doors is open the other could not be unlocked. Windows, if used, must have fixed panels.

Mechanical Requirements

This room must have one-pass air. The room's air pressure must be positive to that of the containment suite corridor and negative to that of the shower stalls and of the vestibule between it and the street clothing change room. It should be noted that the containment suite corridor's air pressure must be positive to that of the containment suite rooms.

Electrical Requirements

No special electrical requirements for this room.

SOILED GARMENT ALCOVE

Location

This alcove is often located in the decontamination vestibule if the containment suite has such a vestibule. It may also be located within the containment laboratory rooms. In any case, this alcove must have walls adjacent to the laboratory clothing change room and to the shower stalls. One of these walls must contain a pass-through that has a special type of bag in which soiled garments are placed before a person takes a shower. Once filled, this bag is sealed and removed by specialized personnel. This type of pass-through is available on the market.

Furnishing

No furnishing is required, except for a bracket to hold the pass-through system.

Architectural, Mechanical, and Electrical Requirements

These requirements are the same as those of the containment laboratory suite, if the alcove is located there. If the alcove is located in the decontamination vestibule, the architectural requirements are the same as those of the vestibule.

DECONTAMINATION VESTIBULE

Location

This vestibule, when used, should be located between the containment laboratory suite and a corridor. If the decontamination vestibule is also the soiled garment alcove, it should also be adjacent to the laboratory clothing change room so that soiled garments can be disposed of as described earlier.

Furnishing

A modified or walk-in enclosure or hood over an area that could contain liquids may be needed in the decontamination vestibule. Large pieces of equipment, instrumentation, or other items should be decontaminated here prior to being moved from the containment laboratory suite. A HEPA filter vacuum cleaner, with attachments allowing the vacuuming of cracks and crevices, should also be available in this vestibule. If the decontamination vestibule is also the soiled garment alcove, it should have the equipment necessary to receive soiled garments as described earlier.

Architectural Requirements

The walls, floor, and ceiling must be finished with a nonporous, washable material and be free of cracks and crevices. Corners between walls, between walls and the floor, and between walls and the ceiling must be coved. Doors connecting the decontamination vestibule with the corridor and doors connecting the vestibule with the containment laboratory suite must be large—at a minimum, 4′-0″ (1.22 m)—and face each other to facilitate the moving of large items to and from the containment laboratory suite. These doors must be interlocked. In case of an emergency, these doors may be forced open. This activates an emergency light and an alarm. It may also activate a silent alarm connected to a 24-hour manned security station. Windows, if used, must have fixed panels.

Mechanical Requirements

One-pass air must be used in this vestibule. The vestibule's air pressure must be negative to that of the laboratory corridor, but must be positive to that of the containment suite corridor and rooms.

Plumbing Requirements

No water should be supplied to this vestibule. No floor drain or any other type of connection to the waste line is allowed out of this room.

Electrical Requirements

Light fixtures must be recessed. Electrical outlets, as required by the decontaminating equipment, must be provided.

OBSERVATION ROOM

Location

The purpose of this room is to allow a select groups of interested people to observe work performed in the containment chambers or the adjacent containment laboratory support rooms without getting contaminated and without interfering with the work. Thus, this room should be located next to the room or chamber being observed.

Furnishing

No special furnishing is required for this room. An elevated countertop located next to the observation window may be useful for people taking notes. Tall chairs may also be provided for better visibility.

Architectural Requirements

The walls, floors, ceilings, and doors may be same as those of an office building.

Mechanical Requirements

Recirculated air may be used in this room.

Electrical requirements

No special requirements.

BIOSAFETY LABORATORIES

Biosafety laboratories are laboratories where biologically hazardous and infectious agents or parts thereof, of known or unknown nature, are handled in a contained manner. The word biohazard, a contraction of the words biological and hazard, is used to identify such agents. Biohazardous materials are defined as infectious agents or materials produced by living organisms that may cause disease in other living organisms. Biosafety laboratories, in order to be biologically safe (as their name suggests), therefore require special consideration with regard to design, construction, and use. This special consideration is necessary in order to safeguard the health and safety of the handler of biohazardous materials and to protect the immediate environment.

The Centers for Disease Control (CDC) and the National Institutes of Health (NIH) have established four biosafety levels used to categorize the potentially pathogenic organisms. These four levels replace the three previous definitions: low-, moderate-, and high-risk. Each of level has special requirements pertaining to the facility, to safety equipment, and to laboratory practices and techniques. The design of a biosafety laboratory should therefore provide the safeguard and containment required by the level of biological hazard of the agents or materials handled in it.

Biosafety Level 1 (BL-1) Laboratories

In BL-1 laboratories, scientists work with agents or materials produced by living organisms and not known to cause disease in healthy adult humans. Such laboratories require the lowest level of safety for biological hazards. They can be designed as a regular laboratory room and as such respond only to functional needs. They do not need containment or safety equipment (primary barriers). However, they must have an open benchtop sink as a secondary barrier.

ARCHITECTURAL REQUIREMENTS

Walls

BL-1 laboratory walls may be constructed of gypsum boards but must be free of holes, cracks, and crevices. Walls should preferably be painted with a glossy, washable paint.

Floors

The floors may be vinyl tile squares with a rubber base as long as the tiles and base are easy to clean, not porous, and resistant to the disinfectants generally used for decontamination.

Casework

BL-1 laboratory casework and other types of furniture should be easy to clean and have finishes resistant to the disinfectants used for cleaning.

Doors

Large BL-1 laboratory rooms (three modules and more) must have at least two doors, one for regular access to the room and the other for emergency exit. These doors should be located so as to always allow an escape on either side of the room should an emergency situation occur. Doors must have gaskets and be gusseted, equipped with panic hardware, and airtight to keep the laboratory's negative pressurization. They also must be equipped with flush view windows. At least one door must be wide enough to allow bulky equipment to be moved in and out of the room.

Windows

Windows (used for light or as view windows) should have fixed panels.

MECHANICAL REQUIREMENTS

All BL-1 laboratory rooms must have one-pass air. The airflow must be directed from the clean areas to the contaminated areas. The air supply diffusers must be equipped with fans directing the airflow away from fume hoods and biosafety cabinets and located to minimize disruptive air currents.

The BL-1 laboratory room's air pressure must be negative to that of adjacent, connecting spaces and corridors. Most BL-1 laboratory rooms should have independent temperature controls. The temperature in these rooms should normally be maintained around 72°F (22.2°C). However, this temperature may vary depending on the requirements of the operation being conducted. Relative humidity in a BL-1 laboratory should be controlled. The range depends on the requirements of the operation being conducted. The recommended maximum relative humidity is 45 percent, plus or minus 10 percent for general applications and 5 percent for humidity-sensitive applications.

SERVICES AND EQUIPMENT

In order to function safely, BL-1 laboratories are generally provided with chemical fume hoods, vacuum systems, flammable liquid storage cabinets and other chemical storage cabinets as may be needed, biosafety cabinets, a fire alarm system, and eye washes and safety showers.

Chemical Fume Hoods and Biosafety Cabinets

Chemical fume hoods and biosafety cabinets, if used, must be located as far away as possible from the laboratory doors to isolate the hood air from the air movement created by traffic.

Vacuum Systems

BL-1 laboratories' vacuum systems must be carefully designed so that the vacuum air is not recirculated. This air must be vented to the outside of the building. The vacuum systems must be protected with filters such as a 0.2 micron hydrophobic filter or equivalent.

Flammable Liquid Storage Cabinets

BL-1 laboratories should be provided with flammable liquid storage cabinets when flammable liquids are used and, if needed, with acid/base storage cabinets. Sufficient and strategically located space should be provided in the laboratory room for the cabinets needed.

Safety Showers, Eye Washes and Fire Alarm System

Eye washes must be located inside the BL-1 laboratory room as close as possible to the location of risk. Safety showers with eye washes may be located inside or outside the door. This room must be protected by a fire alarm system.

Plumbing

The BL-1 laboratory room generally uses cold and hot water, deionized water, and several gases. The output location of these plumbing needs should be at the point of use indicated by the laboratory user.

Biosafety Level 2 (BL-2) Laboratories

In BL-2 laboratories, scientists work with certain infectious agents or materials that are produced by living organisms and are associated with human disease. These infectious agents or materials are acquired through autoinoculation, ingestion, and/or by mucous membrane exposure. The design of BL-2 laboratories is similar to that of regular laboratory rooms. They must respond to functional needs but should also conform to the BL-2 containment requirements. They must have, as a primary barrier, class I or II biosafety cabinets or other physical containment devices used for operations that can cause splashes or aerosols of infectious materials. The personnel entering a BL-2 laboratory may have to use personnel protective devices such as laboratory coats, gloves, and shields to protect their faces. As a secondary barrier, they must have an open benchtop sink as well as access to an autoclave.

ARCHITECTURAL REQUIREMENTS

Walls

BL-2 laboratory walls must be constructed and finished with hard, nonporous materials free of holes, cracks, and crevices. Wall finishes must be smooth and corner joints between walls and floors must be coved. Corner joints between walls and between walls and the ceiling should preferably be coved. Walls must be painted with a glossy, washable paint. Fixtures on the wall must be recessed and flush with the wall surface to avoid lodging infectious materials. Similarly, electrical and other types of outlets should preferably be flush with the wall to eliminate the lodging of infectious materials in cracks, corners, and crevices. Openings may require gusseted covers.

Floors

BL-2 laboratory floors must be covered with a durable, seamless material that is capable of withstanding the wear imposed by the operations that will take place. The floor material should be easy to clean, nonporous, and resistant to the disinfectants generally used for decontamination of biohazardous spills. These disinfectants may include halogen-containing compounds (hypochlorite and iodine), phenols, alcohols, aldehydes, and quaternary ammonium compounds. The junctions between the floor and the walls and between the floor and the fixed cabinets and equipment must be coved.

Ceiling

BL-2 laboratory ceilings should be constructed and finished with hard, nonporous materials. They should be free of holes, cracks, and crevices. A dropped ceiling with cleanable tiles such as those with a smooth Mylar face are acceptable. Although not recommended, open ceilings with minimal ductwork and piping may be acceptable in certain situations if approved by safety personnel familiar with the operations that will be conducted. The junctions between the ceiling and the walls should preferably be coved.

Casework

BL-2 laboratory casework and other types of furniture needed should be constructed of smooth, nonporous materials that are easy to clean and have finishes resistant to the caustic chemical activity of disinfectants and other cleaning materials. These are the same disinfectants and cleaning materials used for the floors. The junctions at the countertops, between the fixed cabinets, and between the fixed cabinets and other equipment also must be smooth and free of cracks. These junctions' seams must be caulked smooth to ensure smooth finishes for easy cleaning. The junctions between the fixed cabinets and the floor must be coved.

Doors

BL-2 laboratory rooms must have at least two doors, one for regular access to the room and the other for emergency exit. These doors should be located so as to allow an escape on either side of the room should an emergency situation occur. Doors must be gusseted and tight, equipped with panic hardware, and airtight to keep the laboratory's negative pressurization. They also must be equipped with flush view windows. Doors at opposite ends of an air lock, if an air-locked vestibule is used, should be interlocked so that only one door can be opened at a time. At least one door must be wide enough to allow bulky equipment to be moved in and out of the room.

Windows

Windows (used for light or as view windows) should be fixed and flush with the inside walls of the laboratory.

MECHANICAL REQUIREMENTS

All BL-2 laboratory rooms must have one-pass air. The air must be supplied to the room at a maximum velocity of 10 inches per second (50 fpm) at a height of 6 feet above the floor level. The airflow must be directed from the clean areas to the contaminated areas and from the low-hazard areas to the high-hazard areas. The air supply diffusers must be equipped with fans directing the airflow away from fume hoods and biosafety cabinets and located to minimize disruptive air currents.

The BL-2 laboratory room's air pressure must be negative to that of adjacent, connecting spaces and corridors so that any leakage that should occur would be from the adjacent rooms or areas to the BL-2 laboratory room. The minimum positive pressure differential between a BL-2 laboratory room and adjacent rooms or areas should be 0.05 inches of water when all the doors are closed. When the doors are open, the capacity of the blower providing air to the adjacent rooms or areas should be enough to maintain a sustained outflow of air that could prevent contamination from the BL-2 laboratory room.

BL-2 laboratory rooms should have independent temperature controls. The temperature of these rooms should normally be maintained around 72°F (22.2°C). However, this temperature may vary depending on the requirements of the operation being conducted. In most instances, the temperature should be 67–77°F (19.4–25°C), plus or minus 5°F (2.8°C) in rooms where a constant temperature is not critical and plus or minus only 0.5°F (0.28°C) in rooms where a constant temperature is critical. Relative humidity in a BL-2 laboratory should be controlled. The range depends on the requirements of the operation. The recommended maximum relative humidity is 45 percent, plus or minus 10 percent for general applications and 5 percent for humidity-sensitive applications.

SERVICES AND EQUIPMENT

In order to function safely, most BL-2 laboratories must be provided with autoclaves, vacuum systems, medical pathological waste containers, flammable liquid storage cabinets and other chemical storage cabinets as may be needed, biosafety cabinets, a fire alarm system, and eye washes and safety showers.

Autoclaves

Autoclaves are used in BL-2 laboratories where microbiological work is performed. Separate autoclaves are used for clean and dirty procedures. Clean autoclaves are used for sterilization of microbiological media and dirty autoclaves are used for decontamination purposes. The number and size of the autoclaves must be decided by the laboratory users. The size of a BL-2 laboratory's autoclave must be sufficient to accommodate carts containing equipment and other materials needing sterilization. The interior of the autoclave must be epoxy-coated, free of cracks and crevices, and caulked and sealed to prevent retention of biological wastes.

Sufficient space must be provided for locating the required number of autoclaves in BL-2 laboratories. One or more canopy hoods with the required exhaust capacity must be located just over the doors of the autoclaves to remove steam and odors. The air pressure of the space where the autoclaves are located must be negative to that of the surrounding spaces so that air will always be directed toward the autoclaves.

Vacuum Systems

BL-2 laboratories must be provided with a special type of vacuum system that is carefully designed by safety-conscious engineers. The system's air must not be recirculated and must be vented to the outside of the building. The system also must be protected with filters such as a 0.2 micron hydrophobic filter or equivalent. In order to minimize the possibility of contaminating the vacuum pumps, the filters must be located as close as possible to the laboratory and must be equipped with a mechanism capable of decontaminating them.

Medical Pathological Waste Containers

BL-2 laboratories must be provided with medical pathological waste containers and other types of waste receptacles as may be needed by the operations conducted in the laboratory rooms. Sufficient and strategically located space should be provided in the laboratory room for these containers and receptacles. The handling and storage of these containers and receptacles requires special design. Collection stations should be located at the end of each laboratory wing and on each floor. Further, a special space should be provided in the general area of the building's loading dock for a cold box large enough to store all the containers and receptacles used in the building.

Flammable Liquid Storage Cabinets

BL-2 laboratories must be provided with flammable liquid storage cabinets and, if needed, with acid/base storage cabinets or with special cabinets to store other chemicals. Sufficient and strategically located space should be provided in the laboratory room for all the cabinets needed.

Biosafety Cabinets

Class I or II biosafety cabinets or other physical containment devices must be used in operations that can cause splashes or aerosols of infectious materials.

Safety Showers, Eye Washes, and Fire Alarm System

Eye washes must be located inside the BL-2 laboratory room as close as possible to the location of high risk. Safety showers with eye washes must be located outside the door of the BL-2 laboratory room. Additionally, this room must be protected by a fire alarm system.

Biosafety Level 3 (BL-3) Laboratories

In BL-3 laboratories, scientists work with indigenous or exotic agents that may, if inhaled, cause serious or lethal disease. BL-3 laboratories should not be designed like regular laboratory rooms, nor should BL-2 laboratories be converted to BL-3 laboratories, except in very rare cases. Because of the agents handled, BL-3 laboratories have specific and restrictive requirements for containment, limited access, air locks, HVAC filter decontamination processes, and space for autoclaves, to name a few. The sum of these requirements makes it necessary to design and build such laboratories using the appropriate criteria for a BL-3 laboratory. Abiding by all these requirements would make it difficult if not impossible to convert a BL-2 laboratory into a BL-3 laboratory. BL-3 laboratories are required to have as primary barrier class II or III biosafety cabinets or other physical containment devices for all manipulations of infectious materials.

The personnel entering a BL-3 laboratory must have specific training in handling pathogenic and potentially lethal agents, must be supervised by experienced scientists, and must use appropriate protective devices such as laboratory coats, gloves, and shields to protect their faces.

LOCATION

A BL-3 laboratory must be separate from areas open to unrestricted traffic flow. Entrance to a BL-3 laboratory must be through a vestibule with two sets of interlocked, self-closing doors. In many instances this is sufficient separation between a BL-3 laboratory and the rest of the facility. There are some situations in which a set of antechambers should separate the BL-3 laboratory from the other areas of the facility. When this is the case, individuals entering the BL-3 laboratory should first go through a personnel clothing change room with a clothing locker area (or alcove). There, they either remove their street clothing and put on biolab garments or put on a positive pressure suit over their street clothing. These are one-piece positive pressure suits that are ventilated by a life support system that includes alarms and emergency backup breathing air tanks. Once individuals have changed into the biolab garments or suits, they can enter the biolab garments room through an air-locked vestibule. After completing their work in the BL-3 laboratory, they must remove the biolab garments, dispose of them in a special autoclave, and shower in an air-locked shower stall. If they used a positive pressure suit, they must take a decontamination shower with the suit on in an air-locked chemical shower stall and, once in the clothing change room, remove the suit and place it in a special container. The two sets of shower stalls and the air-locked vestibule separate the clothing change room and the biolab garments room. These antechambers, as well as others that may be needed, are similar to the BL-4 laboratory antechambers. These antechambers' requirements are described in detail later in this chapter (see figure 11-4).

ARCHITECTURAL REQUIREMENTS

Walls

BL-3 laboratory walls must be constructed and finished with hard, nonporous materials free of holes, cracks, and crevices. Wall finishes must be smooth and corner joints between walls, between walls and floors, and between walls and the ceiling preferably should be coved. Wall finishes must be sealed and water-resistant so that they can be easily cleaned and decontaminated. They may be painted with a washable epoxy paint. All penetrations in walls must be sealed with a smooth finish to facilitate decontamination and cleaning.

Fixtures on the wall must be recessed and flush with the wall surface to avoid lodging infectious materials. Electrical and other types of outlets should be flush mounted with the wall to eliminate the lodging of infectious materials in cracks, corners, and crevices. Openings may require gusseted covers.

Access panels to critical mechanical equipment should be located outside the BL-3 laboratory room. When this is not possible, the access panel must be hinged with a piano-type hinge and gusseted with gastight gaskets to provide appropriate seal. This also assures that contaminants are contained within the BL-3 laboratory and removed when walls are decontaminated.

Floors

BL-3 laboratory floors shall be covered with a durable, seamless material capable of withstanding the wear imposed by the operations that will take place. The floor material should be easy to clean, nonporous, and resistant to the disinfectants generally used for decontamination of biohazardous spills. These disinfectants may include halogen-containing compounds (hypochlorite and iodine), phenols, alcohols, aldehydes, and quaternary ammonium compounds. Junctions between the floor and the walls and between the floor and the fixed cabinets and equipment should be coved.

Ceiling

BL-3 laboratory ceilings should be constructed and finished with hard, nonporous materials and must have a smooth, sealed finish. These ceilings must be free of holes, cracks, and crevices. Junctions between the ceiling and the walls should preferably be coved. Access to mechanical equipment must be provided outside the BL-3 laboratory area. All penetrations in the ceiling must be sealed with a smooth finish to facilitate decontamination and cleaning.

Casework

BL-3 laboratories' casework and other types of furniture should be constructed of smooth, nonporous materials that are easy to clean. Countertop finishes must be resistant to acids, alkalis, and organic solvents. These countertops must be able to sustain moderate heat and be resistant to the materials used for cleaning and disinfection. Junctions at the countertops between the fixed cabinets as well as between the fixed cabinets and other equipment must also be smooth and free of cracks. These junctions' seams must be caulked smooth to ensure easy cleaning. Junctions between the fixed cabinets and the floor must be coved.

Doors

BL-3 laboratory rooms must have at least two doors, one for regular access to the room and the other for emergency exit. These doors should be located so as to always allow an escape on either side of the room should an emergency situation occur. Doors must be gusseted and tight, equipped with panic hardware, and airtight to maintain the BL-3 laboratory's negative pressurization. They also must be equipped with flush view windows. Doors at opposite ends of an air locked space should be interlocked so that only one door can be opened at a time. At least one door must be wide enough to allow bulky equipment to be moved in and out of the room.

Windows

Windows (used either for light or as view windows) should be fixed, sealed, and flush with the inside walls of the BL-3 laboratory.

MECHANICAL REQUIREMENTS

All BL-3 laboratory rooms must have one-pass air. The air must be supplied to the room at a maximum velocity of 10 inches (0.25 m) per second (50 fpm) at a height of 6 feet (1.83 m) above the floor level. The airflow must be directed from the clean areas to the contaminated areas and from the low-hazard areas to the high-hazard areas. The air supply diffusers must be equipped with fans directing the airflow away from fume hoods and biosafety cabinets and located to minimize disruptive air currents.

The minimum positive pressure differential between a BL-3 laboratory room and adjacent rooms or areas should be 0.05 inches (1.27 mm) of water when all the doors are closed. When the doors are open, the capacity of the blower providing air to the adjacent rooms or areas should be enough to maintain a sustained outflow of air sufficient to prevent contamination from the BL-3 laboratory room.

BL-3 laboratory rooms should have independent temperature controls. The temperature of these rooms should normally be maintained around 72°F (22.2°C). However, this temperature may vary depending on the requirements of the operation being conducted. In most instances, this temperature should be 67–77°F (19.4–25°C) , plus or minus 5°F (2.8°C) in rooms where a constant temperature is not critical and plus or minus only 0.5°F (0.28°C) in rooms where constant temperature is critical.

Relative humidity in a BL-3 laboratory should be controlled. The range depends on the requirements of the operation conducted there. The recommended maximum relative humidity is 45 percent, plus or minus 10 percent for general applications and 5 percent for humidity-sensitive applications.

The BL-3 laboratory room's air pressure must be negative to that of adjacent, connecting spaces and corridors so that any leakage that might occur would be from the adjacent rooms or areas to the BL-3 laboratory room. Exhaust ducts must be under negative pressure until the air is discharged outside the building. Supply and exhaust ducts for BL-3 laboratory rooms must be equipped with gastight dampers to allow the decontamination of the BL-3 laboratory room without contaminating other parts of the building.

Exhaust stacks and building air intakes must be located in relation to each other and to the predominant winds so that the exhausted air is dispersed and discharged away from any occupied areas and without being reentrained into the building. When the biosafety cabinets' exhaust system is connected to the building's exhaust system, the connection should be such that the air balance of the cabinets' and building's exhaust systems is maintained.

Normally, HEPA filtration of BL-3 laboratory rooms is not an absolute requirement. However, an evaluation of the need for specific filtration should be performed during the planning and programming process. A panel composed of the facility safety officer or a contracted industrial hygienist and of other safety professionals must conduct an assessment of the level of hazard of the materials and of the procedures that will be used and determine on a case-by-case basis the need for HEPA filtration. If such need is determined, the same panel should be involved in and give final approval to the design of the decontamination mechanism. The filtration system should be isolated from the ventilation system for gas decontamination and testing and the HEPA filter exhaust housing should be constructed to allow easy particulate testing.

The HEPA filtered exhaust air from class II and class III biosafety cabinets must be directly exhausted to the outside. This may be through the building's exhaust system. However, when this is the case, the connection between the safety cabinets' exhaust system and the building's exhaust system should be such as to avoid any interference with the air balance of the cabinets or of the building's exhaust systems. The HEPA filtered exhaust air from class II biosafety cabinets may be recirculated within a BL-3 laboratory room if the cabinets are tested and certified at least once a year.

All aerosol-producing equipment used in a BL-3 laboratory room, including continuous flow centrifuges, must have their exhaust air HEPA filtered prior to discharge. Although this air may be discharged directly in the laboratory room, it is preferable to have it discharged in a contained manner, either through the biosafety cabinets' exhaust system or through the building's exhaust system. BL-3 laboratory rooms must be equipped with an alarm to notify of any failure in directional airflow or failure of air containment. This device should have a visual and/or audible local alarm and must be connected to the building's central alarm system.

SERVICES AND EQUIPMENT

In order to function safely, BL-3 laboratories must be provided with adequate plumbing, autoclaves, vacuum systems, medical pathological waste containers, flammable liquid storage cabinets and other chemical storage cabinets as may be needed, biosafety cabinets, a fire alarm system, and eye washes and safety showers.

Plumbing

BL-3 laboratories must have a sink for hand washing. This sink must be located inside the laboratory, near the exit door. The sink faucet must be foot-, elbow-, or automatically operated.

Autoclaves

Separate autoclaves must be provided for clean and dirty procedures. Clean autoclaves are used for sterilization of microbiological media and dirty autoclaves are used for decontamination purposes. The number and size of the autoclaves must be decided by the laboratory users and be sufficient to accommodate carts containing equipment and other materials needing sterilization. The interior of the autoclave must be epoxy-coated, free of cracks and crevices, and caulked and sealed as needed to prevent retention of biological wastes.

The air pressure of the space where the autoclaves are located must be negative to that of the surrounding spaces so that air will always be direct toward the space containing the autoclaves.

Sufficient space must be provided for locating the required number of autoclaves in the BL-3 laboratories. One or more canopy hoods with the required exhaust capacity must be located just over the doors of the autoclaves to remove steam and odors. It should be noted that the exhausts from the autoclaves contain a significant amount of moisture, so if the exhausts must be filtered, moisture-resistant (hydrophobic) filter must be used. If such exhausts were filtered through a regular paper HEPA filter, the paper would be damaged and the integrity of the filter would be lost.

Vacuum systems

Like BL-2 laboratories, BL-3 laboratories should be provided with a special type of vacuum system carefully designed by safety-conscious engineers. The systems' air must not be recirculated, must be protected with liquid disinfectant traps and with HEPA filters or their equivalent, and must be vented to the outside of the building. In order to minimize the possibility of contaminating the vacuum pumps, the filters must be located as close as possible to the laboratory and be equipped with a mechanism capable of decontaminating them. These filters must be routinely maintained and replaced as needed.

Medical Pathological Waste Containers

Like the BL-2 laboratories, the BL-3 laboratories must be provided with medical pathological waste containers and other types of waste receptacles needed by the operations conducted in the laboratory rooms. Sufficient and strategically located space should be provided in the laboratory room for these containers and receptacles. The handling and storage of these containers and receptacles in laboratory building must be the same as described for the BL-2 laboratories.

Flammable Liquid Storage Cabinets

Also like BL-2 laboratories, BL-3 laboratories must be provided with flammable liquid storage cabinets and, if needed, with acid/base storage cabinets or with special cabinets to store other chemicals as may be needed. Sufficient and strategically located space should be provided in the laboratory room for all these cabinets.

Biosafety Cabinets

Class II or III biosafety cabinets or other physical containment devices must be used for most operations that take place in BL-3 laboratories.

Safety Showers, Eye Washes and Fire Alarm System

Eye washes must be located inside the BL-3 laboratory room as close as possible to the location of high risk. Safety showers with eye washes must be located outside the door of the BL-3 laboratory room. Further, this room must be protected by a fire alarm system.

Biosafety Level 4 (BL-4) Laboratories

In BL-4 laboratories, scientists work with dangerous and exotic agents that pose a high individual risk of aerosol-transmitted laboratory infections. The diseases caused by these agents are life threatening. Agents with antigenic relationship to biosafety level 4 must be handled at this level until the data either confirm the need to work at this level or indicate a lower level of risk.

Because of the agents handled, BL-4 laboratories have specific and restrictive requirements for, among other things, containment, limited access (locked doors and a log book to record entries and exits) air locks, hazard warning signs, HVAC filter decontamination processes, and space for autoclaves. It is difficult, if not impossible, to convert other laboratory space into a BL-4 laboratory.

Within the work area of a BL-4 laboratory, all activities must be confined to class III biosafety cabinets or to class II biosafety cabinets located in a specially designed suite area within the BL-4 laboratory. This suite area must provide personnel protection equivalent to that provided by class III biosafety cabinets. It

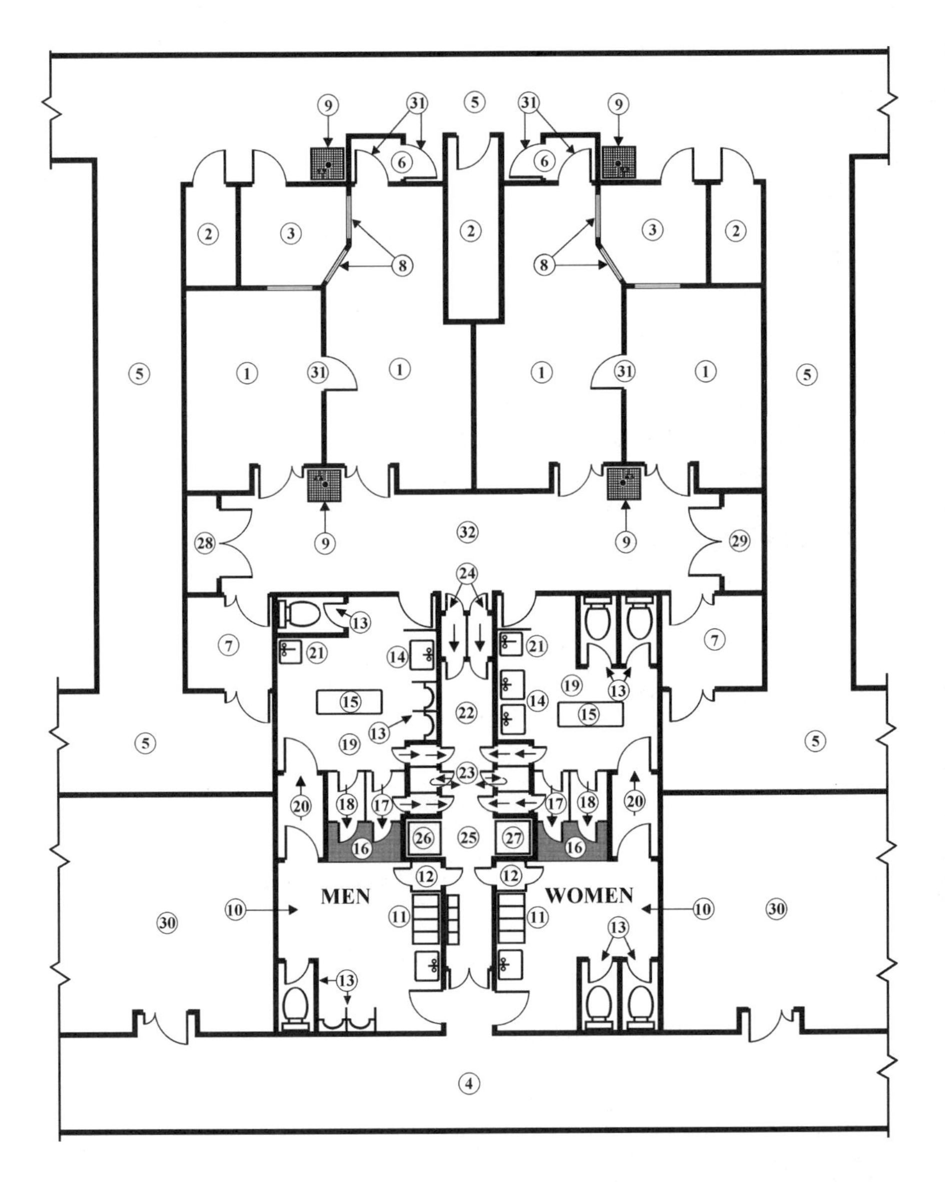

FIGURE 11-4
Schematic plan showing a bio-safety laboratory level 4 suite with related antechambers

1. Bio-safety laboratory level 4 (could also be level 3 2. Utility corridor 3. Visitors' observation room 4. Regular laboratory (access through unrestricted corridor 5. Restricted corridor 6. Air lock emergency exit 7. Decontamination air locked vestibule 8. Observation room window 9. Emergency shower/eye and body wash 10. Personnel clothing change room 11. Clothing lockers 12. Positive pressure lockers 13. Toilet and urinals 14. Lavatory 15. Bench 16. Decontamination pad 17. Wash-up shower stall 18. Chemical shower stall for pressure suit cleaning 19. Personnel bio-laboratory garment room 20. Air lock vestibule leading to personnel bio-laboratory garment room 21. Water cooler 22. Autoclave/fumigation area alcove 23. Autoclave with two sets of interlocked doors 24. Fumigation chamber with two sets of interlocked doors 25. Decontamination/laundry room 26. Washer 27. Dryer 28. Cleaning equipment storage closet 29. Laboratory equipment storage closet 30. Laboratory room non related to the bio- safety laboratory suite 31. Alarmed emergency exit door 32. Internal corridor

must be used only by operators wearing a one-piece positive pressure suit ventilated by a life support system.

Personnel entering a BL-4 laboratory must have specific and thorough training in handling extremely hazardous and infectious agents. They must be supervised by competent scientists who are experienced in working with these agents and they must also understand the primary and secondary containment functions of the standards and special practices, the containment equipment, and the laboratory design.

LOCATION

A BL-4 laboratory must be located in a separate building or in a clearly demarcated and controlled area within a building. This area must be completely isolated from all other areas of the building. Entrance to a BL-4 laboratory must be through a set of antechambers with interlocked, self-closing doors. A clothing change room with a clothing locker area (or alcove) that includes a large locker for the positive pressure suits must be located at the entrance of the BL-4 laboratory suite. This room is separated from the bio-laboratory garments room by an air-locked vestibule, by one or more air-locked shower stalls, and by one or more air-locked chemical shower stalls. These and other antechamber requirements are described later in this chapter. An arrangement of antechambers leading to a BL-4 laboratory is shown in figure 11-4.

Personnel must enter and leave the BL-4 laboratory through the antechambers. Once in the clothing change room, individuals must either remove their street clothing (including undergarments) and put on special laboratory clothing, shoes, and gloves, or put on a positive pressure suit over their street clothing, pressurized as needed and ready to be hooked up. Every time people leave the BL-4 laboratory, they must remove the special laboratory clothing in the biolab garments room, place the clothing in a special autoclave (with two sets of interlocked doors) designed to decontaminate laboratory clothing, and shower in the decontaminating shower stall. Individuals wearing positive pressure suits enter the chemical shower stall (still wearing the suit) and go through the decontamination procedure. Once the procedure is completed, they remove the suit and drop it in a special autoclave located next to the chemical shower stall. They must not use the air-locked vestibule to leave the BL-4 laboratory, except in the case of an extreme emergency.

All supplies and materials needed in the BL-4 laboratory room must be brought in either through a double-door autoclave, a double-door fumigation chamber, or through an air-locked vestibule that is decontaminated. Once the outer doors of the autoclave, the fumigation chamber, or the air-locked vestibule are securely closed, the personnel may open the inner door, retrieve the supplies and materials, and close the inner door.

ARCHITECTURAL REQUIREMENTS

Walls

BL-4 laboratory walls must be constructed and finished with hard, nonporous materials, free of holes, cracks, and crevices. Wall finishes must be smooth and animal- and insect-proof. Corner joints between walls, between walls and floors, and between walls and ceiling must be coved. Wall finishes must be sealed and resistant to liquids and chemicals. The walls, floor, and ceiling must together form a sealed, internal shell with surfaces that can be easily cleaned, fumigated, and decontaminated. They may be painted with a washable epoxy paint. All penetrations in walls must be sealed with a smooth finish to facilitate decontamination and cleaning. Fixtures on the wall must be recessed and flush with the wall surface to avoid lodging infectious materials. Electrical and other types of outlets should be flush with the wall to eliminate the lodging of infectious materials in cracks, corners, and crevices. Openings may require gusseted covers. Access panels to critical mechanical equipment should be located outside the BL-4 laboratory room.

Floors

BL-4 laboratory floors must be covered with a durable, seamless material that is capable of withstanding the wear imposed by the operations that will take place. The floor material should be easy to clean, nonporous, and resistant to the disinfectants generally used for decontamination of biohazardous spills. The junctions between the floor and the walls and between the floor and the fixed cabinets and equipment should be coved.

Ceiling

The BL-4 laboratory ceilings should be constructed and finished with hard, nonporous materials and must have a smooth sealed finish. These ceilings must be free of holes, cracks, and crevices. The junctions between the ceiling and the walls must be coved. Access to any mechanical equipment must be provided outside the BL-4 laboratory area. All penetrations in the ceiling must be sealed with a smooth finish to facilitate decontamination and cleaning.

Casework

BL-4 laboratory casework and other types of furniture should be constructed of smooth, nonporous materials that are easy to clean and have countertop finishes resistant to acids, alkalis, and organic solvents. These countertops must be seamless, able to sustain moderate heat, and resistant to the materials used for cleaning and disinfection. The junctions between the fixed cabinets, as well as between the fixed cabinets and other equipment, must also be smooth and free of cracks. These junctions' seams must be caulked smooth to ensure easy cleaning. The junctions between the fixed cabinets and the floor must be coved. Spaces between cabinets, benches, and other pieces of equipment must be easily accessible for cleaning.

Doors

BL-4 laboratory rooms must have at least two doors, one for regular access to the room and the other for emergency egress. These doors should be located to allow an escape on either side of the room should an emergency situation occur. Doors must be gusseted and have a lock to allow for access from the outside, and they must be equipped with panic hardware and be airtight to keep the BL-4 laboratory's negative pressurization. They also must be equipped with flush view windows. Doors at opposite ends of an air-locked space should be interlocked so that only one door can be opened at a time. At least one door must be wide enough to allow bulky equipment to be moved in and out of the room.

Windows

Windows (used either for light or as view windows) should be fixed, breakage-resistant, sealed, and flush with the inside walls of the BL-4 laboratory.

MECHANICAL REQUIREMENTS

All BL-4 laboratory rooms must have one-pass air. This air must be provided by a dedicated ventilation system. Adequate filtration of this air should be considered, as it will extend the service life of the exhaust HEPA filters. The air must be supplied to the room at a maximum velocity of 10 inches (0.25 m) per second (50 fpm; 15 mpm) at a height of 6 feet (1.83 m) above the floor level. The supply and exhaust components must be located to direct the airflow from the clean areas to the contaminated areas and from the low-hazard areas to the high-hazard areas. Air supply diffusers must be equipped with fans directing the airflow away from fume hoods and biosafety cabinets and must be located to minimize disruptive air currents.

The differential pressure/directional airflow between areas within the room must be monitored. The airflow in the supply and exhaust components must also be monitored. The supply and exhaust components must be interlocked to assure that the inward or zero airflow is maintained. The system must be monitored and alarmed to indicate any malfunction.

BL-4 laboratory rooms must also have independent temperature controls. The temperature of these rooms should normally be maintained around 72°F (22.2°C). However, this temperature may vary depending on the requirements of the operation being conducted. In most instances, this temperature should be 67–77°F (19.4–25°C), plus or minus 5°F (2.8°C) in rooms where a constant temperature is not critical, and plus or minus only 0.5°F (0.28°C) in rooms where constant temperature is critical.

Relative humidity in a BL-4 laboratory must also be controlled. The range depends on the requirements of the operation. The recommended maximum relative humidity is 45 percent, plus or minus 10 percent for general applications and 5 percent for humidity-sensitive applications.

The BL-4 laboratory room's air pressure must be negative to that of adjacent, connecting spaces and corridors so that any leakage that might occur would be from the adjacent rooms or areas to the BL-4 laboratory room.

Exhaust ducts must be under negative pressure until the air is discharged outside the building. Supply and exhaust ducts for BL-4 laboratory rooms must be equipped with gastight dampers to allow the decontamination of the room without contaminating other parts of the building.

Exhaust stacks and building air intakes must be located in relation to each other and to the predominant winds so that the exhausted air is dispersed and discharged away from any occupied areas without being reentrained into the building.

The HEPA filtration system must be located as close as possible to the source of contamination in order to minimize the length of contaminated ductwork. The filtration system must also be isolated from the ventilation system for gas decontamination and testing. The HEPA filter exhaust housing constructed to facilitate filter installation and to allow on-site decontamination of the filters prior to their removal in a sealed, gastight container for decontamination and/or destruction elsewhere.

The HEPA filtered exhaust air from class II and class III biosafety cabinets must be directly exhausted to the outside. However, the air exhausted from class III biosafety cabinets must pass through two HEPA filters systems placed in series prior to being discharged to the outside. Once filtered as required, this air may be discharged through the building's exhaust system. The connection between the safety cabinets' exhaust system and the building's exhaust system should be such as to avoid any interference with the air balance of both systems.

The HEPA filtered exhaust air from class II biological safety cabinets in a BL-4 laboratory room in which workers wear a positive pressure suit may be directly exhausted to the outside through the BL-4 laboratory room's exhaust system. This air may also be recirculated within the BL-4 laboratory room under certain situations, with approval from the facility's industrial hygienist and safety officer. In either case, the cabinets must be tested and certified at least once a year.

All aerosol-producing equipment used in a BL-4 laboratory room, including continuous flow centrifuges, must have their exhaust air HEPA filtered prior to discharge. It is preferable that the air be discharged in a contained manner, either through the biosafety cabinets' exhaust system or through the building's exhaust system.

BL-4 laboratory rooms must be equipped with an alarm to notify of any failure in directional airflow or of failure of air containment. This device should have a visual and/or audible local alarm and must be connected to the building's central alarm system.

SERVICES AND EQUIPMENT

In order to function safely, BL-4 laboratories must be provided with adequate plumbing, autoclaves, fumigation chambers, vacuum systems, medical pathological waste containers, flammable liquid storage cabinets and other chemical storage cabinets as may be needed, biosafety cabinets, a fire detection and alarm system, and eye washes and safety showers.

Plumbing

BL-4 laboratories must have a sink for hand washing. This sink must be located in the laboratory near the exit door. The faucet must be foot-, elbow-, or automatically operated.

If water fountains are needed, they should be located outside the laboratory room in an adjacent corridor or vestibule or in the biolab garments room. The water fountains must be foot-operated and the water service to the fountain must not be connected to the backflow-protected system that supplies water to the laboratory areas.

If floor drains are used, they must be equipped with traps filled with a chemical disinfectant selected for its efficacity against the target agent. The drains must be directly connected to the liquid waste decontamination system. Traps and sewer vent lines must contain HEPA filters.

The liquid effluents from the floor drains as well as from the laboratory sinks and the autoclave chambers must be decontaminated by heat treatment before being discharged to the sanitary sewer. This treatment is in addition to any chemical disinfectant poured in the floor drain or sink. This heat treatment process must be validated physically and biologically with a constantly recording temperature sensor in conjunction with an indicator microorganism that has a defined heat susceptibility profile. Effluents from showers and toilets do not need to be treated prior to discharge to the sanitary sewer. Effluents from the chemical shower stall for cleaning pressure suits also do not need to be treated because of decontaminating agents used in the shower.

Autoclaves

Two sets of autoclaves must be provided for use in a BL-4 laboratory. One set must be exclusively for the supply, decontamination, and transportation of materials, clothing, and other items to and from the laboratory room. This set of autoclaves must be located on an external wall of the laboratory. Each of these autoclaves must have two interlocked doors. One of these doors must open to the area outside the laboratory. It must be sealed to the outer wall of the laboratory and automatically controlled so that it can only be opened after the autoclave has completed its sterilization cycle.

The other set consists of separate autoclaves for clean and dirty procedures. Clean autoclaves are used for sterilization of microbiological media and dirty autoclaves are used for decontamination purposes. The number and size of the autoclaves must be decided by the laboratory users and be sufficient to accommodate carts containing equipment and other materials needing sterilization. The interior of the autoclave should be epoxy-coated, free of cracks and crevices, and caulked and sealed to prevent retention of biological wastes.

The air pressure of the space containing the autoclaves must be negative to that of the surrounding spaces so that air will always be directed toward the space containing the autoclaves.

Sufficient space must be provided for locating the required number of autoclaves in BL-4 laboratories. One or more canopy hoods with the required exhaust capacity must be located just over the doors of the autoclaves to remove steam and odors. Exhausts from the autoclaves and canopy hoods must be filtered. These exhausts contain a significant amount of moisture; therefore, a moisture-resistant (hydrophobic) filter must be used, as the paper filter used in a regular HEPA filter would be damaged by the moisture and the integrity of the filter would be lost.

Fumigation chambers

A pass-through dunk tank fumigation chamber must also be located on an external wall of the BL-4 laboratory for the supply, decontamination, and removal of materials and equipment that cannot be decontaminated in the autoclave. This fumigation chamber must have two interlocked doors. One of these doors must open to the area outside the laboratory. It must be sealed to the outer wall of the laboratory and be automatically controlled so that it can only be opened after the decontamination process has completed its sterilization cycle.

Vacuum systems

If a central vacuum system is needed in a BL-4 laboratory, it must be used only within the laboratory and for the exclusive use of the laboratory' operation. This system must be carefully designed by safety-conscious engineers. The system's air must not be recirculated and must be protected with liquid disinfectant traps and with HEPA filters placed as close as possible to the point of use. The HEPA filter exhaust must be vented to the outside of the building. In order to minimize the possibility of contaminating the vacuum pumps, the filters must be equipped with a mechanism capable of decontaminating them at their service location. These filters must be routinely maintained and replaced as needed.

Medical Pathological Waste Containers

BL-4 laboratories must be provided with medical pathological waste containers and other types of waste receptacles needed for the operations conducted. Sufficient and strategically located space should be provided in the laboratory room for these containers and receptacles. The handling and storage of these containers and receptacles is the same as described for BL-2 laboratories.

Flammable Liquid Storage Cabinets

BL-4 laboratories must be provided with flammable liquid storage cabinets and, if needed, with acid/bases storage cabinets or with special cabinets to store other needed chemicals. Sufficient and strategically located space should be provided in the laboratory room for all cabinets.

Biosafety Cabinets

Class II and III biosafety cabinets must be used in most operations that take place in BL-4 laboratories.

Safety Showers, Eye Washes and Fire Alarm System

Eye washes and safety showers with eye washes must be located inside the BL-4 laboratory room as close as possible to the location of high risk. Another set of safety showers with eye washes should be located outside the door of the BL-4 laboratory room. Further, this room must be protected by a fire alarm system.

Antechambers of Biosafety Laboratory Rooms

There are several types of antechambers used for BL-3 and BL-4 laboratories. Which antechamber should be used depends on code requirements, as well as on the safety and operational requirements dictated by the operations that will take place in the laboratory.

The following are among the antechambers that should be considered: a clothing change room with an area or alcove for clothing lockers and positive pressure suits, shower stalls, chemical shower stalls, a biolaboratory garments room, one or more air-locked vestibules, one or more Visitor Observation Rooms, an Equipment Storage Room, and a Cleaning Equipment Storage Room. A brief explanation of each of these rooms' functions follows.

A Personnel Clothing Change Room is a room where the personnel authorized to enter biosafety laboratory level 3 or 4 will either remove all their clothing, including underwear and shoes, and wear special garments or wear, over their street clothing, a positive pressure suit, pressurized as needed and ready to be hooked up as required. This clothing protection must be completed prior to entering the BL-4 laboratory rooms suite. There must be a Personnel Clothing Change Room for men and one for women.

Toilets, lavatories, clothing lockers, and alcoves for positive pressure suits are generally an extension of the clothing change room.

Street clothing is kept in the lockers in the clothing locker area or alcove while the personnel are working in the biosafety laboratory. The positive pressure suits locker is a large locker where the decontaminated positive pressure suits are stored. After putting on the required garments in the clothing change room, the personnel can move directly to the BL-4 laboratory through an air-locked vestibule.

One or more air-locked vestibules must separate the BL-4 laboratory from connecting rooms or antechambers. The doors of these vestibules must be interlocked so that to open the BL-4 laboratory door, all other doors must be closed. Most air-locked vestibules servicing a BL-4 laboratory suite are one-way, leading to or from the BL-4 laboratory suite, depending on their function. Air-locked vestibules leading from the BL-4 laboratory suite must have decontaminating equipment or be connected to the decontaminating equipment storage room or cabinet so that any equipment removed from the BL-4 laboratory suite is first decontaminated. Air-locked vestibule's air pressure must be positive to the BL-4 laboratory room and negative to all other rooms connected to it. It is good practice to include a glass view in all the doors in the air-locked vestibule.

The air-locked vestibule that gives access from the clothing change room to the BL-4 laboratory must not be used to evacuate the laboratory room(s) except in an extreme emergency, during which local and central alarms are activated.

Personnel must leave the BL-4 laboratory through the biolaboratory garments room. There must be separate biolaboratory garments rooms for men and women. These rooms are equipped with toilets and lavatories and may have foot-activated water fountains. When entering this room, individuals wearing biolaboratory garments remove the garments and insert them into an autoclave with interlocked double doors. Once completely naked, they then go through one of the shower stalls, where they wash as required, and return to the clothing change room to get dressed. Individuals who enter the biolaboratory garments room wearing positive pressure suits must go directly to the chemical shower stall to decontaminate the suit. Once this is done, they remove the suit and insert it in a special autoclave as is done with the biolaboratory garments.

The autoclaves where the biolaboratory garments and the positive pressure suits are inserted must be exclusively used to decontaminate clothing and materials before they are removed from the contained area. The autoclave must be located so that one of its doors opens into the biolaboratory garments room and the other opens outside of the containment area, probably into the laundry room.

The air pressure of the biolaboratory garments room must be positive to that of the BL-4 laboratory suite and negative to that of the shower stalls, the chemical shower stalls, and the air-lock vestibule connecting the biolaboratory garments room and the clothing change room. The doors separating the BL-4 laboratory suite and the biolaboratory garments room must be interlocked with the doors of the shower stalls, the chemical shower stalls, and the air-locked vestibule leading to the biolaboratory garments room.

A set of one or more shower stalls and chemical shower stalls must be provided outside the biolaboratory garments room for men and women. Each shower stall must have double interlocked doors allowing a one-way passage from the biolaboratory garments room to the clothing change room. The air pressure of the shower and chemical shower stalls must be positive to that of the biolaboratory garments room and negative to that of the clothing change room. Each shower stall must also have a standard showerhead, a hand-held shower, as well as decontaminant dispenser. Out of each chemical shower stall is an area containing a decontamination pad. Individuals must step over this decontamination pad in order to return to the clothing change room.

A chemical shower stall with a chemical decontamination shower must be provided outside the biolaboratory garments room for men and women using positive pressure suits. This stall must be large enough to allow the decontamination procedure to take place comfortably. It must have double interlocked doors allowing a one-way passage from the biolaboratory garments room to the clothing change room. The exhaust air from the chemical shower stall must be filtered by two sets of HEPA filters. Each stall must be equipped with a duplicate filtration unit, an exhaust fan, emergency lighting and communication systems, and an emergency power source that is automatically activated. Waste materials removed from the suit area must be inserted into an autoclave with interlocked double doors. This autoclave must be located so that one of its doors opens in the chemical shower stall and the other opens outside the containment area, probably into the laundry room. The air pressure of the chemical shower stall must be positive to that of the biolaboratory garments room and negative to that of the clothing change room.

Some BL-4 laboratory rooms may need either a special equipment storage room or one or more equipment storage closets. These rooms or closets must be located across an air-locked vestibule where the equipment must be cleaned prior to entering the storage room or closet. This vestibule must be equipped with the required equipment to clean the instruments and equipment.

A cleaning equipment storage room or closet must be located close to the special equipment storage room and across an air-locked decontamination vestibule leading to it. The equipment stored in this room is used to clean the BL-4 laboratory room and, if needed, the instruments and equipment stored in the special equipment storage room. This cleaning equipment, once used, must be cleaned prior to entering the cleaning equipment storage room. Thus, the cleaning equipment storage room must also have equipment to clean the cleaning equipment.

A decontamination air-locked vestibule—preferably the vestibule connecting the special equipment storage room and the cleaning equipment storage room—must connect the BL-4 laboratory room to a main corridor, with the interlocked doors of the room and of the corridor located directly in line with each other. This arrangement allows direct access to the laboratory room for the installation or removal of equipment and for unrestricted emergency exit from the room.

Sometimes an observation room is placed adjacent to the BL-4 laboratory room. This room's large glass window permits visitors to observe work done in the room without entering it or disturbing operations. The observation room also provides an easy means of checking on the safety of the laboratory room's occupants.

The walls, floors, and ceilings of clothing change rooms, the clothing locker and areas or alcoves for positive pressure suits, the shower stalls, the chemical shower stalls, the biolaboratory garments room, the air-locked vestibules, the equipment storage room, and the cleaning equipment storage room must be constructed of and finished with hard, nonporous materials, free of holes, cracks, and crevices. Wall finishing must be smooth and animal- and insect-proof. Corner joints between walls, between walls and floors, and between walls and ceiling preferably should be coved. Wall finishes must be sealed and resistant to liquids and chemicals. The walls, floor, and ceiling must together form a sealed, internal shell with surfaces that can be easily cleaned, fumigated, and decontaminated. They may be painted with a washable epoxy paint. All penetrations in walls must be sealed with a smooth finish to facilitate decontamination and cleaning. Fixtures must be recessed and flush with the wall surface to avoid lodging infectious materials. Electrical and other types of outlets also should be flush. Openings may require gusseted covers.

CLEAN ROOMS

A clean room is where contaminant content is controlled. There are several different ways to control contaminant content, and choosing one should depend on the room's requirement for "cleanliness." For instance, some operations take place in a bench that must be protected from particle contaminants. In this case, the aim should be to keep the operation and the immediate area around it free of contaminants. Other operations require that the entire room be kept totally clean. A very different method must be used to deal with this situation. Before defining, classifying, and identifying the different types of clean rooms, a few questions must be asked: "What are exactly the contaminants that we want to control?" "Are they particles only or are they particles and vapors?" And finally, "Why do we want this control?"

There are many operations that require a clean room environment. The miniaturized components for satellites manufactured for the aerospace industry must be assembled in clean rooms. Space vehicles and much of their instrumentation needs to be tested in clean rooms. Medical and pharmaceutical research operations as well as production and research operations in chemistry, electronics, and physics often need to be conducted in clean rooms. The radioactive industry, the photographic industry, the food industry, and the plastic industry, to name a few, all need to conduct some of their operations in some kind of clean room. Different operations require clean rooms of different types and classes and with different requirements.

With this understanding, we can start to define clean rooms: A clean room is a fully enclosed room where the air supplied has its particulate matters—and sometime its vapors and fumes—removed and is temperature-, humidity-, and pressure-controlled as required by the operations that will be conducted in the room. To meet the requirements of a clean room, a particulate count must be kept. The number of particulates must be maintained within the count allowed by the clean room's class. Any rooms or areas with a higher contamination level

FIGURE 11-5
Class limits in particles per cubic foot of size equal to or greater than particles size shown. (The class limit particle concentrations shown are defined for class purposes only and do not necessarily represent the size distribution to be found in any particular situation.)

ROOM CLASS	MEASURED PARTICLE SIZES (MICROMETERS)				
	0.1	0,2	0.3	0.5	5.0
1	3.5	7.5	3	1	N/A
10	350	75	30	10	N/A
100	N/A	750	300	100	N/A
1,000	N/A	N/A	N/A	1,000	7
10,000	N/A	N/A	N/A	10,000	70
100,000	N/A	N/A	N/A	100,000	700

than those defined hereafter are not considered clean rooms. Clean rooms are identified by class and by type.

Clean Room Classification

There are six classes of clean rooms. Each class allows a different maximum number of particles of a given size. In this book, clean room standards are not converted into metric units because there is no single set of standards for all of Europe (Germany and England each have their own standards, for instance). Furthermore, Japan and other Asian countries also have unique sets of standards. All these standards are very close to (and many may have derived from) the U.S. Federal Standard 209D, which is the one used in this book.

- A class 100,000 clean room should not have more than 100,000 particles of 0.5 and larger microns, or 700 particles of 5.0 and larger microns, per cubic foot.
- A class 10,000 clean room should not have more than 10,000 particles of 0.5 and larger microns, or 70 particles of 5.0 and larger microns, per cubic foot.
- A class 1,000 clean room should not have more than 1,000 particles of 0.5 microns per cubic foot.
- A class 100 clean room should not have more than 750 particles of 0.2 microns, or 300 particles of 0.3 microns, or 100 particles of 0.5 microns, per cubic foot.
- A class 10 clean room should not have more than 350 particles of 0.1 microns, or 75 particles of 0.2 microns, or 30 particles of 0.3 microns, or 10 particles of 0.5 microns, per cubic foot.
- A class 1 clean room should not have more than 35 particles of 0.1 microns, or 7.5 particles of 0.2 microns, or 1 particles of 0.5 microns, per cubic foot.

Figure 11-5 shows a table from Federal Standard No. 209D for clean room and work station requirements and for controlled environments. This standard for clean rooms was developed by the federal government. This table illustrates the limits per cubic foot for particles of equal or greater size than the particle sizes listed. Figure 11-6, from the same federal standard, shows the same information in graphic form.

Types of Clean Rooms

There are two types of clean rooms: conventional flow clean rooms and the laminar flow clean rooms. Following is a short description of each of these room types.

CONVENTIONAL FLOW CLEAN ROOMS

Conventional flow clean rooms, also known as mixed flow clean rooms, are where highly filtered and conditioned air is supplied through diffusers located in

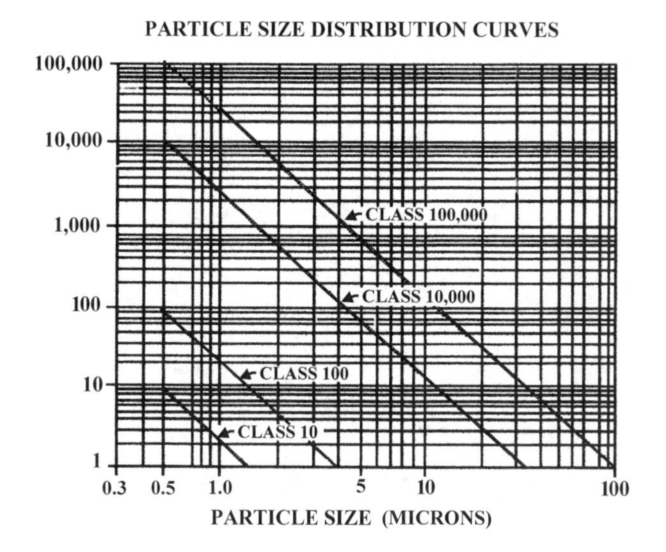

FIGURE 11-6
Particles/cubic foot equal to or greater than stated particle size

the ceiling. The air supplied to these rooms is blown at high speed and, once out of the diffuser, slows down and moves in all directions. This air hits anything in its path, such as people and equipment, is deflected, and becomes highly nondirectional. As a result, eddies are formed in the room's corners and the sweeping effect that would have resulted from a uniform and directional path is lost. This air is then exhausted through grilles located near the floor around the perimeter of the room and into return air ducts. Figure 11-7 shows a diagram of a conventional flow clean room with such diffusers.

Sometimes a clean perforated ceiling is placed under the source of air supply to help diffuse the air more evenly. The area between the perforated ceiling and the source of air supply becomes a plenum that works to equalize the air pressure entering the room and therefore diffuse it. Here, too, the air is exhausted through grilles located near the floor around the perimeter of the room and into return air ducts. Figure 11-8 shows a diagram of a conventional flow clean room with a clean perforated ceiling under the source of air supply.

When one of these two schemes is used, it is important, in order to keep these rooms as clean as possible, to limit the amount of contamination brought into the room by controlling the personnel and materials entering the room. A conventional flow clean room cannot be cleaner than a class 10,000 clean room. Access to this room should be limited to personnel wearing special clothing that

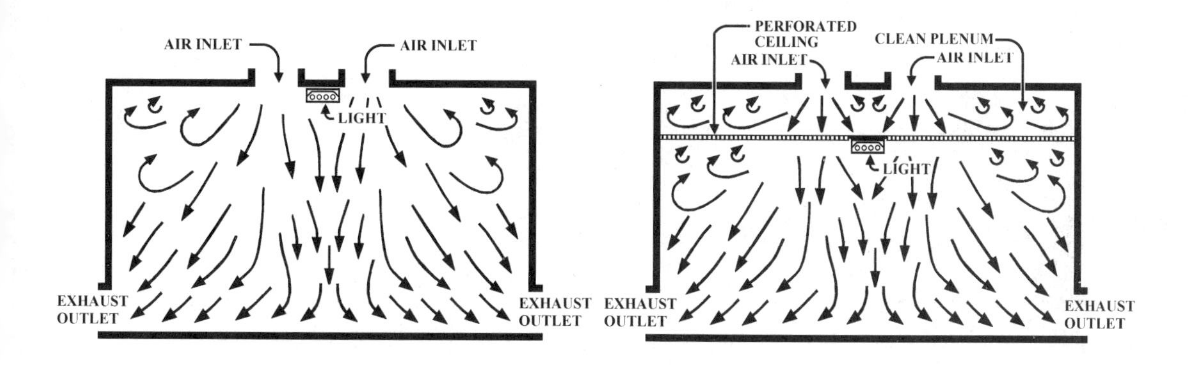

FIGURE 11-7 (left)
Side view of a conventional flow clean room with diffusers

FIGURE 11-8 (right)
Side view of a conventional flow clean room with a clean perforated ceiling under the source of air supply

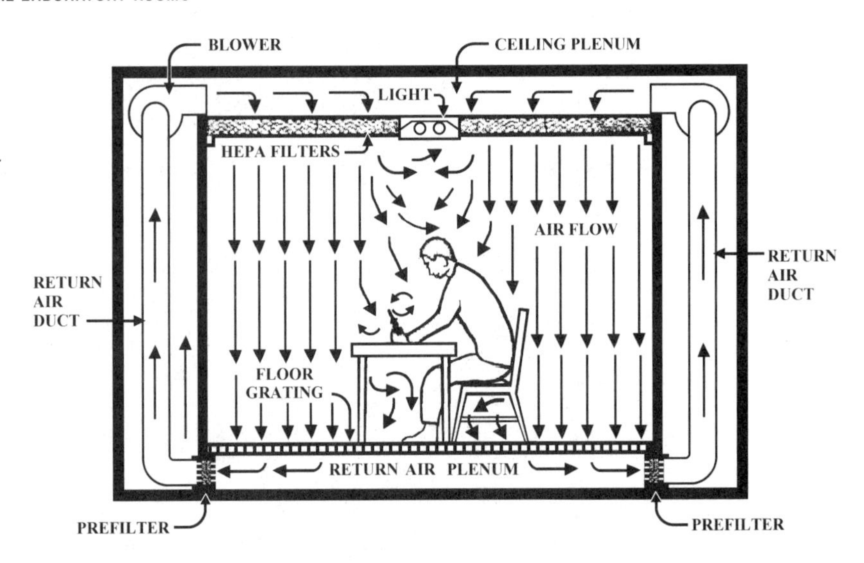

FIGURE 11-9
Side elevation of a conventional flow clean room with diffusers
Ceiling to floor vertical laminar air flow

sheds few particles and people and materials must be "cleaned" before entering the room. Janitorial service, to remove room contaminants, should be specialized and use approved methods for the removal of contaminants.

The design for a conventional flow clean room is much simpler than that for a laminar flow clean room. It is also much less expensive, since the HEPA filtration required is much less sophisticated. Often, a conventional flow clean room is used for rooms in which some areas must be much cleaner than the rest of the room; the cleanliness of these areas is achieved by laminar flow cabinets. There are several kinds of laminar flow cabinets. Among them are cabinets that move air from the room through a prefilter, and cabinets that have a flexible duct that blows preconditioned air through ducts to a prefilter.

LAMINAR FLOW CLEAN ROOMS

Laminar flow clean rooms are where highly filtered and conditioned air is evenly supplied through the entire surface of a wall or of a ceiling. The air is moved through the room in a laminar flow fashion with uniform velocity along parallel flow lines, with minimum eddies, and then is exhausted through a similar entire wall or floor surface. Figure 11-9 shows a cross section through a clean room with a vertical, ceiling-to-floor, laminar airflow. In the cross section, a blower exhausts the room air through the return air plenum under the floor grating and recirculates it first through prefilters at the entrance of the return air ducts then through HEPA filters located throughout the ceiling. From there, the air is introduced back to the room. Because the eddies created above and around the operator sitting at the work bench may carry contaminants (dust, hair, etc.) from the operator's body and clothing to the sample being worked on (which could potentially damage the sample), the operator must be properly gowned.

Figure 11-10 shows the side view of a room with a horizontal, wall-to-wall, laminar airflow. There, a blower sucks the room air from a plenum behind the exhaust wall and recirculates it back to the room through HEPA filters located all over the supply wall. By moving in a laminar way, the air passes only one time in any given area of the room. This pattern of air movement sweeps away any contaminant brought to the room by personnel or equipment or generated by

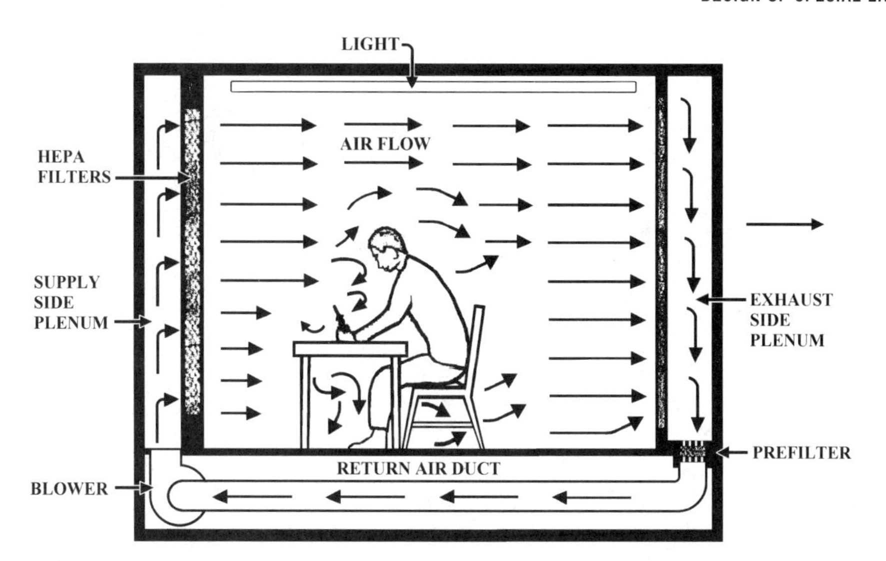

FIGURE 11-10
Side elevation of a conventional flow clean room with wall to wall horizontal laminar airflow

operations and carries it out of the room. In this type of clean room, localized contaminations are isolated by the striations of the laminar airflow and therefore do not, move to other areas of the room. Thus, the work requiring the cleanest air environment should be performed in the path of the undisturbed flow of clean air from the incoming air surface. Personnel's clothing requirements, equipment cleanup, and janitorial restrictions are much less severe in this type of room than in conventional clean rooms.

Design Considerations for Clean Rooms

Clean rooms are almost always designed for specific types of operations, which dictate cleanliness and other types of requirements. Today, available technology permits a high degree of cleanliness. However, as the degree of cleanliness gets higher, the cost of achieving it becomes excessive. A very sophisticated and therefore quite expensive HVAC system is required to adequately control the temperature, humidity, and air pressure of a clean room. A class 100 clean room cost about \$450 to \$550 (\$5,000 to \$6,100/m^2) per square foot to build and over \$30 a day per square foot to run (\$330/m^2).

Often, the requirements that are necessary to conduct operations are in conflict with the cleanliness requirements. For example, the combined cleanliness and operational requirements for electronic, computer chip, and pharmaceutical manufacturing facilities and for microbiological, biomedical, and analytical laboratories are very different from each other. The conflicts between the cleanliness and operational requirements of these facilities are also different. Therefore, it is important that the designer is familiar with the operations for which the clean room is being designed. The designer should work hand in hand with the user, utilizing the guidelines provided here and adhering to safety and operational requirements.

Antechambers for Clean Rooms

There are several antechambers that can or should be incorporated into the design of a clean room. Which antechamber, or group of antechambers, should be used depends on the class and type of the clean room, the level of cleanliness required, and the safety and operational requirements dictated by the activities that will take place in the room. Further, the requirements of each antechamber must be a factor of the antechamber's function and of the clean room it services. These requirements should be worked out with the person or persons who will be using the clean room.

The following are among the antechambers that should be considered: clothing change room with a clothing locker area, one or more air shower stalls, personnel clean room's dressing room, one or more air-locked vestibules, special equipment storage room, cleaning equipment storage room, and observation room. A brief explanation of each of these rooms follows.

A clothing change room is where the personnel authorized to enter the clean rooms may have to remove their street clothing, if this is required, prior to taking an air shower. There should be separate clothing change rooms for men and women. A locker area must be located in the clothing change room to store the street clothing. Depending on how "clean" they should be to enter the clean room, personnel may have to take an air shower in the air shower stall either undressed or with their clothing on. Air shower stalls for men and for women should be located between the clothing change room and the personnel clean room's dressing room. People authorized to enter the clean rooms must move to the personnel clean room's dressing room through the air shower stall and walk over a sticky pad area. Once in the personnel clean room's dressing room, they put on approved garments either over or in place of their clothing. Men and women should have separate personnel clean room's dressing rooms. This room also should have toilet accommodations for the people inside the clean room.

Some clean rooms may need to have a special equipment storage room. These rooms can be located in a clean corridor directly adjacent to the clean room if the equipment to be stored qualifies as clean. The special equipment storage room must be located across the air-locked vestibule if the equipment needs to be cleaned prior to entering the clean room. The equipment room should contain the equipment required to clean whatever instrument or equipment is needed in the clean room.

The cleaning equipment storage room must be located across the air-locked vestibule. The equipment stored in this room is used only to clean the clean room. This cleaning equipment may have to be cleaned prior to entering the clean room. Thus, the cleaning equipment storage room should also be equipped with the equipment required to clean the cleaning equipment before it is moved to the clean room.

One or more air-locked vestibules must separate the clean room from all other connecting rooms or antechambers. The doors of the vestibule must be interlocked so that to open the clean room door, all other doors must be closed. The air-locked vestibule's air pressure must be negative to that of the clean room and positive to all other rooms connected to it. It is good practice to provide all the air-locked vestibule's doors with glass views and to place a sticky mat at the entrance of the clean room. An air-locked vestibule must connect the clean room to a main corridor, with the interlocked doors of the room and of the corridor located directly in line with each other. This arrangement allows direct access to the clean room (for the installation or removal of equipment) and unrestricted exit in case of an emergency.

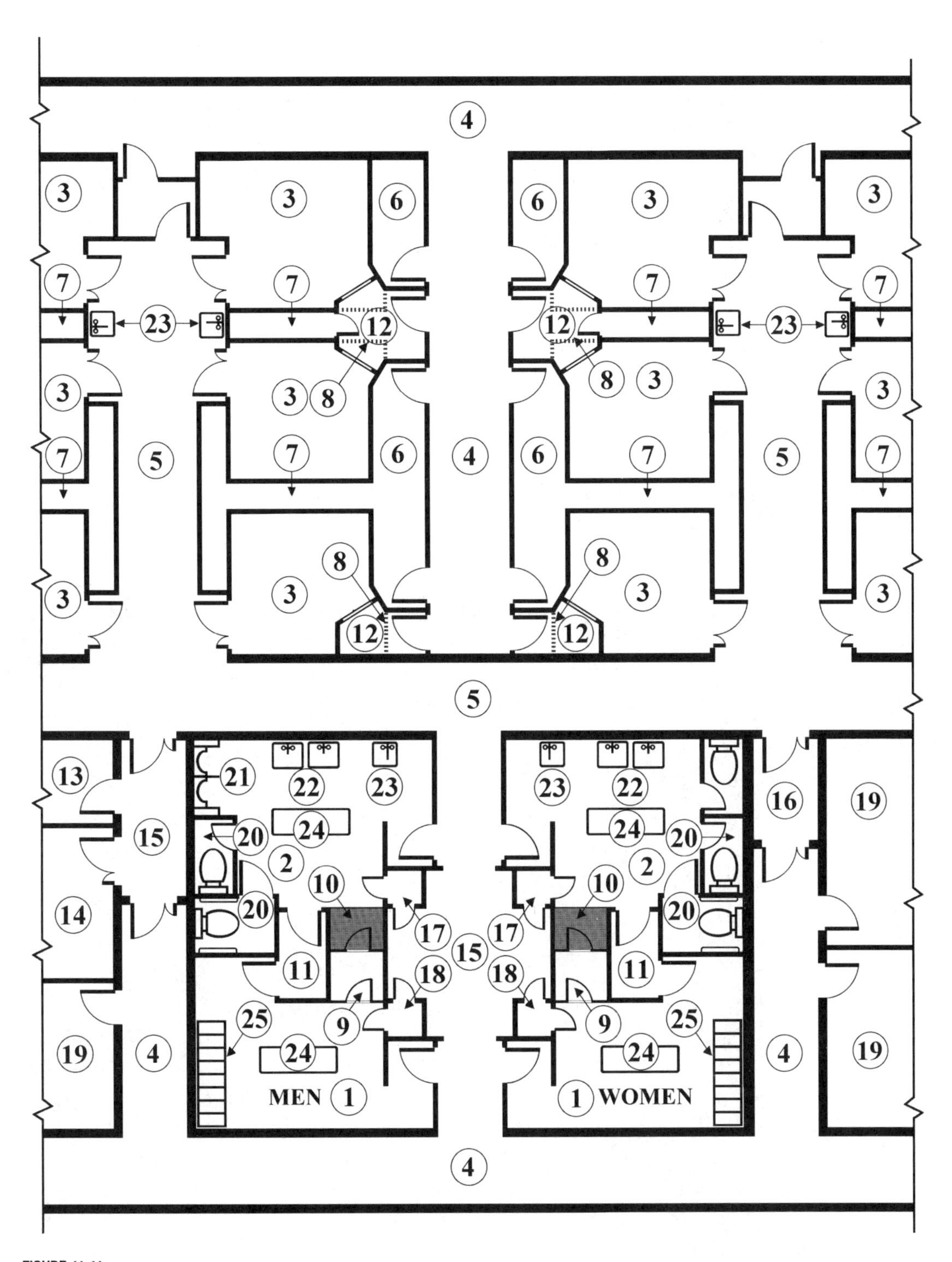

FIGURE 11-11
Plan of a series of clean rooms and related antechambers

1. Clothing change room 2. Personnel clean room dressing room 3. Clean rooms (different classes) 4. Regular staff corridor 6. Service corridor 7. Utility/air duct space 8. Overhead utility/air duct space 9. Air shower stall 10. Sticky pad area 11. Air lock vestibule 12. Observation room 13. Cleaning equipment storage room 14. Special equipment storage room 15. Decontamination/interlocked vestibule 16. Interlocked vestibule 17. Clean rooms' clean garments interlocked double door closet 18. Clean rooms' soiled garments interlocked double door closet 19. Regular laboratory room 20. Toilets 21. Urinals 22. Lavatories 23. Water closets 24. Bench 25. Lockers

Sometimes, an observation room is placed adjacent to the clean room. The observation room has a large glass window overlooking the clean room. This permits visitors to observe the clean room work without entering it or disturbing the operations. This room also provides an easy means of checking on the safety of the clean room's occupants.

Figure 11-11 shows a series of clean rooms serviced by a set of antechambers. It also shows the relationships of these antechambers to each other and to the suite of clean rooms.

Clean Room's Air Pressure, Temperature, and Relative Humidity

All clean rooms, regardless of their class or type, must have a higher air pressure than that of the adjacent rooms or areas with which they connect by doors or windows, so that any leakage that might occur would be from the clean rooms to the adjacent rooms or areas. The minimum positive pressure differential between a clean room and adjacent rooms or areas is 0.05 inches (1.27 mm) of water when all doors and windows are closed. When the doors and windows are open, the capacity of the blower providing air to the clean room should be able to maintain a sustained outflow of air sufficient to prevent contamination from migrating into the clean room.

All clean rooms' temperature normally should be maintained around 72°F (22.2°C). However, this temperature may vary depending on the requirements of the product handled. In most instances, this temperature should be 67–77°F (19.4–25°C), plus or minus 5°F (2.8°C) in rooms where a constant temperature is not critical, and plus or minus only 0.5°F (0.28°C) in rooms where a constant temperature is critical.

Relative humidity in clean rooms should be controlled. The range depends on the requirements of the product handled. The recommended maximum relative humidity is 45 percent, plus or minus 10 percent for general applications and 5 percent for humidity-sensitive applications. It should be noted that if the relative humidity is above 50 percent, metal rusting can become a serious problem, and if the relative humidity is below 30 percent, electric static charges cause particle attraction on dielectric materials.

Architectural Requirements for All Types of Clean Rooms

Clean rooms' floors, walls, and ceilings must be airtight and constructed to prevent air leaks into or from the room. Further, the finishing of the walls, floors, and ceiling must be smooth and easy to clean. The floors must be covered with a durable, seamless material that does not crack or blister and is capable of withstanding the wear imposed by the operations that will take place. This material should shed few or no particles, should be easy to clean, and should be nonporous. The junctions between the floor and the walls and between the floor and the fixed cabinets and equipment must be coved. Perforated floors should be used only when the airflow must go through the floor to the return air plenum or when the floor does not require wiping and mopping. The clean room's ceilings should be constructed of and finished with hard, nonporous materials that do not shed particles. The ceilings should be free of holes, cracks, and crevices. The junctions between the ceiling and the walls must be coved.

The clean room's walls should be constructed of and finished with hard, nonporous materials that do not shed particles and are easy to clean. The walls should be free of holes, cracks, and crevices. Hollow cavity walls are used either as a plenum for air return to the HVAC system or to lodge the air return ducts. The cavity is also used to contain electrical, plumbing, and other required services. Fixtures on the wall must be recessed and flush with the wall surface to avoid accumulation of dirt. Electrical and other types of outlets should be flush with the wall to eliminate cracks, corners, and crevices. Openings may need gusseted covers.

The clean room's casework and other types of furniture should be constructed of materials that are easy to clean and shed few or no particles. The junctions between the fixed cabinets and between the fixed cabinets and other equipment must be smooth and free of cracks. The junctions between the fixed cabinets and the floor must be coved.

Doors, windows, and pass-throughs must be of the double-air-locked type. Doors to and from clean rooms must be tight fitting, have air seals to keep the clean room's pressurization, and be equipped with flush view windows. Doors at opposite ends of an air lock should be interconnected so that only one door can be opened at a time. At least one door must be wide enough to allow bulky equipment to be moved in and out of the room.

Windows (used either for light or as view windows) should be fixed and flush with the inside walls of the clean room.

Air Supply and Filtration for All Types of Clean Rooms

All the air supplied to any class or type of clean room, recirculated or fresh from the outside, must be filtered. Equipment should be large enough to provide a minimum of 20 air changes of filtered air per hour for rooms with ceilings 8–12 feet (2.44–3.66 m) high. This air is also used to pressurize the clean room. Air filtration must remove particles as required by the clean room's class. Air filtration to remove certain vapors or gases may be also be required by certain operations.

HEPA filters should be used to remove particles. A HEPA filter will remove more than 99.97 percent of particles 0.3 microns in diameter. When HEPA filters are used, it is good practice to use prefilters to prolong the life of the HEPA filters. The efficiency of the prefilters should be tailored to the expected level of contamination and to the desired expansion of the HEPA filter's life. When prefilters are used, the HEPA filter must be the final filter for the air entering a clean room. Airflows of 90 fpm (0.46 m/s) or higher may be required to provide sufficient air motion to prevent air particles from settling. This translates into 18–100 air changes per hour.

Some operations required the removal of certain vapors or gases. When that is the case, the removal must take place during prefiltration, before the air is filtered by the HEPA filter. Depending on the vapor or gases to be removed, liquid scrubbers, adsorption train, or carbon filters may be used.

AIR SUPPLY AND FILTRATION IN LAMINAR FLOW CLEAN ROOMS

In a laminar flow clean room, the airflow entering the room from the ceiling must seep through the entire ceiling's surface. Similarly, if the laminar flow enters

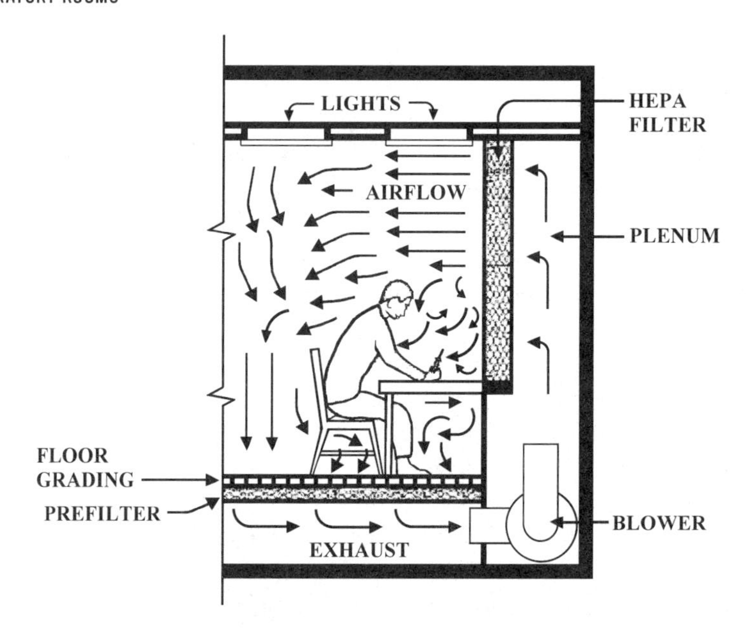

FIGURE 11-12
Section through a clean room with a working bench and with the airflow entering the room through the wall above the work bench and exhausted through the floor

the room from one of the walls, it must seep through the entire wall surface. Therefore, in a laminar flow clean room the bank of HEPA filters should be located to cover the entire surface of the wall or ceiling. If a room has a working bench with cabinets underneath it and has air entering from the wall where the bench is located, the bank of HEPA filters should be located above and around the bench's working surface and should extend to the ceiling.

In a laminar flow clean room, the airflow leaving the room must exit through an entire wall or through an entire grated floor surface. If the air removed from the clean room goes though prefiltration, filters located behind the exit walls or beneath the floor grating should be adjusted to provide an adequate pressure drop across the entire exit surface so that the airflow throughout the room remains uniform. The volume of air exiting the room during any given period should be equal to that of air entering the room through the bank of HEPA filters.

Figure 11-12 illustrates a situation in which the airflow supplied to the room enters from the wall above and around the working bench. This air reaches the sample before reaching the operator working with it. Notice the movement of air around the operator and how this arrangement diminishes the possibility of contaminating the sample. In this example, the air is exhausted through the floor in a manner similar to a ceiling-to-floor airflow.

The airflow velocity through a cross section of a laminar flow clean room should be a minimum average of 90 fpm (0.46 m/s). The velocity can vary plus or minus 20 fpm (0.10 m/s) throughout the undisturbed portions of the clean room. The airflow patterns must also be uniform, with a minimum of turbulent airflow patterns in the undisturbed portions of the clean room.

Clean Air Exhaust Hoods

Certain operations requiring a clean environment may best be conducted in a clean air exhaust hood. This hood may be placed in a clean room, if the operation requires an extreme level of cleanliness, or in a laboratory room, where it

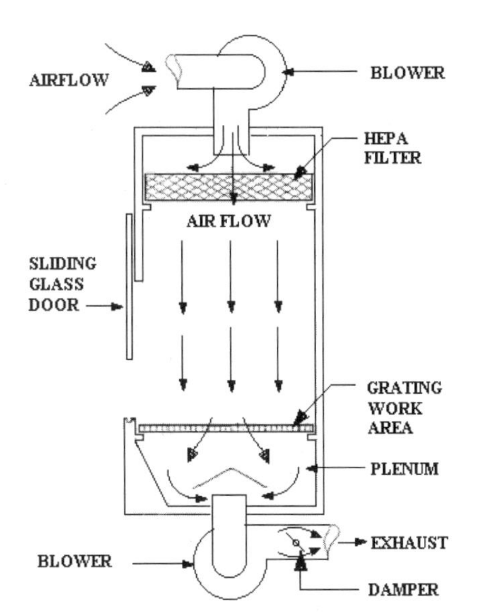

FIGURE 11-13
Section through a clean-air exhaust hood

could be used by operators in street clothes. When used in this manner, the hood still provides a high level of cleanliness within its confines but helps avoid the very high cost of constructing and operating a clean room. A clean air exhaust hood is shown in figure 11-13.

Air Conditioning

The air-conditioning equipment for cooling, heating, and humidification or dehumidification should not only provide the required temperature and relative humidity but also have sufficient blowing power to provide the required airflow velocity across the entire supply areas in a uniform manner.

In most instances, the amount of air supplied to a clean room is much greater than the amount of air required for cooling. Therefore, only a portion of the return air would be passed again through the air conditioner, since it is not necessary to recool the entire volume of circulating air. The recooled air is sent back to the clean air system and mixed with the recirculated air. How much air needs to be recooled is a factor of the clean room's heat load.

If the operation conducted in the clean room requires a lower level of humidity than the air conditioning equipment can provide, a separate dehumidification system should be used. There, too, only a portion of the return air is passed through the dehumidification system, as it may not be necessary to dehumidify the entire volume of circulating air. The dehumidified air is sent back to the clean air system and mixed with the recirculated air. How much air needs to be dehumidified is a factor of the clean room's required humidity level.

Like any other type of laboratory room, a clean room's power consumption, heat generation, and other utility requirements must be listed for each piece of equipment used or for each process taking place in the room. Each piece of equipment requiring emergency power should also be identified. This information is needed to calculate the heat loads of the room and the size of the equipment, uninterruptible power supply (UPS), and emergency generator needed to service the room.

One cannot design the HVAC system required by clean rooms without addressing the question of energy consumption and efficiency. Because of the very high tonnage of cooling they require, clean rooms' energy consumption is extremely high. Generally, owners are much more concerned with the cleanliness of the operation than with the cost of energy, as contamination may be disruptive and production stoppage would cost much more than a high-consumption system. This is reflected by the fact most, if not all, systems available on the market are not yet providing any serious level of efficiency. Further, many clean rooms are not yet able to consistently maintain both the required particle count and the room's temperature and humidity. When an imbalance occurs, work must be interrupted and the problem fixed. The cost to the owner in loss of production is generally very substantial. It is therefore extremely important for an owner to select a mechanical engineering firm highly specialized and experienced in the design of clean rooms.

Lighting Equipment

The clean room's lighting equipment must provide shadowless, uniform lighting at the intensity required by operations. In most instances an intensity of 100–150 foot-candles (1076–1614 lux) is satisfactory. This light intensity should be maintained over the benches and at other work areas. Light fixtures should be flush with the room's ceiling to prevent cracks, crevices, or ledges that accumulate dirt. Further, each fixture must be sealed to prevent air leaks.

Services to the Clean Rooms

Clean rooms often require services and utilities. When this is the case, the services and utilities should be brought in through the inside of the air return walls or through the trenches located under the floor. The number of services and utilities placed in a clean room, including sinks and natural gas valves, should be kept to a minimum and should be placed in the areas where they are needed. This is to reduce the possibility of contaminating the clean room.

The operations that will be performed in the room will determine which service or utility should be brought in. Let's take a clean room designed for trace element chemistry as an example. This room will need cold and hot water, as well as distilled water, in addition to natural gas and vacuum lines. The lines bringing in cold, hot, and distilled water could all be plastic. PVC can be used for the cold and hot water lines. Polyethylene- or Teflon-lined lines may be used to bring in distilled water. PVC lines may also be used to bring in distilled water when the distilled water gets its final polishing in the clean room. PVC faucets can be used for the cold and hot water lines as well as for the vacuum lines. Polyethylene faucets and valves with a high molecular weight may be used for the lines bringing in distilled water. Glass, PVC, or other kinds of approved plastics may be used for drains and waste lines. Polypropylene or polyethylene should be used for sinks, double sinks, cup sinks, and drain boards instead of stainless steel or natural or artificial stone.

Most building codes prevent the replacement of conventional natural gas valves. Therefore, burners using conventional natural gas must be located as close as possible to the gas line's point of entrance into the clean room. This is because use of metal piping should be avoided in a clean room dedicated to trace element chemistry.

Methods for Achieving a Clean Room Class

According to the Federal Standard No. 209D, there are two different ways to design clean rooms: An existing room can be upgraded or a new room that adheres to clean room requirements can be built.

An existing room can be upgraded to meet the requirements of a class 100,000 clean room by simply using HEPA filtered nonlaminar airflow and/or by replacing existing casework with laminar flow work stations. Even with poor quality nonlaminar airflow (and a limited number of laminar flow work stations), these rooms will have areas inside the work stations that meet the requirements of a class 100,000 clean room.

In some cases, an existing room may meet the requirements of a class 10,000 clean room by using HEPA filtered nonlaminar airflow and by replacing existing casework with a small number of laminar flow work stations. This room will have areas inside the work stations that meet the requirements of a class 10,000 clean room. An existing room can be upgraded to meet the requirements of a class 10,000 clean room by using HEPA filtered laminar airflow. Most areas in this room will meet the requirements of a class 10,000 clean room. All rooms with laminar grating airflow and laminar flow work stations will meet the requirements of a class 10,000 clean room.

An existing room can be upgraded to meet the requirements of a class 100 clean room by using HEPA filtered laminar grating airflows. Also, the first work location that faces the cross-flow laminar airflow entering a room will have an area inside the work station that meets the requirements of a class 100 clean room.

In an uncontrolled area of a newly constructed room with laminar flow work stations, the areas inside the work stations will generally meet the requirements of the three classes of clean rooms—100,000, 10,000, and 100. In a controlled area of a newly constructed room, the room area will meet the requirements of a class 100,000 clean room and may meet the requirements of a class 10,000 clean room. But when this room is equipped with laminar flow work stations, the areas inside the work stations may meet the requirements of a class 100 clean room.

In a newly constructed room with laminar cross-flow air, most of the room areas will meet the requirements of a class 100,000 and a class 10,000 clean room. The first work location that faces the laminar cross-flow air will meet the requirements of a class 100 clean room. In a newly constructed room with a laminar air floor grating, the entire room will meet the requirements of all three classes of clean rooms, 100,000, 10,000 and 100.

Operational Guidance For Clean Rooms

Once a room is certified as a clean room, many precautions must be taken to maintain its status as such. The air introduced into the room must be monitored, and regular, scheduled airborne particle counts should be taken at given locations. Further, rooms with laminar flow air must be regularly monitored for non-laminar flow air.

Equipment and personnel may be a major source of contamination of a clean room. Because of this, the enforcement of administrative controls as required by the room's class is necessary. First, all equipment must be thoroughly cleaned before entering a clean room. Depending on what the piece of equipment is, it must be dusted, vacuumed, washed, etc., to ensure compliance with the required specifications of cleanliness. Transport and storage containers should be constructed of materials that shed few or no particles and should be cleaned as required before being brought in a clean room.

Maintenance operations and janitorial cleaning services to the room should be regularly scheduled and properly supervised. After these operations or services are performed, the room's HEPA filters should be activated and the room should be thoroughly cleaned of all airborne particles that may have been generated by those operations or services. Before the room is reused, it should be tested, as required, to assure its compliance to its proper class.

Access to the clean room should be limited to the persons who are required to perform work in the room. Special attention should be paid to the cleanliness of these persons. They should cover, clean, or change their shoes before entering the room. They should wear special lint-free gowns or use a coverall over their street clothing. The type of gowns or coveralls used is based on the level of cleanliness required and must be determined in the planning stage of the project. The persons with access to the clean room should also cover their hair to avoid contamination by loose bits of hair or skin flakes. They should not wear fingernail polish or other cosmetics such as eye makeup, face powder, and hair spray, nor should they use medical or other kinds of ointments on their body. Obviously, they should not be allowed to smoke or eat in a clean room. They should avoid touching solvents, since most solvents remove natural skin oil and cause skin to peel or flake. They should use hand lotions, creams, or soap containing lanolin, as lanolin tends to tighten skin particles. They should use special, nonshedding paper for the paperwork required in the clean room and write with ballpoint pens only, as lead pencils and erasers are not permitted for obvious reasons. When handling products or containers containing products that require a clean room environment, the handler may have to use gloves, finger cots, tweezers, or other adequate equipment to avoid contaminating the product.

Chapter 12

INDOOR AIR QUALITY IN LABORATORY BUILDINGS AND ROOMS

THIS chapter addresses the issue of air quality within laboratory buildings and rooms. This issue is of particular importance, as these buildings require a controlled environment and ventilation that is provided solely by mechanical means. Although laboratory rooms have, in most instances, one-pass air with many air changes per hour, the people who work in these rooms and possibly the operations conducted in them are affected if the air supplied to the room is sufficiently polluted. The same is true for the parts of the laboratory building where the air is recirculated, the problem perhaps being compounded by modern efficiency.

In the last few decades, more and more building materials containing chemicals that may affect people's health or comfort have been used. A rise in the cost of energy during these decades increased the cost of building materials and construction, and led architects to design energy-efficient buildings. More buildings became airtight, with fixed panel windows and centrally controlled HVAC systems. The users of these buildings, in the private sector or in government, increased the density of occupancy to counteract the increase in costs. They reduced the space allocation per person to the minimum required for the task and subdivided the space more efficiently to respond to the new densities. The open space concept was used more frequently in office and administrative spaces, with 60- to 110-square-foot (5.57- to 10.22 m^2) cubicles per person. Simultaneously, a substantial increase in the use of office equipment took place. In addition to a higher density of copiers and fax machines, almost every work station was equipped with a personal computer. As these changes took place, insulated partitions between work stations and wall-to-wall carpeting became a standard to attenuate sound in the space.

People working in the new environment started to feel uncomfortable and had physical and psychological reactions that affected their health and their performance. It was only when the U.S. Environmental Protection Agency employees—scientists and others working in this type of environment—decided to test their working environment that this issue was taken seriously. Pollutants and contaminants were identified and the effect of the indoor environment on people began to be recognized.

To correct or avoid an environmentally unsafe or uncomfortable indoor occupancy, it is helpful to raise the consciousness of the architects and engineers about the problems associated with pollutants: how the problems affect the health, safety, and comfort of the user; how the pollutants reach the indoor environment; and, finally, how the pollutants or their effects can be controlled or eliminated. This chapter is intended to serve only as an introduction to this subject. The following four sources provide more detailed information:

- *Building Air Quality: A Guide for Building Owners and Facility Managers.* Available at Superintendent of Documents, published jointly by EPA and NIOSH. This document provides practical information on how to reduce potential for indoor air quality problems as well as how to deal with such problems. The guide contains flowcharts, forms, checklists, and tables for quick reference managing indoor air quality and diagnosing problems.
- ASHRAE 62-1989, Ventilation for Acceptable Indoor Air Quality.
- *The Inside Story: A Guide to Indoor Air Quality.* EPA 402-K-93-007, April, 1995.
- EPA Sources of Information on Indoor Air Quality (IAQ) Publications (www.epa.gov/iaq/pubs/index.html).

A building's indoor environment is the result of four elements: the outdoor conditions, the indoor building conditions (e.g., pressure differentials, temperature differentials, and pollutants), the building's HVAC system, and the building's user or occupants. The relationship of these elements is illustrated in figure

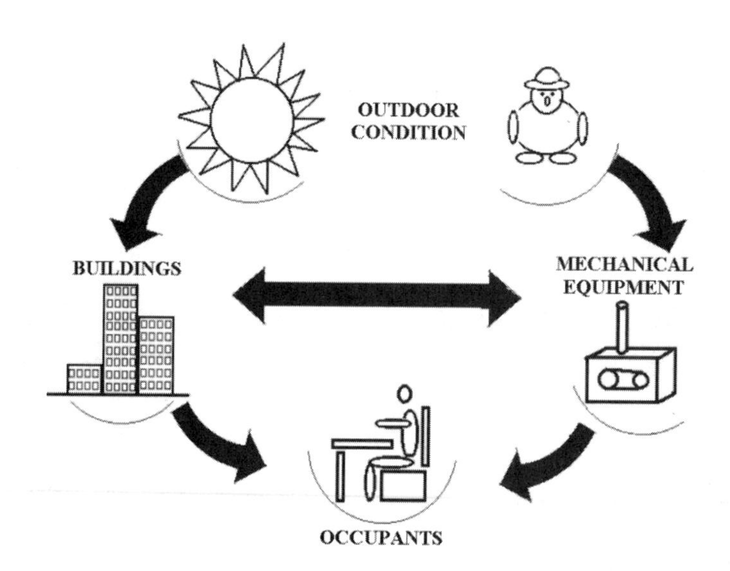

FIGURE 12-1
The elements that create the indoor environment

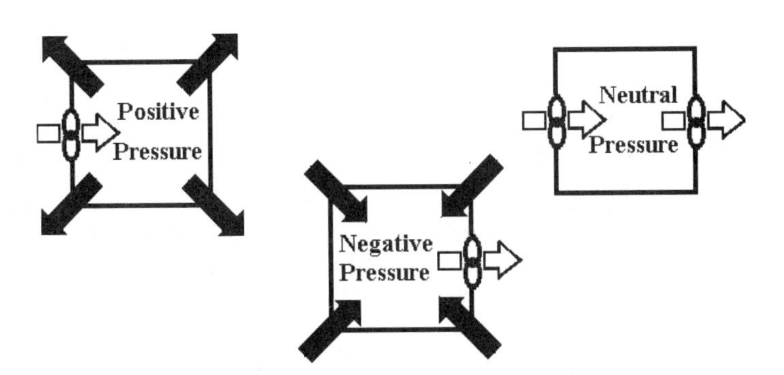

FIGURE 12-2
Indoor/outdoor pressure differences

12-1. All contaminants, once indoors, are spread directly by diffusion, without airflow, or by convection, with airflow, based on pressure differences. Pressure differences are caused by temperature differentials, wind, or mechanically altered indoor/outdoor pressures.

The pressure between two rooms can be positive, negative, or neutral. A room is under positive pressure when more air is supplied to the room than exhausted. A room is under negative pressure when more air is exhausted than supplied, and it is neutral if the volume of air supplied is equal to the volume exhausted. An understanding and application of these principles makes it possible for the HVAC system to play a major role in controlling or eliminating air pollutants in an indoor environment. This relationship is illustrated in figure 12-2.

The eleven major indoor air pollutants are: asbestos, biological contaminants, carbon monoxide, tobacco smoke, formaldehyde, lead, nitrogen dioxide, organic gases, particulates, pesticides, and radon. These pollutants' effects on human beings are as irritants, asphyxiants, neurotoxins, allergens, pathogens, and carcinogens. Irritants cause inflammation and irritation of the skin on contact. Formaldehyde, cleaning agents, particulates, tobacco smoke, biological contaminants, and pesticides are irritants. Asphyxiants interfere with oxygen's attempt to reach tissues and can cause irreversible brain damage. Nitrogen oxide, carbon monoxide, and tobacco smoke are asphyxiants. Neurotoxins act as narcotics; they depress the nervous system, prevent its normal function, and may damage the liver and kidney. Neurotoxins are found in organic gases, pesticides, glues, organic solvents, and in the chemicals in paint. Allergens cause immune-system reactions. Symptoms are from mild irritation to life-threatening fever and illnesses. Mold, indoor dust, and other biological contaminants are among the allergens. Pathogens are disease-producing microorganisms that, if ingested, cause fever, chills, tightness of the chest, and sicknesses such as Legionnaire's disease and hypersensitivity pneumonitis. Biological contaminants are pathogens. Carcinogens are matters that can cause cancer. Asbestos, radon, tobacco smoke, particulates, and organic gases are carcinogens.

It should be noted that tobacco smoke is a mixture of aldehydes, carbon monoxide, benzopyrene, tar, and 4,200 other chemical compounds, of which 43 are carcinogenic. One should also be aware that relatively small quantities of lead reaching the body will cause serious damage to the kidneys, nervous system, and red blood cells. It will decrease the mental ability and impair the normal development of children. The eleven pollutants enumerated above can be ingested by mouth, inhaled to the lungs, or can penetrate the body through the skin. Once absorbed by the body, they distribute themselves to organs and tissues. They can

FIGURE 12-3
Moisture migration in the building shell

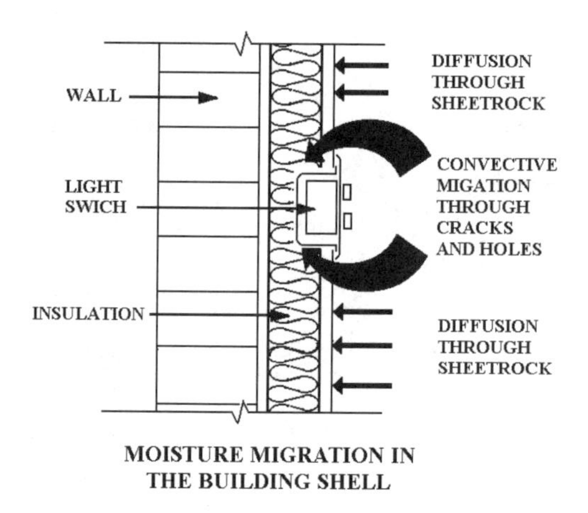

be introduced into a building, or may be part of the building itself. Carbon monoxide, lead, nitrogen, dioxide, and particulates are produced by industrial and vehicular exhausts. Asbestos and other particulates may result from demolishing a building containing them. These pollutants are introduced into a building through the HVAC outdoor air intake or through windows and doors if the building pressure is equal or negative to the outside.

Buildings' equipment and materials will also generate pollutants. Asbestos is found around pipes and furnaces, in certain types of insulation, in ceilings, floor tiles, shingles, and siding. Biological contaminants are found in mold from insulation and carpets not protected by moisture barriers. Insect excreta and animal dander are also contaminants found in insulation, carpets, and other locations. Figure 12-3 shows how contaminants, insect excreta, and animal dander can penetrate most walls and lodge in the wall's insulation. Carbon monoxide is generated by poorly exhausted furnaces, by air entrapment through air supply ducts, and by tobacco smoke. Formaldehyde fumes emanate from particle boards, plywood, urea formaldehyde insulation, and certain types of carpet adhesives. Lead is generated from fumes and flakes of lead-based paint. Organic gases emanate from paints, solvents, and certain kinds of adhesives. Particulates will result from badly vented kerosene, gas, and wood heaters and furnaces. Pesticides are generated from indoor pesticide application and outdoor application drawn in by air intakes. Radon emanates from soil and rock when buildings are erected, through unsealed basements, and from certain kinds of stones and natural cements.

Occupants and users may generate tobacco smoke, lead from soldering and electronic repairs, organic gases from different types of cleaners and air fresheners, and biological contaminants from spills not disinfected and dried immediately. Now that we have identified the major contaminants, their chemical compositions, how they enter and disperse throughout the human body, and their short-and long-term health effects from an industrial hygienist point's of view, let us discuss ways to deal with eliminating, controlling, or reducing their effects. This discussion is deliberately limited, since it is introductory in nature and not meant to be definitive.

The HVAC system can be a pathway, a driving force, and/or a source of contaminants. The system introduces outdoor air, handles and conditions air

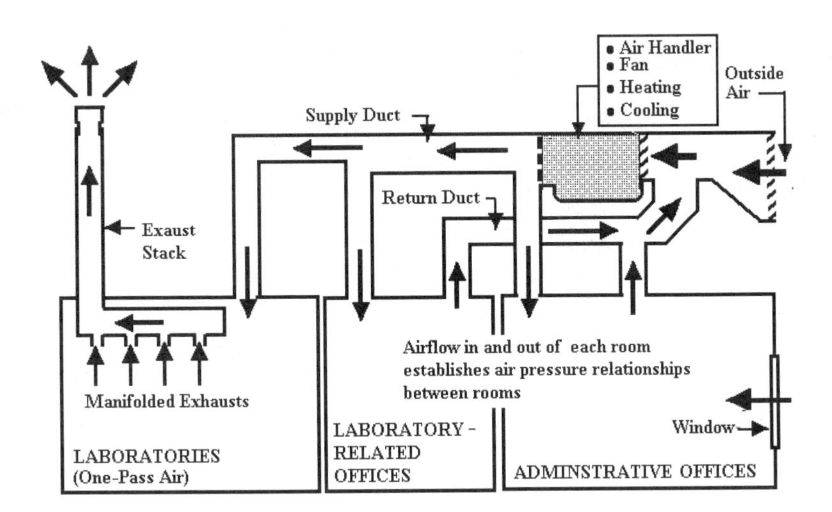

FIGURE 12-4
Schematic overview of the HVAC functions

within the building, controls the building's various pressure relationships, and, finally, exhausts the air. An overview of these functions (for areas where air is recirculated) is illustrated in figure 12-4. There are several ways to use the HVAC system to control air pollutants. The following are among the simplest.

Make sure that outdoor air intake is located so that it does not suck in pollutants generated by vehicular and industrial exhaust. Sometimes this is impossible, either because of the location of the building or because of the air quality index. There are ways to address this situation that can be used together or separately. Depending on the form, source, intensity, and permanency of the pollutants, one can use filters, electrostatic precipitators, and/or water-activated scrubbers. One can also use pressure differentials. There are several kinds of filters and water-activated scrubbers that can be used to filter outdoor air. Filters and scrubbers have different levels of potency. Particulate/dust filters prevent 70–95 percent of particulates from entering. HEPA filters block as much as 99.9 percent of particulates down to the size of 1/10 of a micron. As the potency of a filter increases, the resulting static pressure increases, and the air-handler power required to circulate the air should also increase, as will the cost of upgrading the system.

Electrostatic precipitators, water-activated scrubbers, and charcoal filters can filter intake air separately or in a multifilter arrangement. These filters are also used on the exhaust end of the HVAC system to make sure that no contaminant is released to the environment.

Pressure differential is a good way to assure that no untreated air migrates from one space to another. If the building pressure is higher than the outside pressure, any leak would be from the building to the outside. Only the air sucked in through the outdoor air intake of the HVAC system and treated as needed will enter the building. Pressure differential can also be used inside the building as a tool to diminish indoor pollution spreading and to assure that contaminated air generated in a room does not move to another room or to the corridors.

Buildings where smoking is permitted should have strategically located smoking rooms. These rooms should have one-pass air. Their pressure should be negative to the surrounding rooms and corridor so that when a door opens, the air moves from the corridor to the room, where it is exhausted to the outside.

Finally, the HVAC system should be regularly purged during weekends or days when the building is not occupied and when the outside air is clean or least polluted. In addition to using the HVAC system, there are many other ways that architects can address indoor pollutant control. It is sufficient to say that the architects' and the engineers' most powerful tools are their specifications and construction details. They should specify, as often as possible, contaminant-free materials and verify that these materials are used. When this is not possible, they should specify the encapsulation of some of these materials and/or the ample venting of materials containing fumes before installing them. They should provide good installation and construction details, particularly for moisture barriers and mechanical equipment.

Chapter 13

STRATEGIC MASTER PLANNING FOR LABORATORY FACILITIES

THIS chapter addresses five topics:

- Definition of strategic master planning
- Identification of needs for strategic master planning
- Identification of existing resources
- Development of the strategic master plan and implementation plans
- Institutionalization of the process

DEFINITION OF MASTER PLANNING

One cannot write about strategic master planning without starting by defining the terms. This is necessary because often these terms mean different things to different people, depending on their disciplines.

According to *Webster's Third New International Dictionary,* a master plan is an overall plan into which the details of specific plans are fitted. It is a plan that gives overall guidance. It can also be a graphic or verbal scheme for the development of a city, town, or other building project of an evolutionary nature. According to the *Dictionary of Architecture and Construction,* edited by Cyril H. Harris, the definition is more graphic and physical in nature. It is a plan, usually graphic and drawn at a small scale, that is often supplemented by written material explaining the elements of the project or scheme.

Webster's Third New International Dictionary gives the term *strategic* several definitions, two of which relate to master planning for facilities. The first is concerned with strategy. Again, according to *Webster's, strategy* has many definitions, but the following are most relevant: "the art of devising or employing plans or stratagems toward a goal," and "the science and art of employing the political, economical, psychological, and [other] forces ... to afford the maximum support to adopted policies." If we combine these definitions and relate them to facilities, we conclude that the master planning of facilities should be an evolutionary process leading to the development of an overall plan that gives overall guidance.

The second definition of strategic is "of great or vital importance within an integrated whole, or to the taking place of a planned occurrence." The master plan includes specific subplans, detailed as needed and comprehensively fitted together. In order to be strategic, the master planning process should lead toward one or more goals, defined within an integrated whole. It should employ relevant resources (political, economic, technical and otherwise) to achieve the goals, and it should be implemented in an orderly, planned way. This plan should be updated as needs and are satisfied or changed.

IDENTIFICATION OF NEEDS FOR STRATEGIC MASTER PLANNING

The first question to be asked when identifying the needs is: What are the goals toward which the master planning process should lead? Many architects, engineers, and facility maintenance people tend to orient themselves toward facilities-related goals: the best and most efficient facilities for the functions and tasks to be performed in them. However, having this as a primary goal is a mistake. This is because the functions and tasks performed in a facility are simply the means of achieving higher goals. The primary goal therefore should be to identify the highest, or ultimate, goal, and this should be the goal that drives the planning process.

These guiding goals are generally the program goals of the organization (corporation or agency) for which the strategic master plan for the facility is developed. The *program* can be defined as the activity or activities of the organization. It may be a research program, product development for an industry, or any other activity that requires the use of a laboratory facility. The master plan for a new organization is usually derived from that organization's mission. In most instances, however, the organization is already existing and is projecting the development of new activities or simply wants to know how their current activities could take place under different environments or market situations. These projected activities, new or current, are the organization's programmatic goals, for which a strategy and strategic master plan for the facility are developed. After the goals have been identified, it is necessary to determine how the goals will be achieved, as well as how such achievement can be measured.

The next step is to identify the subgoals or objectives that will be required to achieve each main goal, as well as the tasks and operations that must be undertaken to accomplish those objectives. This exercise is generally performed by or for the corporation or agency's upper management. The management must undertake a strategic analysis of what is needed to reach each objective and look at the program's strategic options. It should then test, refine, evaluate, and analyze these options and then prioritize and select the most promising of them.

FIGURE 13-1
Strategic facility master planning

Simultaneously, a similar set of tasks relating to the laboratory facilities should take place. This overlapping is shown schematically in figure 13-1.

Often, the laboratory facility provides or restrains strategic business solutions. Thus, multidisciplinary, concurrent planning processes lead to the achievement of program goals. One can say, therefore, that program strategy determines facility needs. However, changing facility needs require substantive capital and human resources, which may affect or limit the achievement of program goals. One may conclude, therefore, that the strategic master planning for the facility addresses the interaction between programmatic strategy and the facility's needs and requirements.

STRATEGIC MASTER PLANNING FOR THE FACILITY

In order to perform the tasks and operations to reach each objective, certain resources are required. These resources can be subdivided into three categories: human, instrumentation and equipment, and the facility where the tasks and operations will take place. Acquiring resources from scratch can take a very long time. It normally takes 3 to 6 months to acquire qualified personnel, 6 months to a year to purchase scientific instrumentation and equipment, and 1½ to 3 years for an organization to acquire highly specialized space (3½ to 5 years are required for a government agency to acquire the same type of space). Obviously, securing specialized space causes the greatest delay for organizations, corporations, or government agencies beginning new or additional tasks and operations. In order to avoid or reduce these delays, it is imperative to marry the strategic planning for the program with that for the facility. Additionally, the acquisition process must be started very early in the planning stages, as soon as the needs have been determined. The process of identifying these needs is long and tedious. The steps to follow are described in detail in chapters 1 and 2.

The marriage of the strategic planning for the program and of the strategic planning for the facility requires that the following issues be analyzed:

1. The organizational structure of the different activities requiring laboratory space and the interaction of those activities with the other activities of the organization.
2. The organizational policies on merger, joint venture, subcontracting, co-locating, and centralizing or decentralizing.

3. The organization's acquisition strategies and policies.
4. The market demand, the potential for growth and change.
5. The link between research, development, and production.
6. The organization's approach to quality improvement and technical entrepreneurship.
7. The organization's financial and real estate strategies.
8. The regulatory impact on the facility's acquisition strategies.
9. Human resources.
10. The capabilities and limitations of existing and available facilities and sites.
11. The organization's relocation policies and analysis.

Generally, an organization undertakes a strategic plan for its programs (which is what most nongovernmental organizations refer to as their business plan) when it has a good reason to do so. When this reason is to project for required future changes, it may be wise to develop a complete planning process to determine how and when the contemplated changes could take place and what steps are necessary for the execution of these changes. This planning process includes four (and sometimes five) sets of activities (each being a full process itself):

1. *The program strategic master planning process* determines what the entity intends to do to achieve all or part of its mission or goal. For example, a pharmaceutical organization may have as part of its mission to develop new products. This organization's program strategic master plan would be to identify what product they want to produce and when they want to produce it, and to set a course of action to achieve that production. The program strategic master planning process results in the program strategic master plan (which private sector organizations refer to as their business plan).
2. *The implementation process for the program strategic master plan* describes the steps required to implement the program strategic master plan. These steps identify, analyze, and determine which part of the organization is best equipped to implement all or parts of the program strategic master plan, as well as when and where such implementation should take place. The end result of this process is the implementation plan for the program strategic master plan.
3. *The facility strategic master planning process* determines what type and amount of space is needed to implement the program strategic master plan in accordance with the implementation plan of the program strategic master plan. The facility strategic master planning process will also estimate the cost of acquiring the needed space and determine the most efficient manner of acquisition (build, build-lease, rent or purchase existing space, or modify their own existing space). This process also sets a time schedule to match the required production schedule. The end result of this process is the facility strategic master plan.
4. *The implementation process for the facility strategic master plan* describes the different activities required for the acquisition of space and the order in which these activities should take place according to the budgetary priorities of the organization. The end result of this process is the implementation for the facility strategic master plan.

FIGURE 13-2
Schematic showing transition from strategic planning to implementation/operation planning

5. *The facility physical master planning process* for a given site results in a drawing of the site that shows the location of the structures, the services and utilities, and the opportunities and constraints for building development. It should be a plan showing an ideal configuration of site and building development. The facility physical master planning process is similar to the facility strategic master planning process.

These definitions show the relationships between the different processes. They also emphasize the importance of planning to acquire facility needs when the organization is developing its program strategic plan, as the facilities have a major impact on the capital cost and production schedule. Figure 13-2 shows the sequence of these processes.

The strategic master planning process for laboratory facilities plans for, and ultimately provides, the space required by an organization contemplating changes in its activities. Therefore, master planning should respond to the contemplated changes in a timely manner. When an organization projects for future changes, either to increase or decrease its activities, it must develop a strategic plan for its programs. When such a plan is developed, and the contemplated programmatic changes are identified, the organization should develop a plan to implement the program strategic master plan. This implementation plan identifies what changes should take place and when they should take place. It also identifies the functions and tasks that will be added or subtracted and the personnel, equipment, and space needs of such functions. The development of the facility master plan is based on this information.

The facility master plan identifies the options and tasks requiring space, determines the kind of space required, identifies how the space will be acquired, estimates the cost of acquisition or renting, and determines how the acquisition will respond to the program implementation plan. These findings (supplemented by recommendations) are submitted to management for approval.

Once management selects the option or options that it feels are the most efficient for the tasks at hand, the development of an implementation plan for the facility strategic master plan can begin. This plan encompasses the selection of a planning team, the determination of the space required by each task, the evaluation of existing space, the determination of availability of space within existing space, the identification and determination of the resources needed to acquire or modify the space, and finally the confirmation of consistency between the implementation plan for the facility strategic master plan and the facility strategic master plan.

When an organization has many laboratory facilities in different locations, this organization must develop individual facility master plans (to identify the best use of available resources) for each facility. This process is similar to the facility strategic master planning process.

The sum of the above-described activities constitutes the master planning process for laboratory facilities. Figure 13-3, at the end of the chapter, summarizes these processes and shows how the tasks are intermeshed.

Often it is difficult to determine many of the laboratory facility's requirements during the early stages of the planning process. There is a way of overcoming the problem, particularly when the future laboratory procedures are either an extension of existing ones or are similar to the type of work the corporation or agency is already doing. When this is the case, the program history must be analyzed.

The first step is to look analytically at how the program started and evolved. Evaluate the resources that were needed as the tasks increased and how these resources contributed to meeting objectives, if they did. Also analyze the shortcomings, if any, and how they were handled. Another step is to identify the restraining factors, existing or projected, that must be accounted for. These factors may be scientific, economic, geographic, psychological, market driven, or related to the effort planned in almost any way. This analysis should identify the program trends and use of resources as the mission increases or decreases. It results in a projection curve for each of the resources. This projection curve should show the most optimistic, most likely, or most pessimistic projection.

Future need of resources can be deduced from studying these projections. This can be done as follows: The objectives that must be reached at given times should be identified first. Then they should be defined. The procedures (chemical, biochemical, physical, or otherwise) necessary to reach these objectives should be developed and described in detail. Each procedure, task, subtask, and operation should also be identified and described. The way in which these tasks and operations will flow and evolve in the laboratory space should also be described in a detail.

This whole "package"—from objectives to tasks and subtasks—is called an "operation plan." A more detailed explanation of how to develop an operation plan is provided in chapter 1. Once the operation plan is completed, the projected objectives should be coordinated, as needed, with the procedures that are expected to continue without change. It then becomes possible to quantify and qualify the human resources needed, as well as the required instrumentation and equipment for the combined procedures. The acquisition of these resources can therefore be scheduled.

It should be noted here that certain types of laboratories (such as research laboratories) are not driven by capacity as production laboratories are. In these types of laboratories, the operation itself controls the amount of space required. It also determines if the task should be segregated from other activities and dictates the requirements for the space or room in which the task is performed. Sometimes segregation is also required for the space servicing the operations, such as a glassware washroom and storage space or the unpacking part of a receiving area. Other types of operations may be better conducted in open laboratory space where the operations flow from countertop to countertop and from operator to operator, allowing continuous and direct communication between the operators.

An accurate determination of space requirements for each task becomes possible once the following information is collected:

- The tasks and operations required by the procedures
- The human resources allocated for the procedures
- The equipment and instrumentation needed to perform the procedures
- The date by which the procedures should start (thus, the date at which the space should be available)

The space needed for tasks will dictate the size of the laboratory room. This should be described in the RDS (see chapter 4). Rooms with similar characteristics or needs should be grouped together. The support space needed to directly and indirectly service these rooms should also be identified. Finally, all spaces should be grouped according to their function into blocks, as described in chapter 1.

Once the space needed and the date in which it should be available is identified, the planning process to acquire it can begin. This process uses information on existing resources, which should be collected simultaneously with the previously mentioned information.

IDENTIFICATION OF EXISTING RESOURCES

The search up to now has been focused on the identification of the space needed for the tasks that will help accomplish the corporate or institutional goals and objectives. This need is expressed as so many square feet of a specified type of space needed at a given date for a given sets of tasks. For example, a facility may need 50,000 square feet of class A laboratory, 10,000 square feet of class B laboratory, and 5,000 square feet of laboratory support space by June 2005 for tasks 1, 2 and 3 of the Genetic Engineering Research Program. This stipulation of space and when it is needed should be done for each of the set of tasks to be preformed.

Once the required space is identified and the date by which it will be needed is established, one should evaluate if and how the existing facilities could partially or fully respond to these needs. The existing facilities should be evaluated according to their condition and their adequacy for the present and projected uses. Evaluators should use the same standards, which fall into one of two categories:

- engineering and architectural standards
- health and safety standards

These two categories should encompass and supplement all local, state, and federal codes. It is not sufficient to base one's evaluation of a facility's physical condition only on its response to existing regulations—the facility's efficiency should also be evaluated according to its adequacy for the projected and existing tasks. It should be evaluated according to the efficiency and cost of operation and maintenance during its life cycle (see chapter 2) and according to the way in which its systems are affected by the surrounding environment and climate (see chapter 3). Finally, the facility systems should also be evaluated according to their resistance to the chemicals and to the wear and tear from required tasks and operations. Engineering and architectural standards and health and safety standards should be tailored to the kinds of tasks and operations required by each corporation or government agency.

The primary objective in performing a physical evaluation of existing facilities, in the contest of strategic planning, is to determine the possible use of the facilities for the projected tasks. The physical evaluation consists of four activities:

1. *An analysis of the physical condition of the facility.* Each structure or component of the facility and the grounds surrounding the structures should be looked at during the evaluation of the facility's physical condition. The age and condition of these structures should be identified. The facility's systems (architectural, mechanical, and electrical) and their age, performance, condition, and efficiency should be evaluated. Are they in need of repair or replacement? What is the life expectancy of each of these systems? What services and utilities are available? What level of services and utilities is presently used? How much can these services and utilities be expanded without replacement or addition? (See chapter 2 for more information on this type of evaluation.)
2. *An evaluation of how the facility responds to the present needs.* Once the physical condition of the facility is fully analyzed, one should look at how this facility is responding to the present needs. What operations or procedures are performed in the facility? How are these operations or procedures subdivided into tasks? What tasks are performed in each of the laboratory rooms? Are these rooms adequately equipped and ventilated? Are these rooms adequately provided with services and utilities for the tasks performed in them? What deficiencies do they have? If part of the facility is not adequate for the tasks being performed in it, could this space be upgraded? Once these types of questions are addressed, one can deduce how the facility is responding to the present needs and start analyzing the potential for expansion, construction, or modification of the facility's systems and components.
3. *An evaluation of whether the facility could be upgraded or downgraded to respond to the projected needs.* If the projected needs require expansion, is there enough land to accommodate the additional space needed? Could this space be added onto the existing structure, or should it be a separate structure connected or not connected to the existing facility? Could the additional space be serviced by existing systems and components? If so, do they need to be expanded or modified, or can they be used as they are? What would be the level of efficiency in each situation? How would the new situation affect the operation and maintenance of the system?

   If the projected needs require a contraction of the operations, how could the facility respond? Do the projected tasks and operations require the same amount of space but substantially fewer services? If so, can the services be reduced? Can this reduction be done efficiently? If the projected tasks and operations require much less space than what is available, how can the surplus space be used? Can the services and utilities servicing the surplus space be disconnected or reduced without affecting the rest of the facility? What would the level of efficiency be in each of the described situations? How would the new situation affect the operation and maintenance of the systems?

   Finally, there may be a situation in which the projected uses do not require expansion or contraction but do require types or levels of services and utilities different from the ones presently provided. If that is the case, can the existing facility adapt to the situation? Can the space be reconfigured? Do the mechanical and electrical systems and their components lend themselves to the projected realignment? Is the existing equipment capacity sufficient to carry the additional loads? Is there enough space above ceilings or in service corridors for realignment or reconfiguration of the ductwork, plumbing, and other services and utilities? If the answer to all or most of these question is yes, can the reconfiguration be made efficiently? What would the cost of this work be? What would be the effect of the required construction work be? Would temporary relocation or interruption of operations be necessary? If not, would the dust, noise, or vibration from the construction affect operations?

There are three major questions that should be asked in all of these situations:

- Can the change be made efficiently?
- How disruptive to the ongoing operations will this change be and can the operations cope with these disruptions?
- Could these changes be cost effective in comparison to other options?

4. *An estimate of the costs of upgrading or downgrading the facility.* When looking at the cost of an operation, make sure that you are considering all the costs of that option. This includes not only the demolition and construction costs, but also the relocation, disruption, interruption, and similar costs. Sometimes the advantages or disadvantages of an option may be difficult to describe in terms of cost. When this is the case, the advantages or disadvantages must be described in terms of how they affect the option under consideration and taken into account during the evaluation. For example, an entity may be looking at an option with a location that would be economically perfect for its operations and distribution. However, the neighbors the entity would have at that location would not welcome it, which could attract media attention and greatly tarnish the entity's image. This, of course, would be considered a disadvantage, although the economic impact would be difficult to calculate.

Once all options are fully evaluated and costs have been estimated, a comprehensive analysis of the options becomes possible. This evaluation's objective, as stated earlier, is to determine the options' responses to the projected tasks in the context of strategic planning. A life-cycle cost analysis, which includes not only initial costs but also operation and maintenance costs, should be made for each option, and all costs should be translated into a constant dollar value. Also, each option's evaluation should include a description of its impact on the facility's operations (see chapter 2).

Next, one should look at other possible options that may respond more efficiently to the future programmatic needs. As discussed in chapter 2, the options include:

To lease an existing facility and use it as is.
To lease a facility and modify it to suit needs.
To lease a new building built to suit needs.
To purchase a piece of land and build a new building on it to suit needs.
To build a new building on piece of land already owned.
To add or modify a building already owned.
To purchase a building and then add to it or modify it to suit needs.

Any combination of these options is also possible, as situations rarely present themselves as simply as the list indicates. For instance, a lease option may be offered with the possibility of purchase or with financial arrangements that are very convenient. A building or a lot for sale may have advantages not necessarily related to its condition or location. Although all of these options may be available, certain factors may eliminate the viability of some and narrow the choice to two or three.

If the corporation owns a piece of land in a convenient location, it may not want to consider the options of leasing or purchasing a building located somewhere else. If this corporation, on the other hand, does not own a site or does not want to face a large initial capital outlay, it may decide to lease the needed space. The first task to be undertaken by the board of directors or administrator is, therefore, to eliminate the nonviable options.

In order to make this decision, the strategic planner must answer many managerial questions, including: How soon and for how long is the new space needed? How much funding is available? How much of this funding is already lined up? What is the projected income for the project's life cycle? These and similar questions should be addressed in order to decide on the type of options to consider. One should look at the projections for each situation, consult with accountants, attorneys, and bankers, and then determine the expansion direction. Let's assume, for example, that the future operation requires a type of space not readily available, that the corporation does not own a piece of land, that the income and work projections show a 30-year need for the projected space, and that the corporation has or can secure sufficient funds for a down payment and financing for the required space. The options for this corporation are: to lease a building, either new or modified to suit needs, or to purchase a building. The next step is to determine which is more efficient—to build a new laboratory or to modify an existing one.

The evaluation and analysis of the modification of an existing building was addressed earlier in chapter 2 and in this chapter. If building a new facility is considered, once the space needed is determined and one or more locations for the building are identified, the cost of each location option, including buildings, site development, roads, etc., should be estimated. The advantages and disadvantages of each option, as well as its impact on the operation's efficiency, should be identified. Finally, the costs and benefits if each option should be evaluated.

It then becomes possible to compare the costs/benefits, advantages, and disadvantages for the options, as well as the impact of each of these options on the operations and tasks of the corporation. This information greatly facilitates the prioritization of options and allows the strategic planner to recommend a selection and outline the implementation procedure.

DEVELOPMENT OF THE STRATEGIC MASTER PLANS AND IMPLEMENTATION PLANS

This chapter has shown how to identify a corporation or government agency's objectives and needs and determine existing resources that could be used to reach those objectives. It also discussed the analytical steps necessary for the

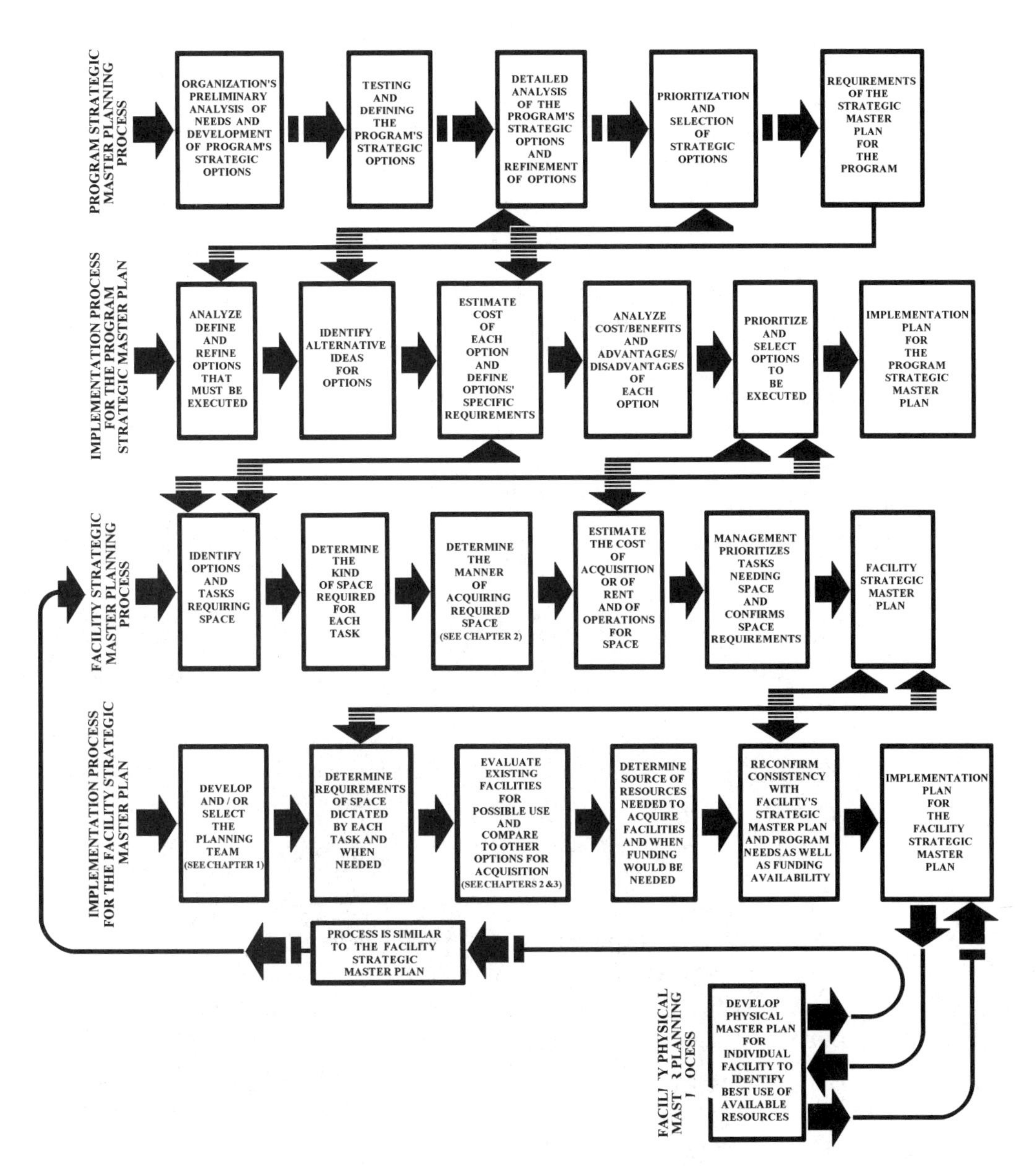

FIGURE 13-3
The facility master planning process

development of a complete planning process and of master plans. With this information in hand, it becomes possible to develop a program strategic master plan for an organization, corporation, or government agency. It also becomes possible to derive from the program strategic master plan long-, intermediate-, and short-term implementation processes for the program strategic master plan.

Once the organization's program strategic master plan is developed and its requirements are identified, these requirements become the goals of the organization's master plans. These goals are then prioritized in terms of what program tasks should take precedence over others and should be approved by the organization's management. Then the required tasks should be analyzed (using the process described earlier and illustrated in figure 13-3), and ultimately, the management should decide which of these tasks require space. Once this determination in made, a strategic master plan for the facility can be developed. The aim of this master plan is to identify what space is needed (including specialized space) and if existing space can be used. Another aim of the strategic master plan for the facility is to determine whether new space must be acquired, how much of this space must be specialized space, and if space can be acquired in increments, as may be required to implement the organization's master plan.

With this as a goal, the facility's personnel (a staff similar in composition to the planning team described in chapter 1) develop a strategic master plan for the facility and an implementation plan to implement the facility strategic master plan. The implementation plan has short-, intermediate-, and long-term plans to realize the organization's needs. The tasks to be preformed by each of these plans will be a factor of available funds and of how some tasks need to have space.

INSTITUTIONALIZATION OF THE PROCESS

As we move towards the execution of the stated goals and objectives, the situation or the aims of the corporation or agency may change as new opportunities and/or difficulties arise. These changes should be expected and continuously addressed. Thus, a genuinely comprehensive and strategic master planning process is constantly evolving.

Once a master plan is developed, it should be continuously updated. That means that when a goal or objective is changed or modified, for whatever reason, the resources should be reevaluated and redirected to respond to the new objectives. When opportunities or difficulties result in a new need, that need should be incorporated into the plan. Further, when a goal is obtained, the resources should be redirected to the objective of next importance. Thus, in order to keep the process continuously up-to-date, it is necessary for the corporation or agency to have a mechanism geared to continuously incorporate the changes. There are many computerized systems available on the market that can help do this.

Chapter 14

LIST OF COMMON LABORATORY INSTRUMENTS

NATURALLY, the type of operations taking place in a laboratory determines which instruments are used. The list provided here cannot include all laboratory instruments, but it should help design professionals familiarize themselves with the most common instruments, what these instruments do, and what type of space each instrument requires.

During the design process, the design professional must consult the instrument user, refer to manufacturer catalogs, and consult the manufacturer's representative before determining which instruments are most fit for the operation at hand. At that time, the designer should also determine the instruments' power and utility needs, footprints on the floor or counter, and any other specific requirements.

In the following list, the terms *analyte* and *constituent* refer to the compound or element being investigated. The term *determine* means to indicate the presence and quantity of the analyte in the compound or element being investigated.

ATOMIC ABSORPTION SPECTROMETER

The Atomic Absorption Spectrometer is a sophisticated instrument used to analyze metal content in water samples or in digested solid samples. Several companies manufacture this type of instrument. Figures 14-1 through 14-3 show three different Atomic Absorption Spectrometers. Each of these instruments may be best suited for different uses. Figure 14-1 is of a Varian Atomic Absorption Spectrometer (AA 220) FS, figure 14-2 shows a Varian Atomic Absorption Spectrometer (AA 50). Figure 14-3 depicts a PerkinElmer

AAnalist 800™ High Performance Atomic Absorption Spectrometer. It should be noted that the AAnalist 800™ Atomic Absorption Spectrometer provides a high analytical performance over the entire AA wavelength range. All three instruments require countertop space

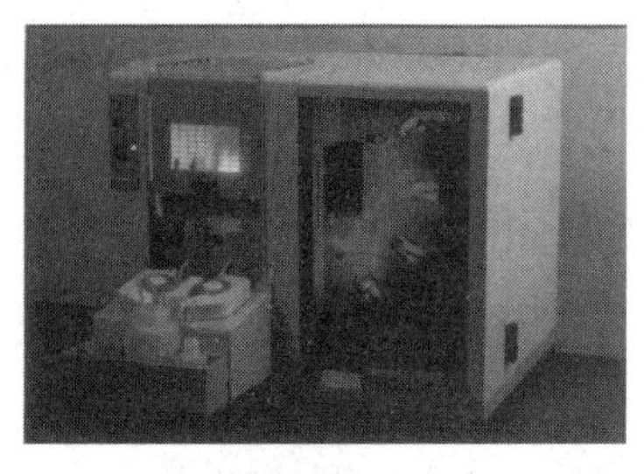

FIGURE 14-1

FIGURE 14-2

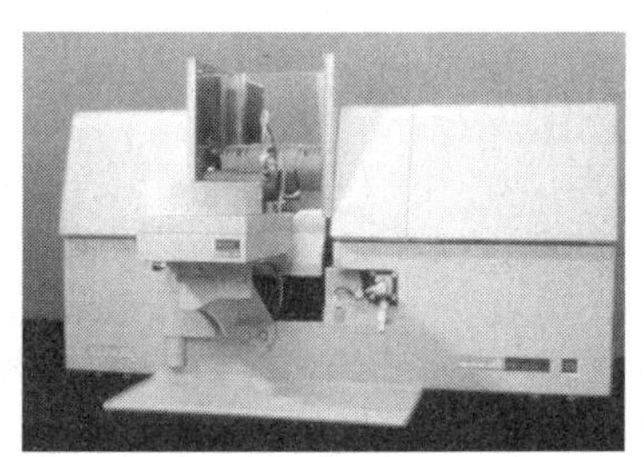

FIGURE 14-3

AUTO ANALYZER

The Auto Analyzer is an automated analyzer that analyzes samples for various water quality parameters. This instrument requires countertop space.

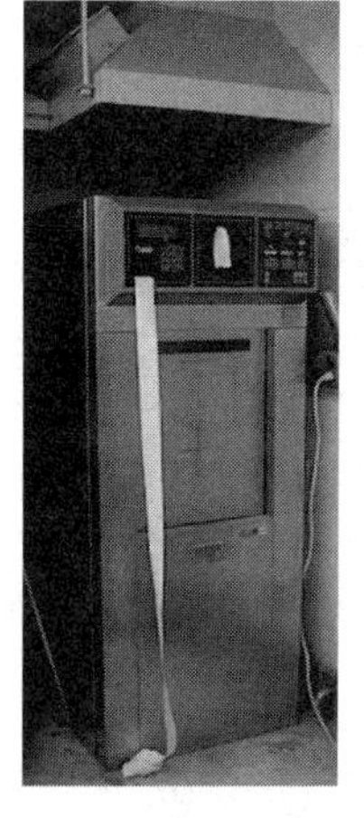

FIGURE 14-4

AUTO ANALYZER CONSOLE

The Auto Analyzer Console is a test-specific, essential component of the Auto Analyzer. It requires countertop space.

AUTOCLAVE

The autoclave (figure 14-4) is a pressurized chamber that uses steam to sterilize laboratory materials and instruments. Several companies manufacture this instrument. Autoclaves usually require floor space.

AUTO SAMPLER

The Auto Sampler is a dedicated robotic instrument used for sample introduction into analytical instruments. This instrument requires countertop space.

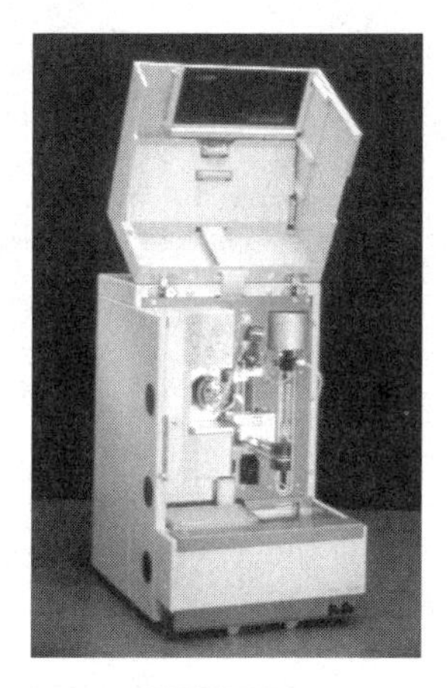

FIGURE 14-5

FIGURE 14-6

Several companies manufacture this type of instrument. Figures 14-5 and 14-6 show different Auto Samplers. Each of these instruments may be best suited for different uses. Figure 14-5 shows a Varian high pressure liquid chromatograph Al-96 Auto Sampler, and figure 14-6 shows a PerkinElmer Automatic Head Space Sampler, Series 200 Diode Array Detector and a Hewlett-Packard 6890 plus GC and associated sampler.

AUTOMATED COULTER COUNTER

The Automated Coulter Counter is an instrument used for bacterial counting. This instrument requires countertop space. Figure 14-7 is a picture of an Automated Coulter Counter.

FIGURE 14-7

FIGURE 14-8

BALANCE

A contemporary laboratory balance is an electronic weighing apparatus (figure 14-8). This apparatus requires countertop space.

BALL MILL

The Ball Mill (figure 14-9) is a high-frequency shaker used to divide large solid samples into smaller pieces. This instrument requires countertop space.

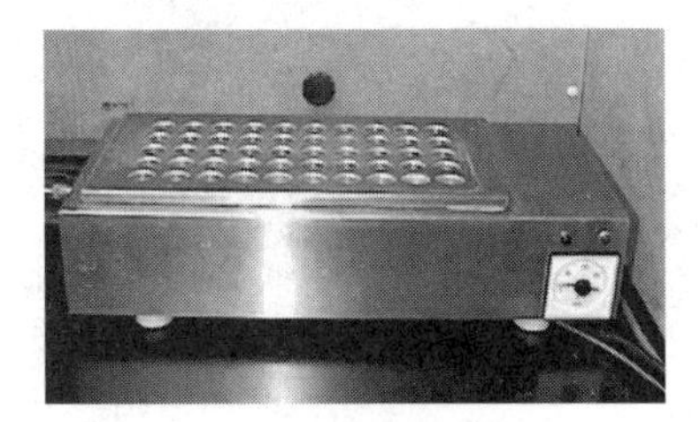

FIGURE 14-9

FIGURE 14-10

BLOCK DIGEST TKN

The Block Digest TKN (figure 14-10) is an apparatus that uses controlled high heat to decompose samples for total nitrogen analysis. This apparatus requires countertop space.

CENTRIFUGE

The Centrifuge (figure 14-11) is a high-speed spinner used to separate liquid/liquid mixtures and liquid/solid mixtures. Depending on the model, this instrument may be located on a counter or the floor.

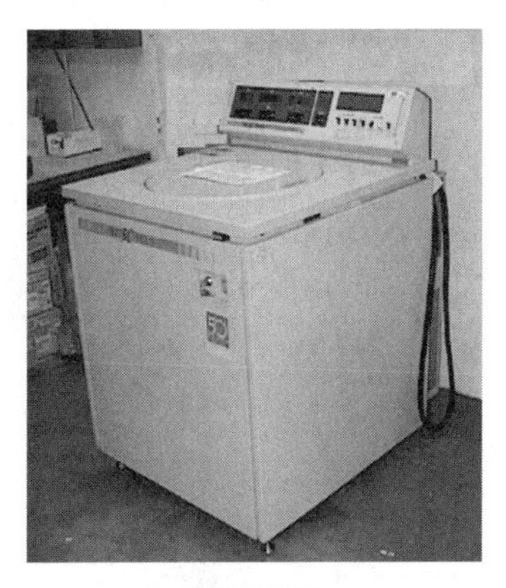

FIGURE 14-11

FIGURE 14-12

CIRCULATED COOLER

The Circulated Cooler (figure 14-12) is a refrigeration system generally used for cooling and recirculating water used by instruments needing cooling. This instrument requires floor space.

COD DIGESTER

The cod digester (figure 14-13) is an apparatus that uses controlled high heat to decompose samples. It is used for determining the chemical oxygen demand (cod). This instrument requires countertop space.

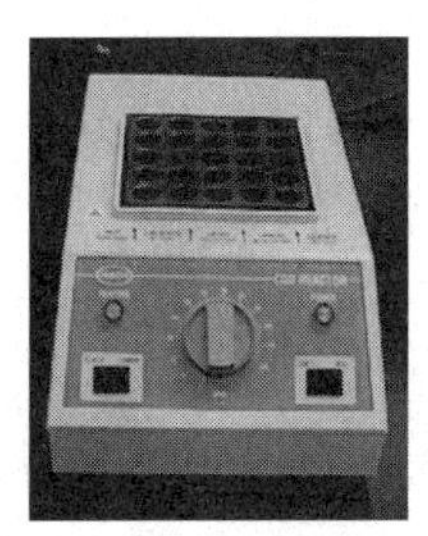

FIGURE 14-13

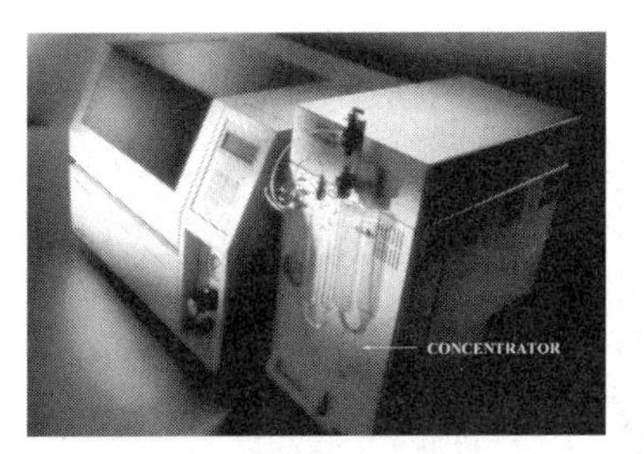

FIGURE 14-14

CONCENTRATOR

The Concentrator, also called a Purge and Trap Apparatus, is an instrument used to bubble gas, usually Helium, through an aqueous sample to get the volatile organic compounds that are in the sample into the gas stream. The compounds are then trapped in a chemical trap, concentrated, and then sent to a gas chromatograph or a gas chromatograph/mass spectrometer. Figure 14-14 shows a Varian LSD 3000 concentrator (on the right) next to and servicing a gas chromatograph.

CONCENTRATOR ORGANOMATION

The Concentrator Organomation, also known as the S-EVAP Evaporator, is a hot water bath used for solvent evaporation and for organic analyses. This instrument requires countertop space. Figure 14-15 pictures two models of an Evaporator Organomation.

FIGURE 14-15

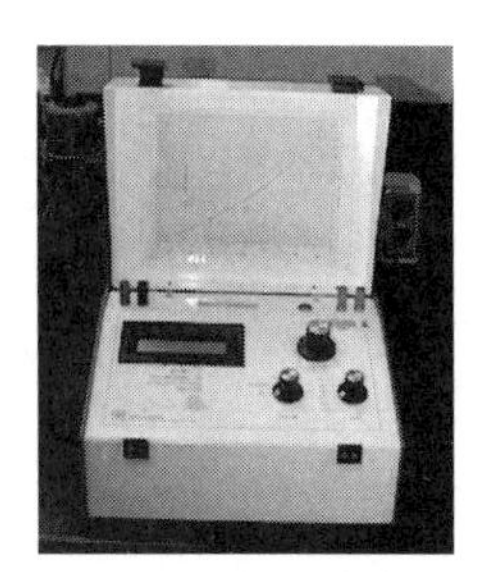

FIGURE 14-16

CONDUCTIVITY METER

The Conductivity Meter (figure 14-16) is an instrument used for measuring the conductivity of liquids. This instrument requires countertop space.

DISHWASHER

The dishwasher (figure 14-17) used in laboratories is an industrial-grade washer used to wash laboratory glassware. This instrument requires under-counter floor space or may be freestanding.

DISTILLATION UNIT

The distillation unit (figure 14-18) is a large electronically controlled apparatus used to purify solvents. This apparatus requires floor space.

FIGURE 14-17

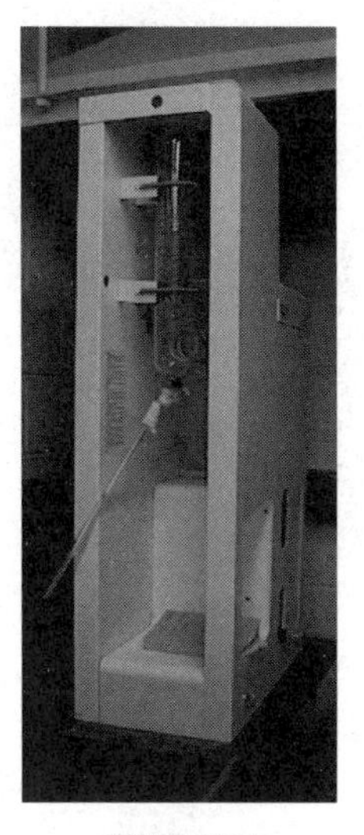

FIGURE 14-18

FIGURE 14-19

ELECTRODE LESS DISCHARGE LAMP (EDL) POWER SUPPLY

The Electrode Less Discharge Lamp is a high-intensity lamp used with the atomic absorption spectrometer. This lamp requires countertop space.

ENVIRONMENTAL SHAKER

The environmental shaker (figure 14-19) is an orbital shaker used for mixing liquids. Depending on the model, this instrument may be located on a counter or the floor.

FERMENTOR

The Fermentor is a controlled heating apparatus used for bacteriological analysis. This apparatus requires countertop space.

FIGURE 14-20

FLOW CONTROL BOX

The Flow Control Box (figure 14-20), also called Auto Dispenser, is a peristaltic pump with flow control used for bacteriological analysis. This pump requires countertop space.

FOURIER TRANSFORM INFRARED SPECTROMETER (FTIR)

The Fourier Transform Infrared Spectrometer, often referred to as the FTIR, is a sophisticated instrument used to determine the presence of petroleum hydrocarbons and other organic constituents. This instrument requires countertop space.

GAS CHROMATOGRAPH (GC)

The Gas Chromatograph, often referred to as GC, is a sophisticated instrument used to detect the presence of organic compounds such as pesticides and PCBs in a sample. This instrument requires countertop space. Figure 14-21 shows a Hewlett-Packard Gas 6890 series Chromatograph. Figure 14-22 shows a Varian 3800 Gas Chromatograph, which is generally used for analysis of organic compounds such as pesticides in environmental laboratories. Figure 14-23 shows a PerkinElmer Gas Chromatograph Analyzer for petrochemicals and food flavor laboratories.

FIGURE 14-21

FIGURE 14-22

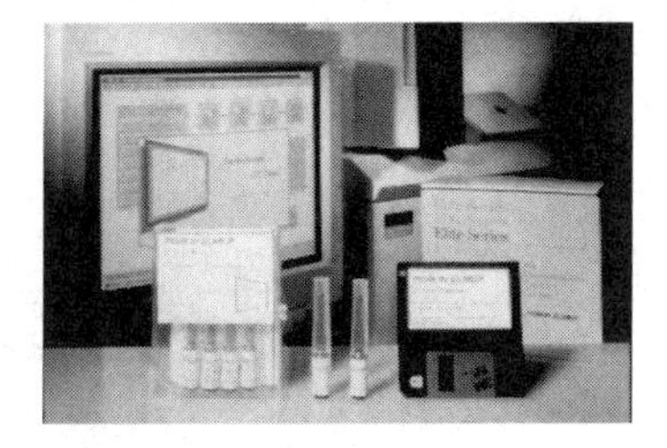

FIGURE 14-23

GAS CHROMATOGRAPH/MASS SPECTROMETER (GC/MS)

The Gas Chromatograph/Mass Spectrometer, often referred to as GC/MS, is a sophisticated instrument used to detect the presence of a wide range of organic compounds in a sample. The Gas Chromatograph separates the organic compounds and the Mass Spectrometer detects the organic compounds' ions and reads the compound's structural information. This instrument requires countertop space. Figure 14-24 shows a Hewlett Packard 6890 Series Gas Chromatograph/Mass Spectrometer, and figure 14-25 shows a Varian 3800GC Gas Chromatograph/Mass Spectrometer, Saturn 2000.

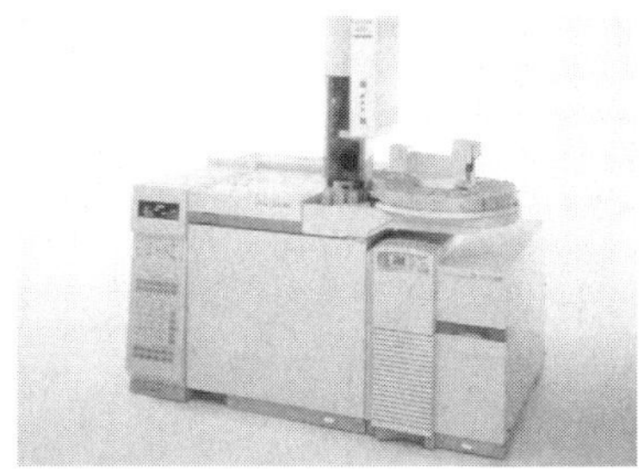

FIGURE 14-24

FIGURE 14-25

GEL PERMEATION CHROMATOGRAPH [GPC] ULTRA WISP

The Gel Permeation Chromatograph (figure 14-26), often referred to as GPC, is an instrument used to purify sample extracts for organic analyses. This instrument requires countertop space.

GRAPHITE FURNACE ATOMIC ABSORPTION (GFAA)

The Graphite Furnace Atomic Absorption, often referred to as the GFAA, is a specialized atomic absorption instrument generally used for low-level metal analysis. This instrument requires countertop space. Figure 14-27 shows a Varian Spectrometer AA 50155 Graphite Furnace System.

H2 GENERATOR

The H2 Generator (figure 14-28) is a sophisticated apparatus that generates and provides hydrogen to various instruments. The instrument shown in the figure requires bench space; other models require floor space.

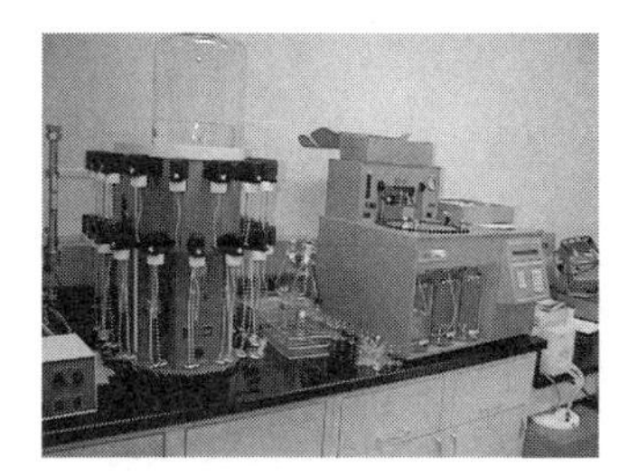

FIGURE 14-26

FIGURE 14-27

FIGURE 14-28

HIGH PRESSURE LIQUID CHROMATOGRAPH (HPLC)

The High Pressure Liquid Chromatograph, also known as HPLC, is a sophisticated instrument used to detect, separate, and identify high boiling organic compounds. This instrument requires countertop space. Figure 14-29 shows a Hewlett-Packard 1100 series High Pressure Liquid Chromatograph, figure 14-30 an analytical Varian High Pressure Liquid Chromatograph, and figure 14-31 a High Pressure Liquid Chromatograph.

FIGURE 14-29

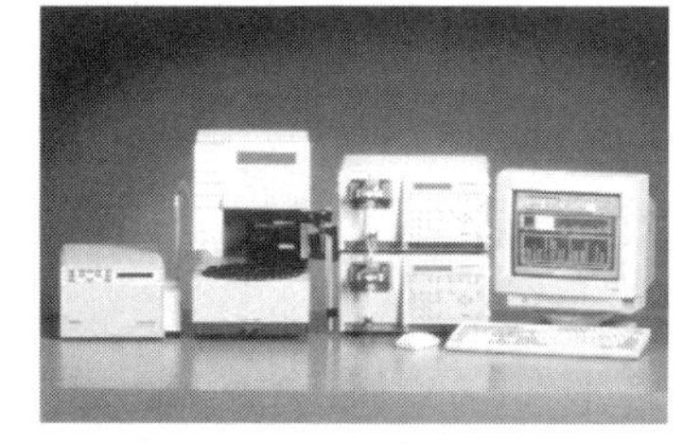

FIGURE 14-30

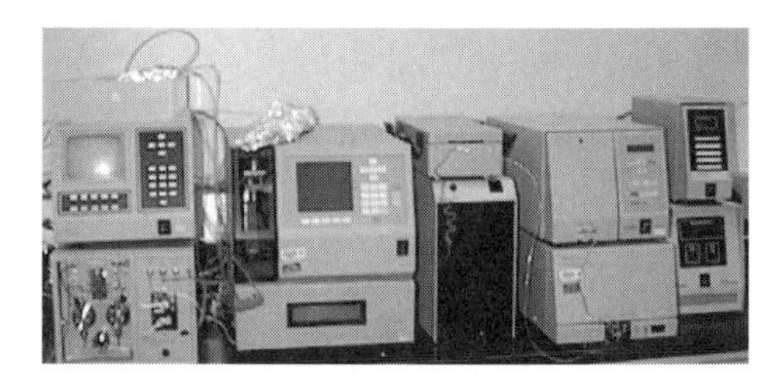

FIGURE 14-31

HIGH PRESSURE LIQUID CHROMATOGRAPH COUPLED WITH A MASS SPECTROMETER (HPLC/MS) AND HIGH PRESSURE LIQUID CHROMATOGRAPH COUPLED WITH TWO MASS SPECTROMETERS (HPLC/MS/MS)

The Liquid Chromatograph Coupled with a Mass Spectrometer, also known as HPLC/MS, is a sophisticated instrument used to detect, separate, and identify high boiling organic compounds. The Liquid Chromatograph/Mass Spectrometer/Mass Spectrometer, also known as HPLC/MS/MS is used for the analysis of organic compounds that require a softer separation procedure than gas chromatography. MS/MS detects daughter ions and is therefore more specific than single MS. Depending on the model, this type of instrument may be located on a counter or the floor. Figure 14-32 shows a PerkinElmer API 365 Liquid Chromatograph/Mass Spectrometer/Mass Spectrometer; figure 14-33 shows a PESCIEX API 3000 Liquid Chromatograph/Mass Spectrometer/Mass Spectrometer; figure 14-34 shows a PESCIEX Triple/Quadruple/Liquid Chromatograph/Mass Spectrometer/Mass Spectrometer; and figure 14-35 shows a Hewlett-Packard Liquid Chromatograph ion trap LC/MS LC/MS/MS.

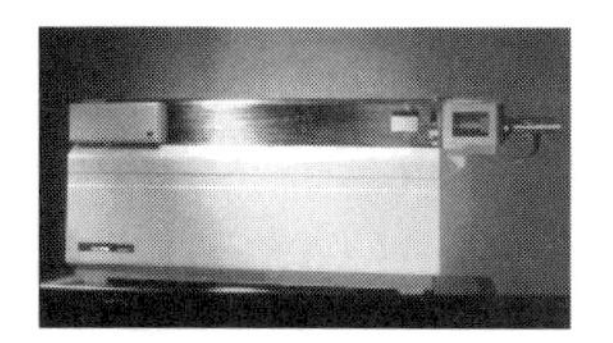

FIGURE 14-32

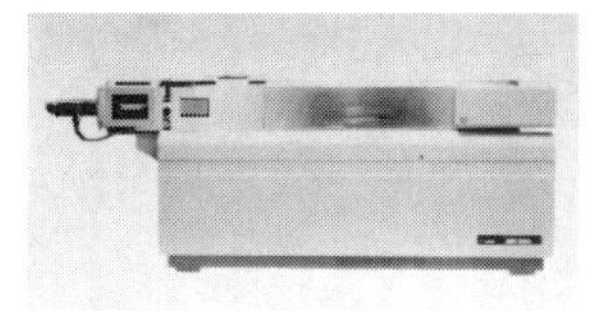

FIGURE 14-33

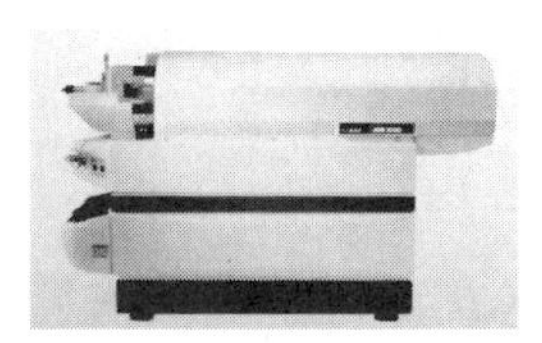

FIGURE 14-34

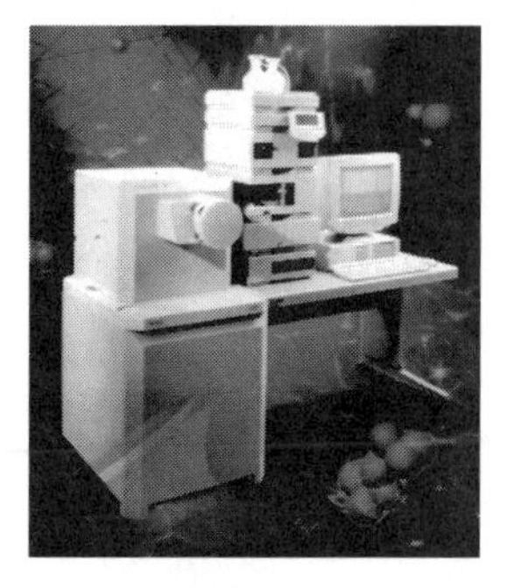

FIGURE 14-35

HOMOGENIZER

This is a sophisticated, high-speed, bladed apparatus used to blend and reduce the size of samples. This instrument requires countertop space.

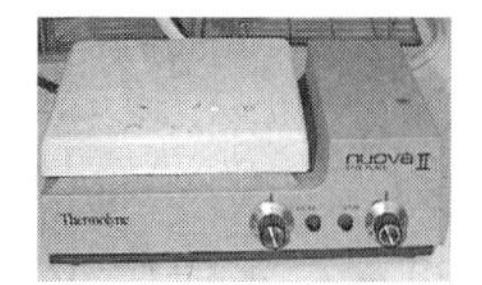

FIGURE 14-36

HOT PLATE

The hot plate (figure 14-36) is a simple apparatus used to heat liquids. This instrument requires countertop space.

INCUBATOR

The Incubator is a controlled temperature chamber used for bacteriological analyses. Depending on the model, this chamber may be located on a counter or the floor.

INDUCTIVELY COUPLED PLASMA MASS SPECTROMETER (ICP-MS) AND INDUCTIVELY COUPLED PLASMA OPTICAL EMISSION SPECTROMETER (ICP-OES)

These are sophisticated instruments used to analyze different levels of low-level multielement metals. These instruments provide simultaneous, multiple detection of metals by measuring the wavelengths of light emitted by the sample when subjected to a plasma flame. They are often used to analyze clean water samples such as drinking water and clean surface waters. Depending on the model, this instrument may be located on a counter or the floor. Figure 14-37 shows a PerkinElmer ElAN® 6000 Inductively Coupled Plasma Mass Spectrometer (ICP-MS); figure 14-38 shows a Hewlett-Packard 4500 plus Inductively Coupled Plasma Mass Spectrometer (ICP-MS); and figure 14-39 shows a PerkinElmer Inductively-Coupled Plasma Optical Emission Spectrometer (ICP-OES).

FIGURE 14-37

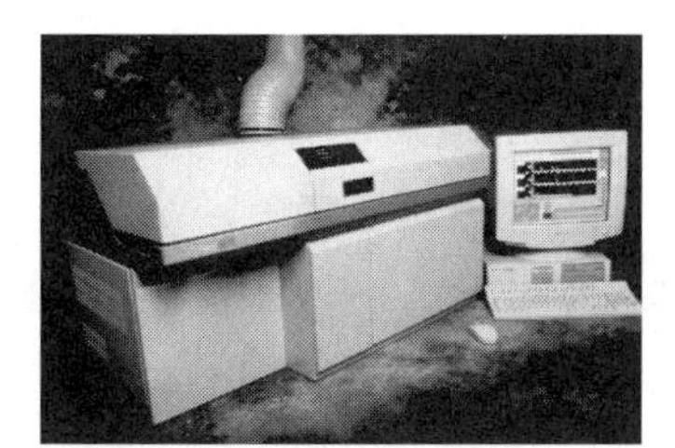

FIGURE 14-38

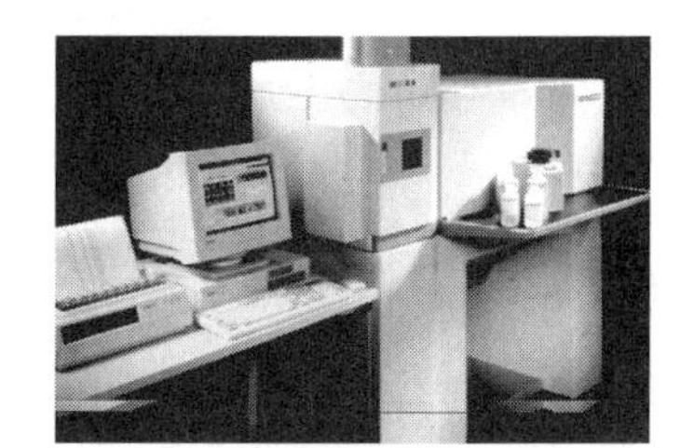

FIGURE 14-39

INFRARED SPECTROMETER (IR- SPEC.)

The Infrared Spectrometer (IR- SPEC.), also known as the Infrared Analyzer, is a sophisticated instrument used for petroleum hydrocarbon analysis. This instrument requires countertop space. Figure 14-40 shows a PerkinElmer Infrared Analyzer.

FIGURE 14-40

INTEGRATED THERMAL CYCLER

The Integrated Thermal Cycler is an automated workstation used for genetic analysis research. This instrument's high-precision robot is capable of pipeting extremely low volumes necessary for both PCR and sequencing reactions. Figure 14-41 shows a PE Biosystems Integrated Thermal Cycler.

FIGURE 14-41

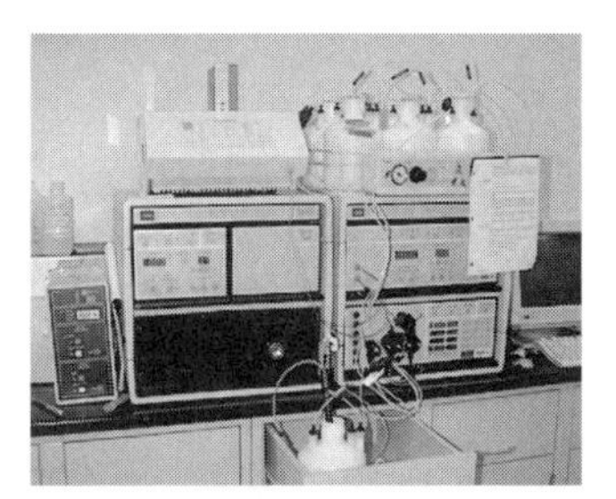

FIGURE 14-42

ION CHROMATOGRAPH (IC), DIONEX PUMPS, AND DIONEX SAMPLERS

The Ion Chromatograph, often referred to as IC, is a sophisticated instrument used to determine various water quality analytes. The Dionex pump is an ancillary apparatus for use with Dionex Ion Chromatographs. The Dionex sampler is a robotic sample delivery system for the Dionex Ion Chromatograph. These three instruments, as well as a Conductivity Detector, are generally connected to one another via a computer interface. This group of instruments, shown in figure 14-42, requires countertop space.

LIQUID CHROMATOGRAPH MASS SPECTROMETER DETECTOR

The Liquid Chromatograph Mass Spectrometer Detector is a sophisticated instrument that separates the chemicals in a mixture. The Mass Spectrometer detects the eluding chemicals and reads chemical structure information. Figure 14-43 shows a Hewlett-Packard Liquid Chromatograph Mass Spectrometer Detector.

LYOPHILIZER

The Lyophilizer (figure 14-44) is a specialized freeze-drying apparatus. This instrument requires countertop space.

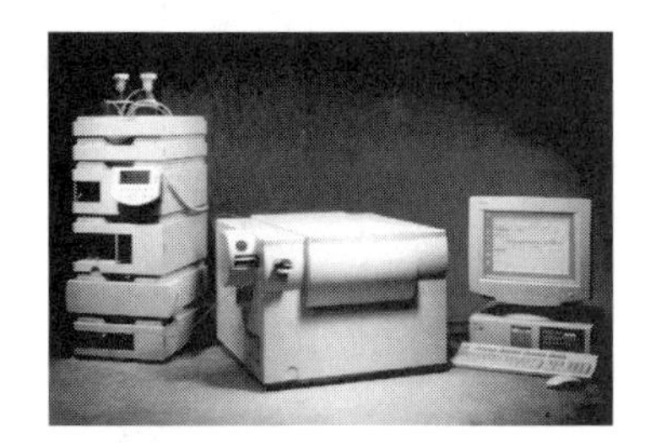

FIGURE 14-43

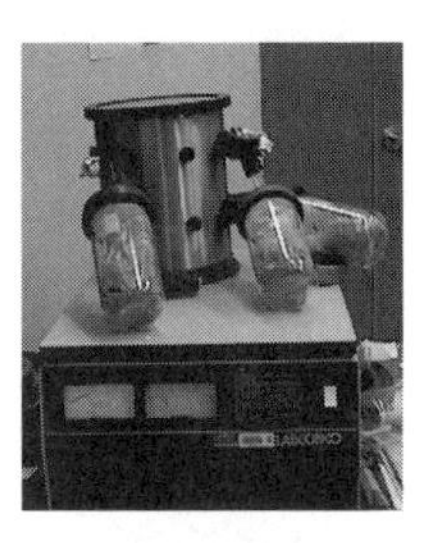

FIGURE 14-44

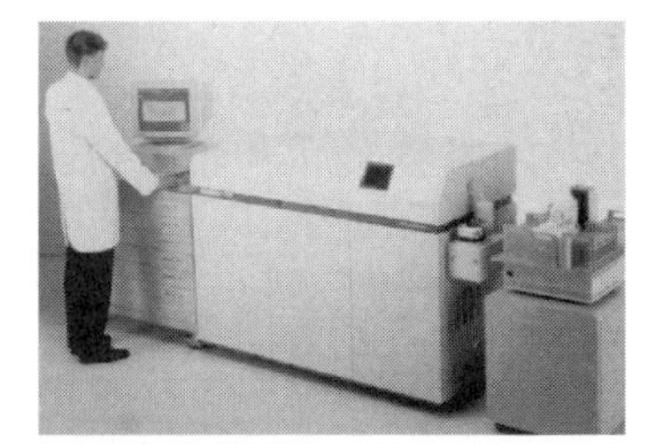

FIGURE 14-45

MASS SPECTROMETER

The Mass Spectrometer is a sophisticated instrument used to detect and identify sample constituents. It is usually coupled with a Gas Chromatograph (GC) or Inductively Coupled Argon Plasma (ICAP). Depending on the model, this instrument may be located on a counter or the floor. Figure 14-45 shows a Varian Ultramass-700 Inductively Coupled Mass Spectrometer.

MASS SPECTROMETER DETECTOR (MSD)

The Mass Spectrometer Detector, often referred to as MSD, is a sophisticated detection system for a wide range of organic compounds. This instrument is generally placed next to a Gas Chromatograph and requires countertop space. Figure 14-46 is a picture of a Mass Spectrometer Detector placed next to a Gas Chromatograph.

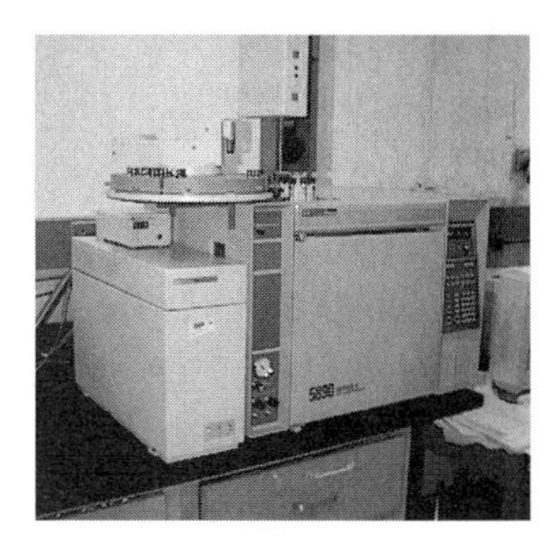

FIGURE 14-46

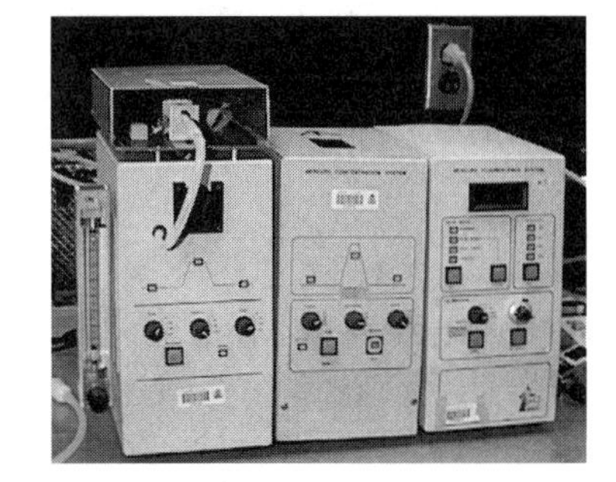

FIGURE 14-47

MERCURY ANALYZER

The Mercury Analyzer (figure 14-47), as its name indicates, is a dedicated instrument used for mercury analysis. This instrument requires countertop space.

MICROSCOPE

The microscope is a piece of equipment used to identify microscopic organisms by enlarging them several thousand times. This instrument requires countertop space. Figure 14-48 shows a PerkinElmer Auto Image Near Infrared (NIR) Microscopy System.

MICROWAVE OVEN

The microwave oven is a laboratory-grade microwave generally used to prepare samples for metal and other types of analyses. This instrument requires countertop space.

MOLECULAR BIOLOGY ASSISTANT (MBA 2000)

The Molecular Biology Assistant MBA 2000 TM is a piece of equipment used for genetic analysis, biotechnology, and pharmaceutical drug discovery. It is used to handle and/or measure samples containing DNA or proteins. Figure 14-49 shows a PerkinElmer Molecular Biology Assistant.

FIGURE 14-48

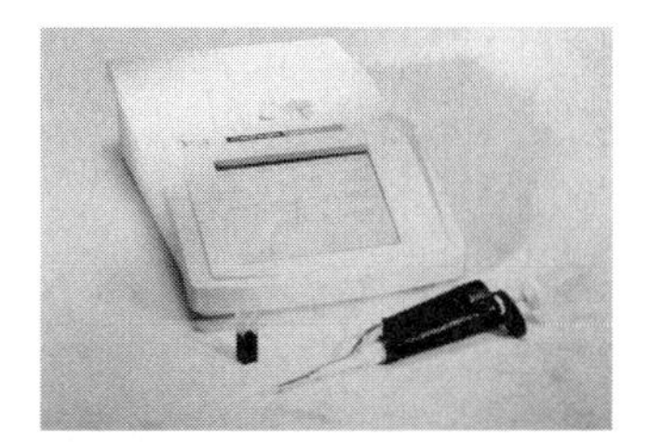

FIGURE 14-49

MUFFLE FURNACE

A Muffle Furnace is a controlled (dry) heat furnace used to decompose samples. Furnaces are different from ovens as they provide much higher temperatures. Depending on the model, this instrument may be located on a counter or the floor. Figure 14-50 shows a PerkinElmer HPA-5 High Pressure Asher (left) and a Muffle Furnace (right).

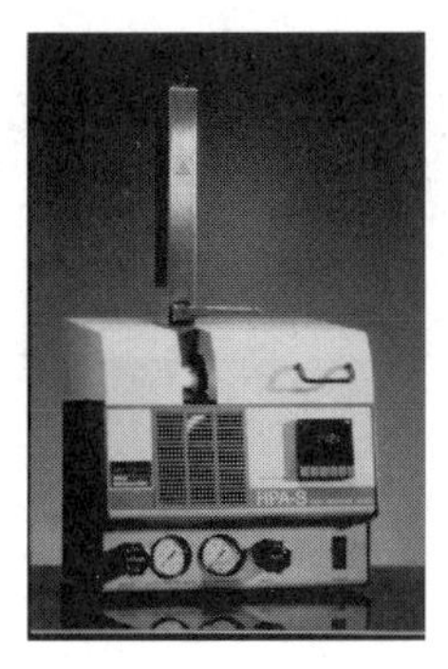

FIGURE 14-50

MULTI-METER

The Multi-meter, also known as a Volt-Ohm-Amp meter, is an apparatus used to measure voltage, ohms, and amps. This instrument requires countertop space.

MULTIWAVE MICROWAVE DIGESTION SYSTEM

The Multiwave Microwave Digestion System is a powerful system used for fast and safe high-pressure acid digestion of all types of organic and inorganic solid samples and sediments for analysis by Atomic Absorption, Inductively Coupled Plasma, and Inductively Coupled Plasma/Mass Spectrometer. Figure 14-51 shows a PerkinElmer Multiwave™ Microwave Digestion System.

FIGURE 14-51

FIGURE 14-52

OVEN

Ovens used in laboratories are controlled heating chambers. Depending on the model, these instruments may be located on a counter or the floor. The oven shown in figure 14-52 is located on the floor.

PENSKY MARTIN PETROTEST

The Pensky Martin Petrotest (figure 14-53) is an instrument used to determine the ignitability of liquid samples. This instrument requires countertop space.

FIGURE 14-53

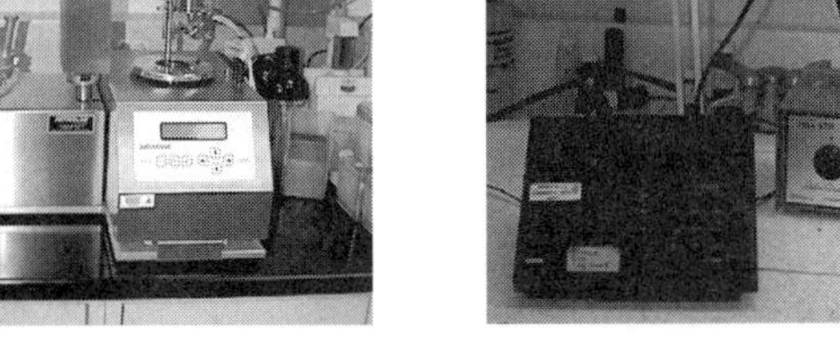

FIGURE 14-54

PH METER

The PH Meter (figure 14-54) is an instrument used for determining the pH of a given sample. This instrument requires countertop space.

RADIO FREQUENCY (RF) GENERATOR

The Radio Frequency (RF) Generator is a piece of ancillary equipment used to provide controlled electrical output for the Inductively Coupled Argon Plasma (ICAP). This piece of equipment is also known as an ICP-AES. Depending on the model, this instrument may be located on a counter or the floor.

REFRIGERATOR FOR FLAMMABLE MATERIALS

The refrigerator for flammable materials (figure 14-55) is a refrigerator designed to be explosion-proof. The refrigerator for flammable materials requires floor space.

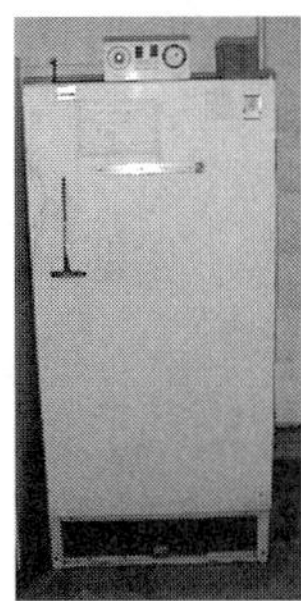

FIGURE 14-55

FIGURE 14-56

ROBOTIC SYSTEM

The Robotic System is a personal computer-controlled system used for robotic manipulation of specific laboratory operations. This instrument requires countertop space. Figure 14-56 shows a Varian Sample Preparation Robotic System.

SIEVE SHAKER

The Sieve Shaker is a set of different-sized sieves mounted on a shaker. This set of sieves is used to determine the distribution of a sample's particle sizes.

Depending on the model, this instrument may be located on a counter or the floor.

SONIC DISRUPTER

The Sonic Disrupter (figure 14-57) is a bladed apparatus coupled with an ultrasonic energy source. This apparatus is used for rending and homogenizing biological samples. This apparatus requires countertop space.

FIGURE 14-57

FIGURE 14-58

SOXHLET EXTRACTOR

A Soxhlet Extractor (figure 14-58) is a specialized piece of glassware used for solvent extraction from solid samples. This instrument requires countertop space, and, for safety reasons, it should be located in a fume hood.

SPECTROPHOTOMETER

A Spectrophotometer is an instrument that uses ultraviolet (UV) and visible (VIS) light to determine various water quality constituents. This instrument requires countertop space. Figure 14-59 shows a Hewlett-Packard 8453 E UV Visible Spectrometry System, and figure 14-60 shows a Varian Cary 50 UV-VIS Scanning Spectrometer.

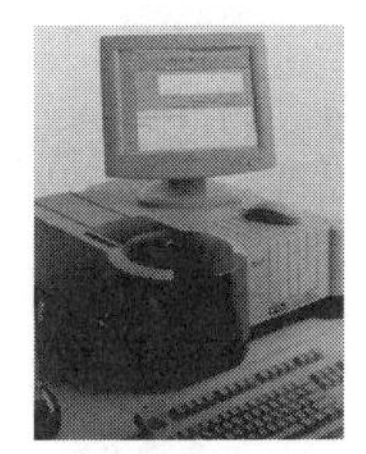

FIGURE 14-59

FIGURE 14-60

STEAM GENERATOR

The Steam Generator is a pressurized boiler that produces steam. Steam has multiple uses in a laboratory. It is often used for cleaning highly soiled laboratory and field apparatus. Depending on the model, this instrument may be located on a counter or the floor.

TEKRAN 2600 HG ANALYSER:

The Tekran 2600 Hg Analyser (figure 14-61) is a specialized type of valving equipment used for sample and reagent introduction to the atomic absorption spectrometer. This instrument requires countertop space.

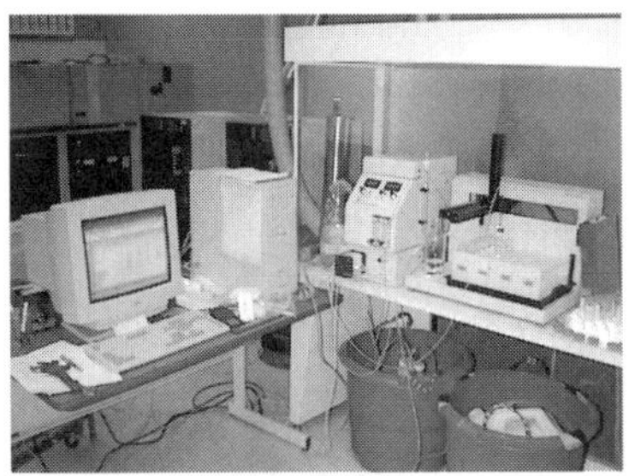

FIGURE 14-61

FIGURE 14-62

THERMOGRAVMETRIC ANALYSER

The Thermogravmetric Analyzer is an instrument used to measure changes in the weight (mass) of a sample as a function of temperature and/or time. The Thermogravmetric Analyzer is commonly used to determine polymer degradation temperatures, residual solvent levels, absorbed moisture content, and the amount of inorganic (noncombustible) filler in polymer or composite material compositions. This instrument requires countertop space. Figure 14-62 shows a PerkinElmer Thermogravimetric Analyzer.

TITRATOR

The Titrator is an apparatus used for electronically sensing titration end points. This instrument requires countertop space.

TOTAL ORGANIC CARBON ANALYZER (TOC)

The Total Organic Carbon Analyzer (TOC) (figure 14-63) is an instrument used to determine the organic carbon content of samples. This instrument requires countertop space.

FIGURE 14-63

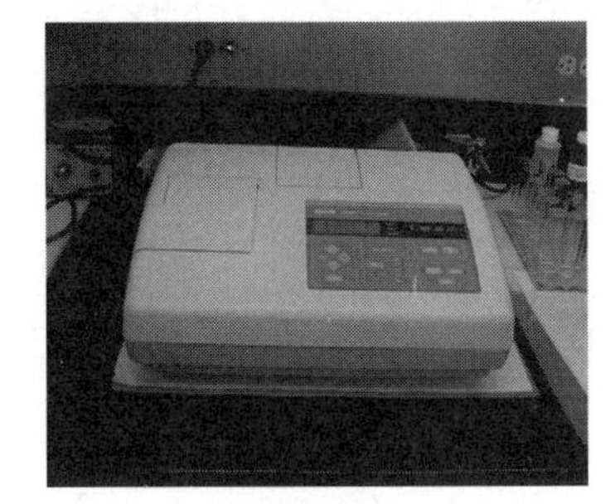

FIGURE 14-64

TURBIDITY METER

The Turbidity Meter (figure 14-64) is an instrument used for measuring a sample's turbidity (opacity). This instrument requires countertop space.

TURBOMASS MASS SPECTROMETER

The Turbomass Mass Spectrometer, in addition to performing routine GC/MS analysis, handles more complex types of analysis and enhances productivity in a wide range of applications. Depending on the model, this instrument may be located on a counter or the floor. Figure 14-65 shows a PerkinElmer Turbomass Mass™ Spectrometer.

FIGURE 14-65

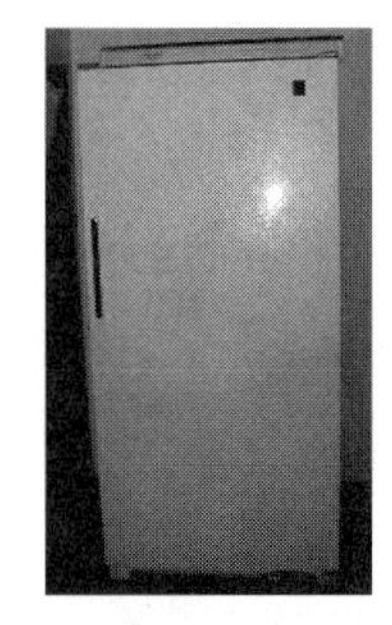

FIGURE 14-66

ULTRA LOW FREEZER

The Ultra Low Freezer (figure 14-66) is a very low-temperature freezer (less than –40°C). This type of freezer is used for storage of unstable biological materials, cultures, and samples. It requires floor space.

ULTRAVIOLET (UV) REACTOR

The UltraViolet (UV) Reactor is a radiant (ultraviolet light) source of energy used for chemical reactions. This instrument requires countertop space.

VACUUM PUMPS

Vacuum Pumps (figure 14-67) are pumps used to provide vacuum for laboratory filtrations. Depending on the model, these instruments may be located on a counter or the floor.

FIGURE 14-67

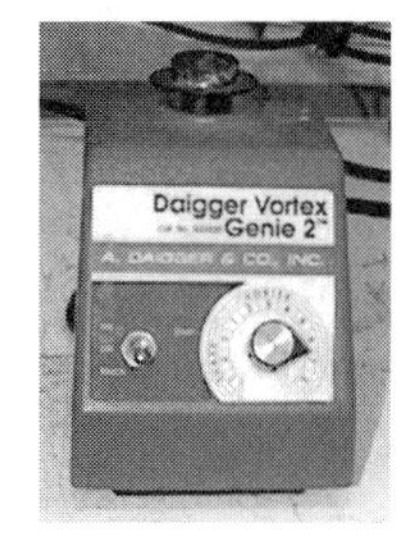

FIGURE 14-68

VORTEX MIXER

The Vortex Mixer (figure 14-68) is an oscillating apparatus used for stirring liquids in test tubes. This instrument requires countertop space.

WATER BATH

The water bath (figure 14-69) is a small apparatus used to heat and control the heating of water for bacteriological analyses. This instrument requires countertop space.

X-RAY FLUORESCENCE (XRF)

The X-Ray Fluorescence (XRF) is a nondestructive metal analyzer that uses X-ray radiation. Depending on the model, this instrument may be located on a counter or the floor.

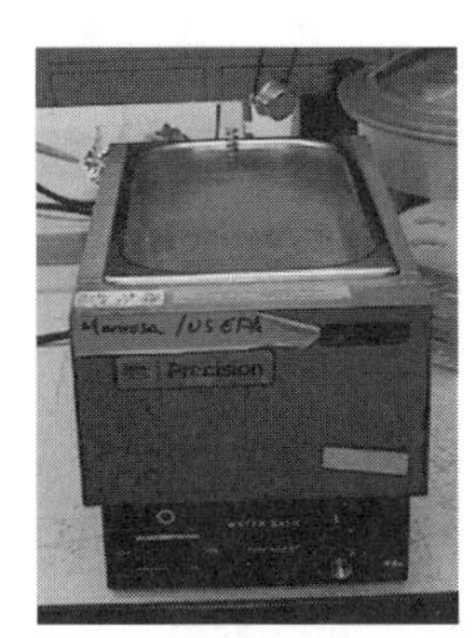

FIGURE 14-69

FIGURE 14-70

YSI DISSOLVED OXYGEN

The Ysi Dissolved Oxygen (figure 14-70) is a specialized instrument used to determine the oxygen content of water samples. This instrument requires countertop space.

Index

A

B

C

D

E

F

G

H

I

K

L

S

T

U

V

W

Y

Z